METHODS IN MOLECULAR BIOLOGY™



*Series Editor*
John M. Walker
School of Life Sciences
University of Hertfordshire
Hatfield, Hertfordshire, AL10 9AB, UK

For further volumes:
http://www.springer.com/series/7651

# Synthetic Gene Networks

## Methods and Protocols


Edited by

**Wilfried Weber**

*BIOSS Centre for Biological Signalling Studies, University of Freiburg, Freiburg, Germany*
*Department of Biosystems Science and Engineering, ETH Zurich, Basel, Switzerland*

**Martin Fussenegger**

*Department of Biosystems Science and Engineering, ETH Zurich,*
*University of Basel, Basel, Switzerland*



Humana Press

*Editors*
Wilfried Weber
BIOSS Centre for Biological
Signalling Studies
University of Freiburg Freiburg, Germany
and
Department of Biosystems Science
and Engineering
ETH Zurich Basel, Switzerland
wilfried.weber@biologie.uni-freiburg.de

Martin Fussenegger
Department of Biosystems Science
and Engineering
ETH Zurich University of Basel
Basel, Switzerland
fussenegger@bsse.ethz.ch

ISSN 1064-3745 e-ISSN 1940-6029
ISBN 978-1-4939-6224-2 ISBN 978-1-61779-412-4 (eBook)
DOI 10.1007/978-1-61779-412-4
Springer New York Dordrecht Heidelberg London



Printed on acid-free paper

Humana Press is part of Springer Science+Business Media (www.springer.com)

# Preface

The rapid expansion of synthetic biology showing exponentially growing publication numbers is based on the design and construction of synthetic gene networks opening many new avenues in fundamental and applied research.

In this book, we provide comprehensive information to design and construct synthetic gene networks in different host backgrounds. In the first section, we focus on design concepts to devise synthetic gene networks and how mathematical models can be applied to the predictable engineering of desired network features while taking into account inherent cellular imponderability like noise and stochastic fluctuations. The second section highlights the construction and validation of biologic tools ranging from engineering the codon usage to the development and validation of sophisticated RNA-based switches. In the third section, strategies are described to optimize and streamline the host cell for optimized network performance. The last section forms a synthesis of the previous ones by describing how optimally designed gene networks can be implemented in a large variety of host cells ranging from bacteria over yeast and insect cells to plant and mammalian cell culture.

This book written by designated experts in synthetic biology represents an encyclopedic resource for biologists, engineers, and computer scientists already established or just entering into the rapidly expanding field of synthetic biology and therefore serves as a positive feedback loop for further promoting the exponential growth of synthetic biology.

*Freiburg, Germany*
*Basel, Switzerland*

*Wilfried Weber*
*Martin Fussenegger*

# Contents

# Contributors


HAL S. ALPER • *Department of Chemical Engineering, The University of Texas at Austin, Austin, TX, USA*
KATJA M. ARNDT • *Institute for Biochemistry and Biology, University of Potsdam, Potsdam, Germany*
YAAKOV BENENSON • *Department of Biosystems Science and Engineering, ETH Zurich, Basel, Switzerland*
LEONIDAS BLERIS • *Department of Bioengineering and Electrical Engineering, University of Texas at Dallas, Richardson, TX, USA*
KATIE BRENNER • *California Institute of Technology, Pasadena, CA, USA*
TONI CATHOMEN • *Institute of Experimental Hematology, Hannover Medical School, Hannover, Germany*
VICTOR DE LORENZO • *Systems Biology Program, Centro Nacional de Biotecnologia, CSIC, Madrid, Spain*
MATTHEW P. DELISA • *School of Chemical and Biomolecular Engineering, Cornell University, Ithaca, NY, USA*
DIEGO DI BERNARDO • *Faculty of Engineering University of Naples "Fecerico II", Naples, Italy*
ANNE-KATHRIN DREYER • *Institute of Experimental Hematology, Hannover Medical School, Hannover, Germany*
MARTIN FUSSENEGGER • *Faculty of Science, Department of Biosystems Science and Engineering, ETH Zurich, University of Basel, Basel, Switzerland*
MARCEL GEERTZ • *Laboratory of Biological Network Characterization, EPFL Lausanne, Lausanne, Switzerland*
ANITA HAJDU • *Institute of Plant Biology, Biological Research Centre, Szeged, Hungary*
DANIELLE JANUS • *Division of Molecular Biotechnology, Helmholtz Centre for Infection Research, Braunschweig, Germany*
ASTRID JUNKER • *Leibniz Institute of Plant Genetics and Crop Plant Research (IPK), Gatersleben, Germany*
BJÖRN H. JUNKER • *Leibniz Institute of Plant Genetics and Crop Plant Research (IPK), Gatersleben, Germany*
MICHAEL M. KÄMPF • *Department of Biosystems Science and Engineering, ETH Zurich, Basel, Switzerland*
MARIA KARLSSON • *Faculty of Biology, BIOSS Centre for Biological Signalling Studies, University of Freiburg, Freiburg, Germany; Department of Biosystems Science and Engineering, ETH Zurich, Basel, Switzerland*
LASZLO KOZMA-BOGNAR • *Institute of Plant Biology, Biological Research Centre, Szeged, Hungary*
AMANDA M. LANZA • *Department of Chemical Engineering, The University of Texas at Austin, Austin, TX, USA*

DIRK LEBIEDZ • *Faculty of Biology, Centre for Biological Systems Analysis, University of Freiburg, Freiburg, Germany; Faculty of Mathematics and Physics, Department of Applied Mathematics, University of Freiburg, Freiburg, Germany*

MADELEINE LEISNER • *Arnold Sommerfeld Centre for Theoretical Physics, Ludwigs-Maximilians-University, Munich, Germany*

AUDREY LEPRINCE • *Division of Molecular Biotechnology, Helmholtz Centre for Infection Research, Braunschweig, Germany*

JASON LOHMUELLER • *Department of Biological and Biomedical Sciences, Harvard Medical School, Boston, MA, USA*

SEBASTIAN J. MAERKL • *Laboratory of Biological Network Characterization, EPFL Lausanne, Lausanne, Switzerland*

MARIO ANDREA MARCHISIO • *Department of Biosystems Science and Engineering, ETH Zurich, Basel, Switzerland*

ESTEBAN MARTINEZ-GARCIA • *Systems Biology Program, Centro Nacional de Biotecnologia, CSIC, Madrid, Spain*

VITOR MARTINS DOS SANTOS • *Division of Molecular Biotechnology, Helmholtz Centre for Infection Research, Braunschweig, Germany; Chair for Systems and Synthetic Biology, Wageningen University, Wageningen, The Netherlands*

LUCIA MARUCCI • *Faculty of Engineering, University of Naples "Fecerico II", Naples, Italy*

FILIPPO MENOLASCINA • *Faculty of Engineering, University of Naples "Fecerico II", Naples, Italy*

KRISTIAN M. MÜLLER • *Institute for Biochemistry and Biology, University of Potsdam, Potsdam, Germany*

FERENC NAGY • *Institute of Plant Biology, Biological Research Centre, Szeged, Hungary*

ALEXANDER J. NINFA • *Department of Biological Chemistry, Cellular Biotechnology Training Program, University of Michigan Medical School, Ann Arbor, MI, USA*

FRANK NOTKA • *Life Technologies Corp/GeneArt AG, Regensburg, Germany*

STEPHEN PAYNE • *Department of Biomedical Engineering, Duke University, Durham, NC, USA*

NICOLAS PERRY • *Department of Biological Chemistry, Cellular Biotechnology Training Program, University of Michigan Medical School, Ann Arbor, MI, USA*

MARCEL REHBERG • *Faculty of Biology, Centre for Biological Systems Analysis, University of Freiburg, Freiburg, Germany; Faculty of Mathematics and Physics, Department of Applied Mathematics, University of Freiburg, Freiburg, Germany*

MICHAEL RETH • *Faculty of Biology, BIOSS Centre for Biological Signalling Studies, University of Freiburg, Max-Planck-Institute of Immunobiology and Epigenetics, Freiburg, Germany*

SYLVIE ROCKEL • *Laboratory of Biological Network Characterization, EPFL Lausanne, Lausanne, Switzerland*

VELIA SICILIANO • *Faculty of Engineering, University of Naples "Fecerico II", Naples, Italy*

DOMINIK SKANDA • *Faculty of Biology, Centre for Biological Systems Analysis, University of Freiburg, Freiburg, Germany; Faculty of Mathematics and Physics, Department of Applied Mathematics, University of Freiburg, Freiburg, Germany*

ROBERT PHILLIP SMITH • *Department of Biomedical Engineering, Duke University, Durham, NC, USA*

BEATRIX SUESS • *Institute of Molecular Biosciences, Johann Wolfgang Goethe-University Frankfurt, Frankfurt, Germany*

RALF WAGNER • *Life Technologies Corp/GeneArt AG, Regensburg, Germany; Molecular Microbiology and Gene Therapy, University of Regensburg, Regensburg, Germany*

DUJDUAN WARAHO • *School of Chemical and Biomolecular Engineering, Cornell University, Ithaca, NY, USA; Biomedical Engineering Program, King Mongkut's University of Technology Thonburi, Bangmod, Toongkru, Thailand*

WILFRIED WEBER • *Faculty of Biology, BIOSS Centre for Biological Signalling Studies, University of Freiburg, Freiburg, Germany; Department of Biosystems Science and Engineering, ETH Zurich, Basel, Switzerland*

JULIA E. WEIGAND • *Institute of Molecular Biosciences, Johann Wolfgang Goethe-University Frankfurt, Frankfurt, Germany*

BIRGIT WILTSCHI • *Faculty of Biology, University of Freiburg, Freiburg, Germany*

ALEXANDER WITTMANN • *Institute of Molecular Biosciences, Johann Wolfgang Goethe-University Frankfurt, Frankfurt, Germany*

G. ANDREW WOOLLEY • *Department of Chemistry, University of Toronto, Toronto, Canada*

ZHEN XIE • *Department of Biological Engineering, Massachusetts Institute of Technology, Cambridge, MA, USA*

JIANYING YANG • *Faculty of Biology, BIOSS Centre for Biological Signalling Studies, University of Freiburg, Max-Planck-Institute of Immunobiology and Epigenetics, Freiburg, Germany*

LINGCHONG YOU • *Department of Biomedical Engineering, Institute for Genome Sciences and Policy, Center for Systems Biology, Duke University, Durham, NC, USA*

FUZHONG ZHANG • *BIOSS Centre for Biological Signalling Studies, University of Freiburg, Freiburg, Germany*

# Part I

## Design Strategies for Synthetic Gene Network

# Chapter 1

## In Silico Implementation of Synthetic Gene Networks

**Mario Andrea Marchisio**


### Abstract

Computational synthetic biology has borrowed methods, concepts, and techniques from systems biology and electrical engineering. Features of tools for the analysis of biochemical networks and the design of electric circuits have been combined to develop new software, where Standard Biological Parts (physically stored at the MIT Registry) have a mathematical description, based on mass action or Hill kinetics, and can be assembled into genetic networks in a visual, "drag & drop" fashion. Recent tools provide the user with databases, simulation environments, formal languages, and even algorithms for circuit automatic design to refine and speed up gene network construction. Moreover, advances in automation of DNA assembly indicate that synthetic biology software soon will drive the wet-lab implementation of DNA sequences.


**Key words:** Standard Biological Parts, Gene circuits, Part composability, Visual design, Formal language, Automation, DNA assembly

## 1. Introduction

In every branch of engineering, the practical implementation of a new, complex product is normally preceded by an accurate computational procedure. In this phase of the project, the system under study is divided into modules that are made, in their turn, of different smaller components. Computer simulations permit to test several possible designs and to identify the components that optimize the performance of the final product.

To be more specific, we can think of electric circuits. Software, like PSpice (1), permit to design them on the computer screen by, first, choosing among standard electric components (e.g., batteries, resistors, capacitors, and solenoids) and devices (e.g., diodes and transistors) and, secondly, connecting them through lines representing wires. Since the temporal behavior of each electric component is well-known and described by physics laws, the circuit dynamics

Wilfried Weber and Martin Fussenegger (eds.), *Synthetic Gene Networks: Methods and Protocols*,
Methods in Molecular Biology, vol. 813, DOI 10.1007/978-1-61779-412-4_1, © Springer Science+Business Media, LLC 2012

is faithfully reproduced via computer simulations that permit to estimate how far the current circuit configuration is from the one that achieves the desired performance. Then, improvements on the circuit are carried out both by modifying its design with the addition, the removal or the spatial rearrangement of some electric units, and by changing the value of a set of parameters (e.g., resistance and capacity) that characterize one or more components. Further simulations and changes on the circuit bring to a satisfactory design that is finally printed on a board.

In the last years, this working procedure has become familiar even to natural sciences after the advent of synthetic biology, the discipline that is commonly regarded as "life engineering" (see Fig. 1). According to the MIT Registry of Standard Biological Parts (http://partsregistry.org), basic modules are DNA traits associated with well-defined functions in transcription and translation process like promoters, ribosome-binding sites (RBSs), coding regions both for proteins and small RNAs (sRNAs), and terminators. Standard parts share the same input/output, i.e., fluxes of molecules like RNA polymerases and ribosomes that are necessary for protein synthesis. Thanks to this common interface, parts are said *composable* and can be assembled into genetic devices and circuits to accomplish specific tasks (2). Furthermore, different mathematical models that successfully describe the dynamics of cell systems have been proposed (3) and software for the visualization, analysis, and simulation of biological systems has been developed (see Table 1).

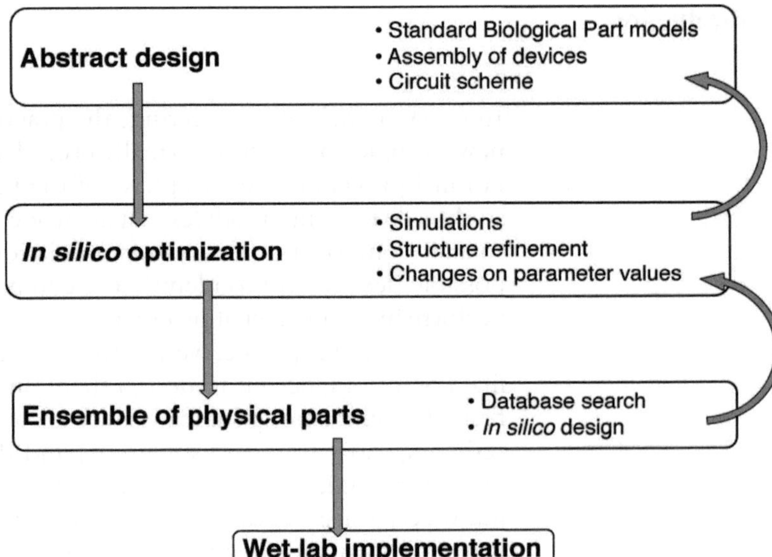


Fig. 1. Schematic workflow for synthetic gene circuit implementation. The procedure in silico starts with the design of a circuit scheme with abstract parts (models) and successively demands refinements to match the theoretical circuit characteristics with the physical parts retrieved from databases or de novo engineered. Ideally, computational design drives the wet-lab implementation of new synthetic gene networks.

**Table 1**
**URLs of the computational tools referenced throughout the chapter**

| | |
|---|---|
| *Visual design* (*Systems biology*) | |
| CellDesigner | http://www.celldesigner.org |
| Cytoscape | http://www.cytoscape.org |
| BisoGenet | http://bio.cigb.edu.cu/bisogenet |
| CADLIVE | http://www.cadlive.jp/ |
| Virtual Cell | http://www.nrcam.uchc.edu/ |
| JDesigner | http://www.sys-bio.org/software/jdesigner.htm |
| BioTapestry | http://www.biotapestry.org/ |
| *Visual design* (*Synthetic biology*) | |
| Biojade | http://web.mit.edu/jagoler/www/biojade/ |
| ProMoT | http://www.mpi-magdeburg.mpg.de/projects/promot/ |
| Asmparts | http://soft.synth-bio.org/asmparts.html |
| Virtual Parts | http://models.cellml.org/ |
| TinkerCell | http://www.tinkercell.com/ |
| SynBioSS | http://synbioss.sourceforge.net/ |
| Clotho | http://www.clothocad.org/ |
| *Formal languages* (*Systems biology*) | |
| BioNetGen | http://bionetgen.org/index.php/BioNetGen_Distributions |
| Kappa Language | http://www.kappalanguage.org/ |
| BIOCHAM | http://contraintes.inria.fr/BIOCHAM/ |
| π-Calculus | http://www.wisdom.weizmann.ac.il/~biospi/index_main.html |
| Bio-PEPA | http://homepages.inf.ed.ac.uk/jeh/Bio-PEPA/biopepa.html |
| little b | http://www.littleb.org/ |
| Jarnac | http://www.sys-bio.org/software/jarnac.htm |
| *Formal languages* (*Synthetic biology*) | |
| Antimony | http://antimony.sourceforge.net/ |
| GEC | http://research.microsoft.com/en-us/projects/gec/ |
| GenoCAD | http://www.genocad.org |
| MDL | See ProMoT |
| Proto | http://groups.csail.mit.edu/stpg/proto.html |
| *Automatic design* | |
| Genetdes | http://soft.synth-bio.org/genetdes.html |
| *Simulations and optimization* | |
| TABASCO | http://openwetware.org/wiki/TABASCO#TabascoSimulator |
| COPASI | http://copasi.org |
| OpenCell | http://www.cellml.org/tools/opencell/ |
| Hy3S | http://hysss.sourceforge.net/index.shtml |
| STOCHSIM | http://www.ebi.ac.uk/~lenov/stochsim.html |
| RoVerGeNe | http://iasi.bu.edu/~batt/rovergene/rovergene.htm |
| *DNA and RNA design* | |
| GeneDesign 3.0 | http://www.genedesign.org |
| Gene designer | https://www.dna20.com/genedesigner2/ |
| DNAWorks | http://helixweb.nih.gov/dnaworks/ |
| Oligo Cuts | http://gcat.davidson.edu/IGEM06/oligo.html |
| Gene2Oligo | http://berry.engin.umich.edu/gene2oligo/ |

(continued)

**Table 1 (continued)**

| | |
|---|---|
| Zinc Finger Tools | http://www.scripps.edu/mb/barbas/zfdesign/zfdesignhome.php |
| NEBcutter | http://tools.neb.com/NEBcutter2/index.php |
| CloneQC | http://cloneqc.thruhere.net/ |
| UNAFold/DINAMelt | http://dinamelt.bioinfo.rpi.edu/ |
| Vienna RNA | http://rna.tbi.univie.ac.at/ |
| RNA FRABASE 2.0 | http://rnafrabase.cs.put.poznan.pl/ |
| *Protein design* | |
| Rosetta | http://www.rosettacommons.org/main.html |
| Raptor | http://www.bioinformaticssolutions.com/products/raptor |
| HHpred | http://toolkit.lmb.uni-muenchen.de/hhpred |
| Modeller | http://salilab.org/modeller/ |
| EGAD | http://egad.ucsd.edu/EGAD_manual/index.html |
| PROTDES | http://soft.synth-bio.org/protdes.html |
| PFP | http://dragon.bio.purdue.edu/pfp/ |
| Autodock 4.2 | http://autodock.scripps.edu/ |
| HexServer | http://www.loria.fr/~ritchied/hex_server/ |
| Biskit | http://biskit.sf.net |

The aim of this chapter is to give a review of the computational tools currently available for the design of synthetic gene circuits with their link to previously developed tools in systems biology. Following their time evolution, I first describe the software for the visual design of synthetic gene networks, and afterward I introduce the concepts of formal languages and the ideas at the basis of automatic circuit design. Finally, before the conclusion, a brief report on the existing tools for DNA, RNA, and protein manipulation is given.

## 2. Materials

### 2.1. Tools for Visual Design of Synthetic Gene Circuits

If synthetic biology requires the definition of modules to use them – in a bottom-up approach – as bricks for the construction of more complex systems, systems biology makes use of the concept of modularity from a reverse point of view (top down) to decompose a biological system into smaller components whose analysis can shed light on the properties of the whole network. Therefore, computational tools based on a visual, modular representation of cellular pathways have been extensively used in systems biology. One of the most popular is CellDesigner (4, 5). Here, a biochemical network is visually depicted via a *process diagram*, where nodes represent states of the molecules and arrows (connecting nodes) transitions between states. Beside the graphic user interface (GUI) for network design, CellDesigner wraps an ordinary differential equation (ODE)-based simulator and offers connectivity to several databases. Furthermore, networks are encoded in System Biology Markup Language (SBML) (6), a standard format for biological system representation as witnessed by the BioModels (7) database, a huge repository of SBML-based biochemical network models. A similar graph-based representation of biochemical networks is accomplished by several other tools like Cytoscape (8) (recently expanded with BisoGenet (9), a plug-in devoted to gene networks); CADLIVE (10, 11); Virtual Cell (12), which turns out to be particularly adapted for the analysis of signaling networks since it permits a precise definition of cell compartments and contains a simulation environment that can solve both ODE and partial differential equation (PDE) systems; JDesigner (13); and BioTapestry (14), which provides a suite of tools for the visualization and modeling of development gene networks. These tools can be used in an engineering perspective as well. In particular, they own all the necessary requisites for the de novo modeling and building of transduction and metabolic pathways. However, they cannot be applied directly to the design of synthetic gene circuits since, among their modules, Standard Biological Parts are ignored. Software products for synthetic gene circuit design appear strongly linked to the framework and the ideas illustrated in the MIT Registry, although they also make use of the notion of modular visual design borrowed from systems biology and electronics tools.

Historically, BioJade (15, 16) can be considered the first tool for the computational design of synthetic gene networks based on Standard Biological Parts. Remarkably, BioJade presents a complete structure that combines a GUI (for circuit design) with a simulation environment and supplies connectivity to the MIT Registry. Furthermore, BioJade can invoke external software, like the simulator TABASCO (17) that runs simulations of transcription and translation at single base-pair resolution. However, beside a nice computational framework, the mathematical description adopted for gene parts is rather simple. In particular, part input/output is defined in translation initiations per seconds (TIPS), a tentative unit to model part interface successively abandoned.

As already explained in the Subheading 1, biological parts are required to be composable, i.e., they must share the same input/output in order to be freely interchangeable as it happens in electronics. In particular, electric components communicate thanks to the exchange of an electrons' flux, the electric current that is measured in Ampere (or Coulomb per second, where Coulomb is the unit of electric charge). Following the MIT Registry, the molecules that represent the biological counterpart of the electrons are called *common signal carriers* since they are shared among parts and bring biological information along a gene network. The choice of common signal carriers has fallen on RNA polymerases and ribosomes since they have to scan the entire sequences of DNA and mRNA in order to accomplish transcription and translation, respectively. Moreover, their fluxes represent biological currents and can be theoretically quantified in polymerases per second (PoPS) and ribosome per second (RiPS). According to this framework, two biological parts are composable if and only if the output of the first part coincides with the input of the second one. For instance, a promoter produces only a polymerase flux; hence, it can be connected to an RBS, an sRNA, or a terminator but not to a coding region, which expects, as an input, a flux of ribosomes as well. Hence, biological composability is subjected to some constraints, differently from electronics. Furthermore, in this overall picture proposed by Endy et al. (18), promoters are regarded as PoPS generators and terminators as PoPS sink.

Starting from these ideas, Marchisio and Stelling (19, 20) developed a more elaborate model that describes thoroughly the exchange of information in a synthetic gene network. Beside RNA polymerases and ribosomes, three other kinds of molecules have been acknowledged the status of signal carriers due to the role they play in protein synthesis. They are transcription factors, sRNAs, and environmental signals. Transcription factor proteins control transcription by binding promoters at specific operators and preventing (repressors) or enhancing (activators) RNA polymerase binding; sRNAs play the same role in translation by binding the mRNA in correspondence to the RBS or the coding region. Environmental signals, finally, are small molecules (chemicals) that can influence both transcription and translation by, for instance, binding and modifying the structure of transcription factors or acting directly on the mRNA at structures, like riboswitches and ribozymes (21). Moreover, both transcription factors and sRNAs are exchanged between transcription units (genetic devices) and their currents can be hypothetically measured as factor per second (FaPS) and RNA per second (RNAPS), respectively. In contrast, chemicals carry information from the cellular (or extracellular) environment to the whole circuit and their flux's unit is signal per second (SiPS). Overall, this new framework demands, at least, five different common signal carriers. Furthermore, each of them is associated with a *pool* that represents the ideal place, inside the cell, where free molecules of signal carriers are stored. The meaning of pools is double. On one hand, in the gene circuit scheme, they are visual interfaces among devices or between the cell and the circuit (signal pools). On the other hand, their content is a biological *potential* that determines the circuit working and performance. In a certain sense, they are like batteries since a circuit is "closed" only once all the pools have been connected to the circuit devices. In this picture, a promoter is no longer a PoPS generator per se and a terminator does not "destroy" any PoPS signal since RNA polymerases return to their pool after leaving the DNA chain. Furthermore, pools can be used just to host biochemical reactions whose reactants are produced by different genes (e.g., enzymatic reactions). This whole framework has been translated into a computational tool as an add-on of Process Modeling Tool (ProMoT) (22). Here, every part and pool are modeled according to mass action kinetics and a circuit can be designed in a drag & drop fashion on the ProMoT GUI. Circuits – and their components – are encoded in Model Definition Language (MDL) (23), the Lisp-based language of ProMoT, and MDL files can be exported into Matlab and SBML format and simulated by several tools (e.g., COPASI (24)).

More conformed to the original MIT ideas is Asmparts (25). Here, standard parts are encoded in SBML and can be assembled into a circuit directly from the command line. Moreover, their mathematical description follows Hill kinetics, which implies a reduction in the number of parameters necessary to specify the whole network.

A rather different approach is followed by Virtual Parts (26) that, essentially, realizes a repository of abstract, mathematical models (templates) for Standard Biological Parts (27). Virtual parts are encoded in CellML 1.1 format (28) that differs from SBML for a better, inherent support to modularity. However, as a drawback, CellML models are ODE based and preclude stochastic simulations that might be relevant for systems characterized by a low number of molecules. Beside the Registry parts, Virtual parts introduce “bioenvironment models” that serve to represent a large range of biochemical reactions (e.g., protein degradation, enzyme–substrate interaction) recalling, in this way, a possible use of the ProMoT pools. However, Virtual parts are modeled according to Hill kinetics and, at their interface, exchange only RNA polymerase and ribosome fluxes. Furthermore, Virtual parts are not assembled in a drag & drop way, but they are specified one after the other, in a similar way to Asmparts. Nevertheless, this operation can be performed via the software OpenCell (29) that, in addition, provides a simulation environment.

Another way for the in silico design of gene networks is proposed in TinkerCell (30). This tool can be thought as a powerful front-end application for Python and C/C++ programs that supply, for instance, analysis and simulation capabilities. Thanks to this modular structure, TinkerCell is easily extendable with new functionalities. Currently, standard parts are encoded in XML files and gene circuits are built either in a visual drag & drop way or by specifying parts and their connections in an Antimony (31) script file. Remarkably, visual design in TinkerCell offers the possibility to choose the mathematical representation of standard parts (Hill or mass action kinetics). Furthermore, a user can also opt for a simpler circuit design, where parts are lumped into transcription units. In addition, TinkerCell allows to design and analyze generic biochemical networks beyond the gene circuit realm. This explains why three different interfaces are available for module connection, namely, reactions (e.g., promoter control by transcription factors), species in common to two modules (e.g., kinases and phosphatases in a signaling pathway), and PoPS between Standard Biological Parts. Finally, information on and sequence of standard parts can be obtained directly from the RegulonDB database (http://regulondb.ccg.unam.mx/).

Overall, TinkerCell resembles BioJade in its attempt to construct a general framework that combines the visual design of gene circuits via abstract parts with analysis and optimization tools and the connection to databases of physical parts to generate DNA sequences for the whole circuit at the end of the design process.

The same philosophy is followed by Synthetic Biology Software Suite (SynBioSS) (32). This suite of software is composed of three environments: the Designer (33), the Wiki, and the Desktop Simulator. The Designer requires, as input, a set of standard parts that can be chosen directly from the Standard Biological Parts Web Service (http://openwetware.org/wiki/Endy:Notebook/BioBrick_Studio). This database can be considered as a mirror of the MIT Registry, where, however, information is reorganized in a machine-readable format. Proteins and chemicals involved in the system – together with both their reciprocal interactions and the proteins' action on promoters – have to be specified as input as well. As a result, the Designer returns a model that describes all the system reactions and can be straightforwardly encoded in SBML format. Afterward, circuits can be analyzed inside the Desktop Simulator – which also incorporates the Hy3S (34) algorithm – whereas the Wiki provides details about biochemical reactions and their kinetic rates.

Even broader appears to be the scope of the Clotho infrastructure under development at University of California, Berkeley (http://www.clothocad.org/). Actually, Clotho is the name of the core application that manages both the exchange of data among the internal tools and the communication with several databases. Gene circuit configurations are computed via a formal language, Eugene, and can be modified within a visualization tool, Spectacles. Furthermore, as a novelty with respect to all other previously described tools, circuit schemes are the input for Kepler, a software that guides a liquid-handling robot in the physical implementation of the DNA sequences realized in silico (see Table 2 for an outline of the principal characteristics of design tools for synthetic gene circuits).

### 2.2. Formal Languages

One of the main hurdles in modeling biochemical networks is the so-called *combinatorial explosion* of the species and the interactions required to properly reproduce the system dynamics. For instance, a protein showing $n$ docking sites for its ligand exists, theoretically, in at least $2^n$ states. Thus, the exhaustive description of the interaction between two such proteins might require $2^{2n}$ reactions. Moreover, the whole system might contain several other proteins of the same kind, and the overall network would be merely intractable. Hence,

**Table 2**
**Comparison of the synthetic biology tools for visual design**

| Tool | Drag & Drop design | Database connection | Simulation environment | Formal language | Export formats |
|---|---|---|---|---|---|
| Biojade | Yes | Yes | Yes | – | – |
| ProMoT | Yes | No | No | MDL | SBML, Matlab |
| Asmparts | No | No | No | – | SBML |
| Virtual Parts | No | No | No | – | CellML |
| TinkerCell | Yes | Yes | Yes | Antimony | XML |
| SynBioSS | No | Yes | Yes | – | SBML, NetCDF |
| Clotho | Yes | Yes | No | Eugene | XML |

one is forced to make some assumptions that permit to reduce the system complexity drastically even though the model predictions become valid only for the particular cases, where all these assumptions are true. In order to cope with high complexity without recurring to rather strict assumptions, some new methods have been developed, originally, for the study of natural occurring systems and have been exported, in a second time, to synthetic (gene) networks. In particular, diverse formal languages have been extensively used because they permit a straightforward definition of biological modules as well.

Formal languages for *rule-based* modeling, for instance, represent protein–protein interactions as rules from which chemical reactions are derived. Importantly, the specification of only few rules generates large reaction networks. Most of the tools developed on these kind of languages have been applied, so far, to the study of signaling pathways. An example is BioNetGen (35, 36). This tool used, first, a text language, where rules were written in the same way as chemical reactions – though more general, since they allowed "wild cards" whenever a reactant or a product was not unequivocally defined. Afterward, BioNetGen has introduced a different notation, where proteins are graphs (whose nodes are specific protein domains) and interactions between proteins are rules for rewriting graphs. Within this framework, modules can be individual species (e.g., ligands and signaling proteins), multistate species (receptors), and complexes of multistate species (a two-receptor system). Interestingly, scripts for BioNetGen (and other formal languages as well) produce a complete list of all species and reactions only for ODE-based simulations. On the contrary, both species and reactions are added "on the fly", i.e., when they effectively appear during a system simulation, if a stochastic simulator has been chosen.

A similar rule-based approach is implemented in the Kappa Language (37). Here, however, proteins – as basic modules – are associated with *agents* and rules express the conditions under which agents can interact (notice that an analogous, agent-compliant modeling can be found in STOCHSIM (38)). Remarkably, the Kappa Language allows model reduction (39) after identifying particular complexes – called fragments – that interact independently of each other.

Another interesting tool, Biochemical Abstract Machine (BIOCHAM) (40), flanks – as a novelty – a rule-based approach with a Computational Tree Logic (CTL)-based query language that permits to get, rapidly, qualitative information about the system without the need for dynamic simulations. Finally, similar formal languages have been realized by adapting *process algebras* (normally used to analyze computer performances) to biological systems: it is the case of $\pi$-Calculus (41) and Bio-PEPA (42).

Currently, none of these languages is used for the design and modeling of synthetic gene networks. The module description they provide hardly matches the requirement of part composability. Moreover, gene circuits engineered so far are too small to show any "combinatorial explosion" and, anyway, they can be rather faithfully described by Hill kinetics that reduces the number of species and interactions to be considered. Hence, new formal languages for synthetic biology have been developed de novo.

We have already encountered Antimony, the language of TinkerCell. Thanks to a rather simple syntax, Antimony allows an intuitive specification of both biochemical reactions and interactions between regulatory factors and their DNA targets. Furthermore, modules – with their own input/output – can be easily defined and interconnected and important features like cell compartments, events, differential equations, kinetic rates, and assignment rules are also available. Importantly, Standard Biological Parts, once defined, can be gathered into transcription units just by writing their names (together with the link operator) in the desired order. Finally, Antimony handles SBML format both to import and export models.

A slightly different syntax is provided in Genetic Engineering of living Cells (GEC) (43). This language allows a highly abstract description of gene circuits, where parts are associated with *properties* that specify, for instance, the kind of regulation (positive or negative) on a promoter, binding and unbinding rate of a transcription factor, protein produced by a gene, and translation rate of an RBS. Modules are made of sequences of parts and contain – if necessary – the definition of new reactions and constraints on some kinetic rates. Parts and modules are assembled together by means of link operators (like in Antimony), but modules need compatible input/output to interact with one another. Importantly, GEC demands connections to two local databases, one for genetic parts and the other for protein–protein interactions. In this way, an abstract model is converted into all the possible combinations of parts and reactions (stored into the databases) that match the model requirements. It should be noted that a simulator – developed ad hoc – is provided with the language as well. Otherwise, circuits can be exported into SBML and simulated by a third-party tool.

An alternative application of a formal language to gene circuit design is realized in GenoCAD (44, 45). This Web-based tool has introduced, in synthetic biology, the concept of *grammar* as a mean to (visually) assemble DNA sequences. Basically, a grammar specifies a set of rules for the composition of gene parts. In GenoCAD, several grammars are available (e.g., one is for *E. coli* and another for yeast) and associated with different part libraries. At each step, the grammar defines the types of parts that can be added to the existing sequence and a set of physical components retrieved from the library. Furthermore, grammars can parse and check the validity of preexisting DNA sequences as well (see (46) for a recent Web service devoted to formal verification of gene regulatory networks). Overall, GenoCAD realizes a very elegant approach that – thanks to the Web interface – can be easily handled by the users.

In the end of this section, I want to point out that, instead of developing languages ad hoc for synthetic biology, a modular, formal description of synthetic gene networks can be achieved via tools and formats proper for engineering. As we have previously seen, Marchisio and Stelling (19) modeled synthetic systems in MDL, the language of ProMoT. MDL was developed for process engineering and successively applied to biochemical systems (47) since it allows a clear definition of independent, interacting modules. Other examples of Lisp-based languages currently used to model biological systems are Proto (48) and Little b (49). The former comes from *amorphous* computation (used in simulation of emergent phenomena) and can be applied to biochemical networks as long as they have a large diameter and a small neighborhood size; the latter, in contrast, has been set up for computational system biology and could potentially be useful also for synthetic system modeling.

### 2.3. Automatic Design

Small gene networks, which are made of a handful of transcription units, can be easily designed in a drag & drop way. This procedure, however, becomes too slow for more intricate circuits that are preferably modeled with the syntax of a formal language, whereas circuit visualization (or direct optimization via tools, like RoVerGeNe (50)) gets useful only in a second time to arrange network details. Nevertheless, as we have previously seen, formal languages can build only variants of the same circuits, i.e., they enumerate all the possible combinations of real parts available in a given database but do not provide the modeler with any algorithm to compute alternative solutions or optimize the current one. A drastic improvement may be achieved through methods and tools for gene circuit automatic design, where circuit structure – and kinetic parameter values – is inferred directly form a target behavior. To this purpose, François and Hakim (51) applied an evolutionary algorithm on a set of small, independent, randomly chosen networks in order to build oscillators and toggle switches. In this work, basic network components are genes and proteins (but not Standard Biological Parts). Translation is modeled as a single-step event in compliance with mass action kinetics and proteins either activate or repress gene expression. Moreover, they can dimerize and are degraded in different ways. At each evolutionary step, different mutations are introduced in the network collection: modification of a kinetic constant or a degradation rate; insertion of a new gene; addition of a new gene–protein interaction, or a new posttranslational reaction. Then, a score, which quantifies how close the network is to the target behavior, is assigned to every circuit and selection is performed. Finally, at the end of the whole evolutionary procedure networks, which contain redundant genes and proteins, undergo a *pruning* operation that returns core circuits that, for oscillators and toggle switches, are made of just two/three genes.

Paladugu et al. (52) chose, as well, an evolutionary procedure to automatically build the same families of small gene circuits with the inclusion of filters. Here, beside mass action, networks are described also according to Michelis–Menten and Hill kinetics. Furthermore, new interactions and mutations (like gene and reaction deletion) are taken into account and pruning is performed with Jarnac (13). Overall, this work points out how the efficiency of an evolutionary algorithm can change by considering both different kinetics (for the circuit modeling) and object functions (to determine the network fitness).

An analogous, theoretical framework is set up with Genetdes (53), where, however, networks evolve via Simulated Annealing (54). As in the previous works, gene expression is entirely regulated by transcription factors and mutations act on kinetic parameters (more probable), genes, and gene regulations. Interestingly, thanks to a library of logic promoters, Genetdes was used to compute new configurations (up to three/four genes) of basic logic gates and the latch bistable circuit (55).

Slightly different appears OptCircuit (56), a tool either for the de novo generation or the only redesign of genetic networks. Here, circuit components are no longer abstract transcription units but real bacterial promoters and coding regions. Furthermore, circuit design is treated as a Mixed Integer Dynamic Optimization problem (MIDO), where the input (small networks) is chosen every time ad hoc – in dependence on the target behavior – and successively "evolved" into a family of properly working circuits via changes in genes' number and connectivity and in the kinetic parameter values.

OptCircuit has been tested on rather complex circuits (requiring at least five genes) like a decoder and a band detector.

Overall, these methods lie on a brute-force approach that employs an optimization procedure both on circuit structure and kinetic parameter values. Since the CPU time required by optimization algorithms grows exponentially with the number of input variables, biochemical models for automatically generated gene circuits have to be very simple. This implies coarse-grained modularity, where Standard Biological Parts are lumped into genes and important reactions and species are omitted, which is apparent when translation is reduced to a single-step event.

In contrast to this approach, Marchisio and Stelling (57) showed that gene *digital* circuits can be automatically designed without the need for any optimization algorithm on the circuit structure. In their method, they employ the Karnaugh map algorithm (58) – commonly used in electronics – to convert a truth table, which completely describes the circuit input/output relation, into a Boolean formula, from which the circuit scheme follows straight-forwardly.

Furthermore, a set of parameter values, which assure a proper reproduction of the truth table entries, can be derived on the basis of simple biological considerations and the optimization of the circuit performance is confined to the single gate that produces the circuit outputs, a fluorescent protein. In addition, Boolean gates are no longer designed only on promoters, but also on RBSs by controlling translation both with sRNAs, which bind their mRNA target sequences, and chemicals inducing or inhibiting riboswitches.

On the whole, this new procedure – implemented as an add-on for ProMoT – requires as an input only the circuit truth table and then generates several circuit schemes (the most complex of which can be made of more than ten transcription units). They are ranked according to a *complexity score* that depends on the number of regulatory factors employed and reflects the effort for a wet-lab circuit implementation. A solution, chosen by the user, is finally encoded in an MDL file after assembling Standard Biological Parts into Boolean gates and linking gates (and pools) to close the circuit. Since no stochastic algorithm is invoked, the overall CPU time is normally lower than ten seconds.

### 2.4. Part and Protein Design

As we have seen, the first step for the in silico gene circuit design makes use of abstract parts, where it is possible to change at will kinetic parameter values until the network reaches the desired performance. Afterward, DNA sequences can be either chosen from a database or de novo designed, if none of the available parts appears to be satisfactory. Synthetic RBSs, for instance, can be engineered via the Web server application RBS calculator (http://voigtlab.ucsf.edu/software). This software estimates the RBS translation initiation rate and, at present, is the only example of computational tool that specifically targets a Standard Biological Part category.

Other software allows various in silico DNA manipulations. Gene Design 3.0 (59), for instance, is a suite of Web applications that range from simple sequence analysis to restriction site addition and subtraction, codon juggling, random sequence generation, and reverse translation. Gene Designer (60), in contrast, permits to combine genetic elements in a drag & drop fashion. Furthermore, codon optimization for protein expression, calculation of the annealing temperature of DNA short fragments, and primer generations are some of the operations provided by this program. Similar capabilities are supplied in Gene Morphing System (GeMS) (61), another suite of tools that can be used, for instance, for stem loop determination and design verification. On the contrary, DNAWorks (62), Oligo Cuts (http://gcat.davidson.edu/IGEM06/oligo.html), and Gene2Oligo (63) furnish different approaches to the design and the analysis of oligonucleotides, whereas Zinc Finger Tools (64) find, on a DNA sequence, special domains that can be exploited as a binding site for Zinc Finger Proteins acting either as transcription factors or as nucleases. Finally, NEBcutter (65) is a Web server that performs a detailed research of restriction sites along a DNA sequence and CloneQC (66) may be used for the quality control of sequenced clones.

Folding, hybridization, and melting pathways of single- and double-stranded nucleic acids are other extremely complicate tasks and require dedicated tools. UNAFold (67) and the corresponding Web server, DINAMelt (68), perform this kind of analysis both on DNA and RNA sequences, whereas the only target of the VIENNA RNA package (69, 70) is the ribonucleic acid. Notice that data on RNA structures and functionalities can be retrieved via RNA FRABASE 2.0 (71).

Beside DNA and RNA, proteins are other fundamental components of gene networks since they regulate transcription by promoter binding. A remarkable suite of tools for computational protein engineering is Rosetta (72). Here, protein three-dimensional (3D) structure is derived via ab initio calculation. This permits to address the design of specific transcription factors and enzymes. Other software programs employ different methods to determine protein 3D structures. Raptor (73), for instance, makes use of protein threading with linear programming; HHpred (74), in contrast, exploits homology detection and structure comparison via hidden Markov models. In a similar way, comparative modeling is adopted in Modeller (75), whereas protein design is based on fixed backbones in EGAD (76) and requires molecular dynamics within PROTDES (77). It should be noted that not only the structure, but also the functions of a protein can be computationally derived: as examples, we indicate the Web servers Protein Function Predictions (PFP) (78, 79), Phyre (80), and I-TASSER (81).

Since chemicals are often necessary to either activate or inactivate transcription factors, simulations of the docking of small molecules to protein may be of extreme importance for synthetic gene circuit design. Ligand–protein docking can be carried out either with Rosetta or Autodock 4.2 (82). Furthermore, one can recur to the HexServer (83) to get estimations on the location of protein-docking sites. Interestingly, simulations with some of the previously described tools may be integrated by using Biskit (84), a Python library that works as a software platform to define workflows for structural bioinformatics.

## 3. Conclusion

In very few years, computational synthetic biology has made rapid progress and appears to be in continuous evolution. Tools provide not only intuitive GUIs for the visual assembly of (abstract) Standard Biological Parts into gene circuits, but also simulation environments and connection to part databases. Furthermore, diverse formal languages have been proposed: they permit, together with new algorithms for automatic circuit design, to face in the near future the in silico implementation of much more complex networks.

However, a commonly accepted part standardization is still missing even though interesting ideas have been put forward by Canton et al. (85) and the Synthetic Biology Open Language (SBOL) project (http://openwetware.org/wiki/The_BioBricks_Foundation:Standards/Technical/Exchange), aimed at defining new conventions for part description an characterization, appears promising.

Moreover, synthetic biology is no longer confined to bacterial transcription networks. For instance, Boolean gates (86), a memory device (87), and a tunable oscillator (88) have been built in mammalian cells. In addition, new RNA-based parts and devices come from natural (89) and synthetic translational control mechanisms, like riboswitches (90), ribozymes (91), and antiswitches (92), as pointed out by Win et al. (93). Furthermore, new results in the reengineering of signaling pathways (94) suggest that a new branch of synthetic biology may be de novo founded on proteins. This would require the definition of corresponding new modules and intramodule interactions (95) as well as the construction of an infrastructure similar to the MIT Registry (96).

Hence, in the near future, in silico synthetic biology will be asked to provide new concepts and methods to cope with the expansion of experimental synthetic biology into translational and signaling networks. Moreover, thanks to recent advance both in computational (97) and experimental (98) DNA assembly automation techniques, current available tools will be soon connected to robot devices – as already initiated with Clotho – in order to drive directly the synthesis of physical gene circuits.

## References


1. Nagel, L. W., and Pederson, D. O. (1973). SPICE (Simulation Program with Integrated Circuit Emphasis). *Memorandum No. ERL-M382*, University of California, Berkley.
2. Endy, D. (2005) Foundations for engineering biology, *Nature* **438**, 449–453.
3. Szallasi, Z., Stelling, J., and Periwal (2006) *System Modeling in Cellular Biology*, The MIT Press, Cambridge, MA.
4. Funahashi, A., Tanimura, N., Morohashi, M., and Kitano, H. (2003) CellDesigner: a process diagram editor for gene-regulatory and biochemical networks, *BIOSILICO* **1**, 159–162.
5. Funahashi, A., Matsuoka, Y., Jourako, A., Morohashi, M., Kikuchi, N., and Kitano, H. CellDesigner 3.5: A Versatile Modeling Tool for Biochemical Networks, In *Proceedings of the IEEE 8* ed., pp 1254–1265.
6. Hucka, M., and others. (2003) The systems biology markup language (SBML): a medium for representation and exchange of biochemical network models, *Bioinformatics* **19**, 524–531.
7. Novere, N. L., Bornstein, B., Broicher, A., Courtot, M. l., Donizelli, M., Dharuri, H., Li, L., Sauro, H., Schilstra, M., Shapiro, B., Snoep, J. L., and Hucka, M. (2006) BioModels Database: a free, centralized database of curated, published, quantitative kinetic models of biochemical and cellular systems, *Nucleic Acids Res* **34**, D689-D691.
8. Shannon, P., Markiel, A., Ozier, O., Baliga, N. S., Wang, J. T., Ramage, D., Amin, N., Schwikowski, B., and Ideker, T. (2003) Cytoscape: a software environment for integrated models of biomolecular interaction networks, *Genome Res* **13**, 2498–2504.
9. Martin, A., Ochagavia, M. E., Rabasa, L. C., Miranda, J., de Cossio, J. F., and Bringas, R. BisoGenet: a new tool for gene network building, visualization and analysis, *BMC Bioinformatics* **11**:91.
10. Kurata, H., Matoba, N., and Shimizu, N. (2003) CADLIVE for constructing a large-scale biochemical network based on a simulation-directed notation and its application to yeast cell cycle, *Nucleic Acids Res* **31**, 4071–4084.
11. Kurata, H., Masaki, K., Sumida, Y., and Iwasaki, R. (2005) CADLIVE dynamic simulator: direct link of biochemical networks to dynamic models, *Genome Res* **15**, 590–600.
12. Loew, L. M., and Schaff, J. C. (2001) The Virtual Cell: a software environment for computational cell biology, *Trends Biotechnol* **19**, 401–406.
13. Sauro, H. M., Hucka, M., Finney, A., Wellock, C., Bolouri, H., Doyle, J., and Kitano, H. (2003) Next generation simulation tools: the Systems Biology Workbench and BioSPICE integration, *OMICS* **7**, 355–372.
14. Longabaugh, W. J. R., Davidson, E. H., and Bolouri, H. (2005) Computational representation of developmental genetic regulatory networks, *Dev Biol* **283**, 1–16.
15. Goler, J. A. (2004) BioJADE: A Design and Simulation Tool for Synthetic Biological Systems. AITR-2004-003. MIT, Cambridge, MA.
16. Goler, J. A., Bramlett, B. W., and Peccoud, J. (2008) Genetic design: rising above the sequence, *Trends Biotechnol* **26**, 538–544.
17. Kosuri, S., Kelly, J., and Endy, D. (2007) TABASCO: A single molecule, base-pair resolved gene expression simulator, *BMC Bioinformatics* **8**:480.
18. Endy, D., Deese, I., and the, M. I. T. S. B. W. G. (2005). Adventures in synthetic biology. Appeared in Foundations for engineering biology p449, *Nature*, **438**, 449–453.
19. Marchisio, M. A., and Stelling, J. (2008) Computational design of synthetic gene circuits with composable parts, *Bioinformatics* **24**, 1903–1910.
20. Marchisio, M. A., and Stelling, J. Synthetic gene network computational design, In *Proc. IEEE International Symposium on Circuits and Systems, ISCAS 2009*, pp 309–312.
21. Serganov, A., and Patel, D. J. (2007) Ribozymes, riboswitches and beyond: regulation of gene expression without proteins, *Nat Rev Genet* **8**, 776–790.
22. Mirschel, S., Steinmetz, K., Rempel, M., Ginkel, M., and Gilles, E. D. (2009) PROMOT: modular modeling for systems biology, *Bioinformatics* **25**, 687–689.
23. Ginkel, M., Kremling, A., Nutsch, T., Rehner, R., and Gilles, E. D. (2003) Modular modeling of cellular systems with ProMoT/Diva, *Bioinformatics* **19**, 1169–1176.
24. Hoops, S., Sahle, S., Gauges, R., Lee, C., Pahle, Jr., Simus, N., Singhal, M., Xu, L., Mendes, P., and Kummer, U. (2006) COPASI – a COmplex PAthway SImulator, *Bioinformatics* **22**, 3067–3074.
25. Rodrigo, G., Carrera, J., and Jaramillo, A. (2008) Asmparts: assembly of biological model parts, *Syst Synth Biol* **1**, 167–170.
26. Cooling, M. T., Rouilly, V., Misirli, G., Lawson, J., Yu, T., Hallinan, J., and Wipat, A. Standard virtual biological parts: a repository of modular modeling components for synthetic biology, *Bioinformatics* **26**, 925–931.

27. Rouilly, V., Canton, B., Nielsen, P., and Kitney, R. (2007) Registry of BioBrick models using CellML, *BMC System Biology* **1** *Suppl 1*, 79–80.

28. Cuellar, A. A., Lloyd, C. M., Nielsen, P. F., Bullivant, D. P., Nickerson, D. P., and Hunter, P. J. (2003) An Overview of CellML 1.1, a Biological Model Description Language, *Simulation* **79**, 740–747.

29. Garny, A., Nickerson, D. P., Cooper, J., dos Santos, R. W., Miller, A. K., McKeever, S., Nielsen, P. M. F., and Hunter, P. J. (2008) CellML and associated tools and techniques, *Philos Transact A Math Phys Eng Sci* **366**, 3017–3043.

30. Chandran, D., Bergmann, F. T., and Sauro, H. M. (2009) TinkerCell: modular CAD tool for synthetic biology, *J Biol Eng* **3**:19.

31. Smith, L. P., Bergmann, F. T., Chandran, D., and Sauro, H. M. (2009) Antimony: a modular model definition language, *Bioinformatics* **25**, 2452–2454.

32. Hill, A. D., Tomshine, J. R., Weeding, E. M. B., Sotiropoulos, V., and Kaznessis, Y. N. (2008) SynBioSS: the synthetic biology modeling suite, *Bioinformatics* **24**, 2551–2553.

33. Weeding, E., Houle, J., and Kaznessis, Y. N. SynBioSS designer: a web-based tool for the automated generation of kinetic models for synthetic biological constructs, *Brief Bioinform* **11**, 394–402.

34. Salis, H., Sotiropoulos, V., and Kaznessis, Y. N. (2006) Multiscale Hy3S: hybrid stochastic simulation for supercomputers, *BMC Bioinformatics* 7:93.

35. Blinov, M. L., Faeder, J. R., Goldstein, B., and Hlavacek, W. S. (2004) BioNetGen: software for rule-based modeling of signal transduction based on the interactions of molecular domains, *Bioinformatics* **20**, 3289–3291.

36. Faeder, J. R., Blinov, M. L., and Hlavacek, W. S. (2009) Rule-based modeling of biochemical systems with BioNetGen, *Methods Mol Biol* **500**, 113–167.

37. Danos, V., and Laneve, C. (2004) Formal molecular biology, *Theoretical Computational Science* **325**, 69–110.

38. Novere, N. L., and Shimizu, T. S. (2001) STOCHSIM: modelling of stochastic biomolecular processes, *Bioinformatics* **17**, 575–576.

39. Feret, Jrm., Danos, V., Krivine, J., Harmer, R., and Fontana, W. (2009) Internal coarse-graining of molecular systems, *Proc Natl Acad Sci USA* **106**, 6453–6458.

40. Fages, F., Soliman, S., and Chabrier-Rivier, N. (2004) Modelling and querying interaction networks in the biochemical abstract machine BIOCHAM, *Journal of Biological Physics and Chemistry* **4**, 64–73.

41. Regev, A., Silverman, W., and Shapiro, E. (2001) Representation and simulation of biochemical processes using the pi-calculus process algebra, *Pac Symp Biocomput*, 459–470.

42. Ciocchetta, F., and Hillston, J. (2008) Bio-PEPA: An Extension of the Process Algebra PEPA for Biochemical Networks, *Electronic Notes in Theoretical Computational Science* **194**, 103–117.

43. Pedersen, M., and Phillips, A. (2009) Towards programming languages for genetic engineering of living cells, *J R Soc Interface* **6** *Suppl 4*, S437–S450.

44. Cai, Y., Hartnett, B., Gustafsson, C., and Peccoud, J. (2007) A syntactic model to design and verify synthetic genetic constructs derived from standard biological parts, *Bioinformatics* **23**, 2760–2767.

45. Czar, M. J., Cai, Y., and Peccoud, J. (2009) Writing DNA with GenoCAD, *Nucleic Acids Res* **37**, W40-W47.

46. Monteiro, P. T., Dumas, E., Besson, B., Mateescu, R., Page, M., Freitas, A. T., and de Jong, H. (2009) A service-oriented architecture for integrating the modeling and formal verification of genetic regulatory networks, *BMC Bioinformatics* **10**:450.

47. Saez-Rodriguez, J., Gayer, S., Ginkel, M., and Gilles, E. D. (2008) Automatic decomposition of kinetic models of signaling networks minimizing the retroactivity among modules, *Bioinformatics* **24**, 213–219.

48. Beal, J., and Bachrach, J. (2006) Infrastructure for Engineered Emergence on Sensor/Actuator Networks, *IEEE Intelligent Systems* **21** *(2)*, 10–19.

49. Mallavarapu, A., Thomson, M., Ullian, B., and Gunawardena, J. (2009) Programming with models: modularity and abstraction provide powerful capabilities for systems biology, *J R Soc Interface* **6**, 257–270.

50. Batt, Gg., Yordanov, B., Weiss, R., and Belta, C. (2007) Robustness analysis and tuning of synthetic gene networks, *Bioinformatics* **23**, 2415–2422.

51. Francois, P., and Hakim, V. (2004) Design of genetic networks with specified functions by evolution in silico, *Proc Natl Acad Sci USA* **101**, 580–585.

52. Paladugu, S. R., Chickarmane, V., Deckard, A., Frumkin, J. P., McCormack, M., and Sauro, H. M. (2006) In silico evolution of functional modules in biochemical networks, *Syst Biol (Stevenage)* **153**, 223–235.

53. Rodrigo, G., Carrera, J., and Jaramillo, A. (2007) Genetdes: automatic design of

transcriptional networks, *Bioinformatics* **23**, 1857–1858.

54. Kirkpatrick, S., Gelatt, C. D., and Vecchi, M. P. (1983) Optimization by Simulated Annealing, *Science* **220**, 671–680.
55. Rodrigo, G., and Jaramillo, A. (2007) Computational design of digital and memory biological devices, *Syst Synth Biol* **1**, 183–195.
56. Dasika, M. S., and Maranas, C. D. (2008) OptCircuit: an optimization based method for computational design of genetic circuits, *BMC Syst Biol* **2**:24.
57. Marchisio, M. A., and Stelling, Jr. (2011) Automatic Design of Digital Synthetic Gene Circuits, *PLoS Comput Biol* 7, e1001083.
58. Karnaugh, M. (1953) The Map Method for Synthesis of Combinational Logic Circuits, *Transactions of the American Institute of Electrical Engineers* **72** *(9)*, 593–599.
59. Richardson, S. M., Nunley, P. W., Yarrington, R. M., Boeke, J. D., and Bader, J. S. GeneDesign 3.0 is an updated synthetic biology toolkit, *Nucleic Acids Res* **38**, 2603–2606.
60. Villalobos, A., Ness, J. E., Gustafsson, C., Minshull, J., and Govindarajan, S. (2006) Gene Designer: a synthetic biology tool for constructing artificial DNA segments, *BMC Bioinformatics* 7:285.
61. Jayaraj, S., Reid, R., and Santi, D. V. (2005) GeMS: an advanced software package for designing synthetic genes, *Nucleic Acids Res* **33**, 3011–3016.
62. Hoover, D. M., and Lubkowski, J. (2002) DNAWorks: an automated method for designing oligonucleotides for PCR-based gene synthesis, *Nucleic Acids Res* **30** *(10)*:e43.
63. Rouillard, J.-M., Lee, W., Truan, G., Gao, X., Zhou, X., and Gulari, E. (2004) Gene2Oligo: oligonucleotide design for in vitro gene synthesis, *Nucleic Acids Res* **32**, W176-W180.
64. Mandell, J. G., and Barbas, C. F. (2006) Zinc Finger Tools: custom DNA-binding domains for transcription factors and nucleases, *Nucleic Acids Res* **34**, W516-W523.
65. Vincze, T., Posfai, J., and Roberts, R. J. (2003) NEBcutter: A program to cleave DNA with restriction enzymes, *Nucleic Acids Res* **31**, 3688–3691.
66. Lee, P. A., Dymond, J. S., Scheifele, L. Z., Richardson, S. M., Foelber, K. J., Boeke, J. D., and Bader, J. S. CLONEQC: lightweight sequence verification for synthetic biology, *Nucleic Acids Res* **38**, 2617–2623.
67. Markham, N. R., and Zuker, M. (2008) UNAFold: software for nucleic acid folding and hybridization, *Methods Mol Biol* **453**, 3–31.
68. Markham, N. R., and Zuker, M. (2005) DINAMelt web server for nucleic acid melting prediction, *Nucleic Acids Res* **33**, W577-W581.
69. Hofacker, I. L., Fontana, W., Stadler, P. F., Bonhoeffer, L. S., Tacker, M., and Schuster, P. (1994) Fast Folding and Comparision of RNA Secondary Strucutre, *Monatshefte fuer Chemie* **125**, 167–188.
70. Hofacker, I. L. (2003) Vienna RNA secondary structure server, *Nucleic Acids Res* **31**, 3429–3431.
71. Popenda, M., Szachniuk, M., Blazewicz, M., Wasik, S., Burke, E. K., Blazewicz, J., and Adamiak, R. W. RNA FRABASE 2.0: an advanced web-accessible database with the capacity to search the three-dimensional fragments within RNA structures, *BMC Bioinformatics* **11**:231.
72. Simons, K. T., Bonneau, R., Ruczinski, I., and Baker, D. (1999) Ab initio protein structure prediction of CASP III targets using ROSETTA, *Proteins Suppl* **3**, 171–176.
73. Xu, J., Li, M., Kim, D., and Xu, Y. (2003) RAPTOR: optimal protein threading by linear programming, *J Bioinform Comput Biol* **1**, 95–117.
74. Soeding, J., Biegert, A., and Lupas, A. N. (2005) The HHpred interactive server for protein homology detection and structure prediction, *Nucleic Acids Res* **33**, W244-W248.
75. Eswar, N., Webb, B., Marti-Renom, M. A., Madhusudhan, M. S., Eramian, D., Shen, M.-Y., Pieper, U., and Sali, A. (2007) Comparative protein structure modeling using MODELLER, *Curr Protoc Protein Sci Chapter 2:Unit 2.9.*
76. Pokala, N., and Handel, T. M. (2005) Energy functions for protein design: adjustment with protein-protein complex affinities, models for the unfolded state, and negative design of solubility and specificity, *J Mol Biol* **347**, 203–227.
77. Suarez, M., Tortosa, P., and Jaramillo, A. (2008) PROTDES: CHARMM toolbox for computational protein design, *Syst Synth Biol* **2**, 105–113.
78. Hawkins, T., Luban, S., and Kihara, D. (2006) Enhanced automated function prediction using distantly related sequences and contextual association by PFP, *Protein Sci* **15**, 1550–1556.
79. Hawkins, T., Chitale, M., Luban, S., and Kihara, D. (2009) PFP: Automated prediction of gene ontology functional annotations with confidence scores using protein sequence data, *Proteins* **74**, 566–582.
80. Kelley, L. A., and Sternberg, M. J. E. (2009) Protein structure prediction on the Web: a case study using the Phyre server, *Nat Protoc* **4**, 363–371.

81. Roy, A., Kucukural, A., and Zhang, Y. I-TASSER: a unified platform for automated protein structure and function prediction, *Nat Protoc* **5**, 725–738.
82. Morris, G. M., Huey, R., Lindstrom, W., Sanner, M. F., Belew, R. K., Goodsell, D. S., and Olson, A. J. (2009) AutoDock4 and AutoDockTools4: Automated docking with selective receptor flexibility, *J Comput Chem* **30**, 2785–2791.
83. Macindoe, G., Mavridis, L., Venkatraman, V., Devignes, M.-D., and Ritchie, D. W. HexServer: an FFT-based protein docking server powered by graphics processors, *Nucleic Acids Res* **38** *Suppl*, W445-W449.
84. Gruenberg, R., Nilges, M., and Leckner, J. (2007) Biskit-a software platform for structural bioinformatics, *Bioinformatics* **23**, 769–770.
85. Canton, B., Labno, A., and Endy, D. (2008) Refinement and standardization of synthetic biological parts and devices, *Nat Biotechnol* **26**, 787–793.
86. Kramer, B. P., Fischer, C., and Fussenegger, M. (2004) BioLogic gates enable logical transcription control in mammalian cells, *Biotechnol Bioeng* **87**, 478–484.
87. Ajo-Franklin, C. M., Drubin, D. A., Eskin, J. A., Gee, E. P. S., Landgraf, D., Phillips, I., and Silver, P. A. (2007) Rational design of memory in eukaryotic cells, *Genes Dev* **21**, 2271–2276.
88. Tigges, M., Marquez-Lago, T. T., Stelling, J., and Fussenegger, M. (2009) A tunable synthetic mammalian oscillator, *Nature* **457**, 309–312.
89. Isaacs, F. J., Dwyer, D. J., and Collins, J. J. (2006) RNA synthetic biology, *Nat Biotechnol* **24**, 545–554.
90. Breaker, R. R. (2008) Complex riboswitches, *Science* **319**, 1795–1797.
91. Win, M. N., and Smolke, C. D. (2008) Higher-order cellular information processing with synthetic RNA devices, *Science* **322**, 456–460.
92. Bayer, T. S., and Smolke, C. D. (2005) Programmable ligand-controlled riboregulators of eukaryotic gene expression, *Nat Biotechnol* **23**, 337–343.
93. Win, M. N., Liang, J. C., and Smolke, C. D. (2009) Frameworks for programming biological function through RNA parts and devices, *Chem Biol* **16**, 298–310.
94. Marchisio, M. A., and Rudolf, F. (2011) Synthetic biosensing systems, *Int J Biochem Cell Biol* **43**, 310–319.
95. Gruenberg, R., and Serrano, L. (2010) Strategies for protein synthetic biology, *Nucleic Acids Res* **38**, 2663–2675.
96. Rialle, S. p., Felicori, L., Dias-Lopes, C., Peres, S., Atia, S. E., Thierry, A. R., Amar, P., and Molina, F. BioNetCAD: design, simulation and experimental validation of synthetic biochemical networks, *Bioinformatics* **26***(18)*, 2298–2304.
97. Densmore, D., Hsiau, T. H. C., Kittleson, J. T., DeLoache, W., Batten, C., and Anderson, J. C. Algorithms for automated DNA assembly, *Nucleic Acids Res* **38**, 2607–2616.
98. Linshiz, G., Yehezkel, T. B., Kaplan, S., Gronau, I., Ravid, S., Adar, R., and Shapiro, E. (2008) Recursive construction of perfect DNA molecules from imperfect oligonucleotides, *Mol Syst Biol* **4**:191.

# Chapter 2

# Standardization in Synthetic Biology


## Kristian M. Müller and Katja M. Arndt



## Abstract

Synthetic Biology is founded on the idea that complex biological systems are built most effectively when the task is divided in abstracted layers and all required components are readily available and well-described. This requires interdisciplinary collaboration at several levels and a common understanding of the functioning of each component. Standardization of the physical composition and the description of each part is required as well as a controlled vocabulary to aid design and ensure interoperability. Here, we describe standardization initiatives from several disciplines, which can contribute to Synthetic Biology. We provide examples of the concerted standardization efforts of the BioBricks Foundation comprising the request for comments (RFC) and the Registry of Standardized Biological parts as well as the international Genetically Engineered Machine (iGEM) competition.



**Key words:** Biobricks, Biological parts, iGEM, Registry, Standard


## 1. Introduction

### *1.1. Development and Scope of Synthetic Biology*

A short look into history provides a basis for the understanding of the scope of standardization in Synthetic Biology. The term Synthetic Biology has been phrased several times and evolved along the development of natural sciences. In the early twentieth century molecular disassembly of life took off and biochemists, such as Jaques Loeb, declared the "abiogenesis" as research goal. In 1912, Stéphane Leduc from France published a book titled "La Biologie Synthétique" (1). In the 1970s, the genetic revolution started and an editorial of Gene stated in 1978: "into the new era of 'Synthetic Biology' where not only existing genes are described and analyzed but also new gene arrangements can be constructed" (2). Yet, in both historic events, the term Synthetic Biology did not make it into scientific mainstream. The terms "biochemistry" or "physiologic chemistry" and later "molecular biology" and "genetic engineering" better described


Wilfried Weber and Martin Fussenegger (eds.), *Synthetic Gene Networks: Methods and Protocols*, Methods in Molecular Biology, vol. 813, DOI 10.1007/978-1-61779-412-4_2, © Springer Science+Business Media, LLC 2012

the research and technology of that time. Finally, at the end of the twentieth century, biologists and engineers again were searching for a new name for their technological ambitions to engineer biological systems. Many aspects of bioscience had matured, and after about 10 years the term Synthetic Biology is finding widespread acceptance. It describes an interdisciplinary approach, which is nurtured from many scientific sources. One uniting theme is a quote attributed to Richard Feynman "What I cannot create, I do not understand," which appeals to researchers and engineers alike. A workgroup initiated by the European Commission developed the definition: "Synthetic Biology aims to engineer and study biological systems that do not exist as such in nature, and use this approach for (i) achieving better understanding of life processes, (ii) generating and assembling functional modular components, and (iii) developing novel applications or processes" (3).

This open view appeals many protagonists. At the same time, it precludes an exhaustive description of all potential standards. This chapter touches standardization as seen in nature (Subheading 2), biology (Subheading 3.1), non-clinical and clinical research (Subheading 3.2), engineering (Subheading 4), and last but not least Synthetic Biology (Subheading 5). Strategies enabling language control and electronic processing are covered (Subheading 6).

### *1.2. Standards and Standard Setting*

Setting standards can serve several purposes. Standards can improve communication, compatibility, interchangeability, reproducibility, effective use, fitness for use, safety, quality assurance, and ultimately consumer protection and environmental protection. Standards can be grouped by type. Technical standards provide specifications, which could be input and output values, procedural standards may describe a workflow, which can be standard operating procedures, and classification standards may provide definitions, which can be a controlled vocabulary. Standards can be grouped by accessibility. Proprietary standards are not freely available, whereas open standards are freely available. In a more narrow definition, a "real" standard has to be freely available. The connotation open standard emphasizes this point and describes a standard which is easily accessible and can be further developed by a wider community. Standards can be described by their development and requirements for use. Standards can be voluntary or mandatory. A de facto standard is a convention accepted by a wider public. A de jure standard is enforced by governments.

A standard is typically described in a document and a standardized material adheres to production or specification standards. In contrast, a standard reference material is a controlled artifact used for calibration.

The standard setting process can be manifold. Standards can emerge unintended as a result of usage habits. They can be set by standards organizations or bodies often affiliated with a state or by standard-developing organizations. Rules and goals of standards vary widely with respect to stakeholders, beneficiaries, formal process, and voting process. In basic research, standards are typically a community effort and users vote with their feet. In clinic studies, governmental regulation prevails.

## 2. Standardization in Nature

### *2.1. Scale of Natural Standards*

Although one might think of nature as wild, untamed, and chaotic, it should be noted that nature evolved standards and solved compatibility issues over space, scale, and time, which humans never will surpass. From a single-cell organism, such as *E. coli*, with a length of about 1 μm to a blue whale with a length of about $3 \times 10^7$ μm or the human brain with more than $5 \times 10^{10}$ cells, nature uses the same set of building blocks of (deoxy-)ribonucleic acids, amino acids, sugars, and lipids. The basics of this standard set emerged about 1.7–2.5 billion years ago when the common ancestor of cellular life diverged into the kingdoms of life. Many building blocks and metabolic processes as we know them today are stably implemented since hundreds of millions of years. Building blocks enabling self-assembling over billion-fold size and data storage for the construction of these building blocks over more than billion years are at present beyond imagination for human technology and provide the gold standard for any sustainable technology. The compatibility between organisms is nicely demonstrated when a gene from a jelly fish is plugged into a bacterium or human cell resulting in the expression of a fluorescent protein. A more natural but less-desired example of compatibility is the rapid spread of antibiotic resistance among microbial pathogens. This admiration for the results of evolution does not preclude attempts to further standardize biological components and systems for current engineering needs in Synthetic Biology.

### *2.2. Reasons for Natural Standards*

Standardization seems to be a human idea rooted in intelligent perception and technical mastery. However, when it comes to biology, standardization can be seen as a necessity of life. Sustaining life on earth requires the endless recycling of all chemical compounds required for life. It is fair to assume that recycling is most energy efficient if all life forms use a similar or even the same set of components and, as a consequence, life that deviates too much from the common chemical basis will sooner or later become extinct. A second argument for the observed standardization in biology is the low likelihood of invention of new functional entities. The latter argument was overcome by humans, who began to create "electronic life" and run the large-scale experiment of how biologic and "electronic life" can coexist.

### *2.3. Abstraction in Nature*

One major goal of standardization is to provide components with defined input and output characteristics that can be used without hassling with the details of each subsystem. At the organismic level, this works perfectly well and even stone-age people were able to handle organisms in an abstracted fashion. They added soil and water as input to crop seeds, which are the gray box, and harvested the output. Furthermore, no biological education is required to combine two interspecies abstraction layers, for example by using the "crop-system" output as "cow-system" input to obtain milk. To some extent, biology's intrinsic compatibility can be used at the genetic level by plugging a gene from one organism into another. However, typically above the layer of basic building blocks and the very similar use of the genetic code, compatibility breaks down. This is even true when organisms use the same principles for metabolism and signaling. At the long-known organismic level, this is seen by the fact that breeding a horse with a horse gives propagable offspring, breeding a horse and a donkey gives almost exclusively infertile offspring, and breeding a horse with a cow does not work. This is why setting standards at the molecular biology level is a major task.

## 3. Standardization in Biological Research and Development

### *3.1. Standardization in Basic Biological Research*

With increasing amounts of biological data and even more with the advent of large-scale data generation experiments, it became apparent that the diversity and incompleteness of classical method publishing provides a roadblock for data comparison and data processing. Electronic databases curated by teams also functioned as nucleation factor for standards in reporting methods and experimental results. A discipline, which early adopted rules of data reporting and submission to a database, was structural biology, namely, crystallography. A formal guideline from 1989 was widely adopted also by publishers (4). DNA sequence data were also among the first deposited in databases (5). In this case, only the final sequence was stored and recently new reporting standards in genomics were demanded (6).

In the meantime, several life science subdisciplines realized the need for traceable experiments and standardized sharing of information and consequently established rules to report experiments. One umbrella effort is the "Minimum Information about a Biomedical or Biological Investigation" (MIBBI) foundry (http://mibbi.org/) (7). An early effort was aimed at the comparability of microarray experiments (MIAME) (8).

A prominent set of rules with implications for Synthetic Biology has been established by the Human Proteome Organization (HUPO, http://www.hupo.org). These guidelines specifically include the proteomics aspect of an experiment and are, thus, named minimal information about a proteomics experiment (MIAPE) (9). For standardization, the HUPO's Proteomics Standards Initiative (HUPO-PSI) has developed a formal process for setting guidelines (10). Within the MIAPE guidelines, there are further detailed guidelines, such as for the reporting of the use of gel electrophoresis (11), for reporting a molecular interaction experiment (MIMIx) (12), or reporting the use of mass spectrometry (13). Further rules are established for experiments, such as chromatography (14), proteomics images (15), and proteomics capillary electrophoresis (16).

### 3.2. Standardization of Research in Non-clinical and Clinical Research

Human health is a major concern for governments, and thus in research and development aiming at substance release or human therapy adherence to standards is required by law. A well-known standard is the "good laboratory practice" (GLP) applying to regulated non-clinical research and development. Requirements are coded, e.g., in the USA in the Code of Federal Regulations (CFR), by European Union directives, and by the OECD. According to the OECD GLP principles, which acquired wide international acceptance, GLP is a "a quality system concerned with the organizational process and the conditions under which non-clinical health and environmental safety studies are planned, performed, monitored, recorded, archived and reported" (17).

A further standard widespread in pharmacological research is the pharmacopeia, which is a reference work composed of monographs describing substances, materials, and dosage forms. For example, the European Pharmacopoeia (Ph. Eur.) is maintained by the European Directorate for the Quality of Medicines and HealthCare (EDQM), which is a Directorate of the Council of Europe (http://www.edqm.eu). EDQM also provides the International Standards for Antibiotics (ISA) and is responsible for the establishment and distribution of WHO's International Chemical Reference Standards (ICRS). Unfortunately, online access has to be purchased.

Standardization efforts have also been developed for clinical experiments. The "Consolidated Standards of Reporting Trials" (CONSORT, http://www.consort-statement.org/) comprises several initiatives to avoid problems arising from inadequate reporting of randomized, controlled trials (18).

#### 3.2.1. Recording Data According to GLP

The following protocol is adapted from ref. 17 with minor modifications to ensure correctness. Data are necessary for the reconstruction of the study after completion.

The following items have to be recorded:

1. WHAT was done: A description of the conduct of the technique, including the results of the observation or measurement, and demonstrating that the actions required by the protocol were carried out.

2. HOW it was done: The data should indicate that they were collected and recorded in accordance with the methods set out in the standard operating procedures (SOPs) and protocol or indicate where there were deviations from these instructions.
3. WHEN the work was performed: A demonstration of compliance with the schedule defined in the protocol. This is done by recording the date and, if necessary, the time that the procedure was conducted (see Note 1).
4. WHO performed the work: The data should clearly identify who was responsible for carrying out the procedure and recording the data. If more than one person was involved in a procedure, this should be recorded in the data, along with an identification of the responsibilities of each.

The generated data should be recorded as follows:

1. Directly: The first written records are considered to constitute the raw data and must be retained. Records should not be made on scraps of paper and then transcribed into a final form. When data are acquired directly by computer, the raw data are considered to be the electronic medium. For data derived from equipment, the raw data may be a direct printout (or trace) or in an electronic form.
2. Promptly: Data must be recorded as the operation is done. It is not acceptable to make the record sometime after the task has been completed.
3. Accurately: This is most important as the accuracy underpins the scientific interpretation and integrity of the study.
4. Legibly: Data that cannot be read are useless, and records that are difficult to decipher raise doubts as to their credibility.
5. Indelibly: Only indelible and waterproof pens and robust machine printouts are permitted. In case of question, additional authorized photocopies might be used for storage.

*In addition, the data need to be*:

6. Identified: The study number, animal number, etc. must be recorded with the data in order to ensure that data mix-ups do not occur. The parameter evaluated must be identified.
7. Signed: Accountability is one of the basic tenets of GLP, hence the need for a record of who did every job on a study.
8. Dated: The date of each signature demonstrates that the procedure was conducted and recorded at the correct point in the study.
9. Corrected only with given reasons: Records may require alteration from time to time, but a clear audit trail is needed showing why a change was carried out, when and by whom.

Data should be recorded and organized in a way that facilitates both the making of the record and the performance of subsequent processes (e.g., data entry, reporting, audit, archiving). Data should be recorded in a logical way, and duplication should be avoided wherever possible. Preformatted documents assist in this by encouraging staff to record all the data required, without forgetting any. A clear structure for the study file, defined upfront, helps to organize and archive the documents as they are produced in real time, preventing loss and facilitating reference between records.

If computerized systems are used to generate or process data, correct operation of data handling and safe storage of the system have to be validated.

## 4. Standardization in Technical Disciplines

### 4.1. Comparison of Synthetic Biology with Mechanic and Electrical Engineering

In the realm of Synthetic Biology, the commercial success of mechanical engineering and electronics has often attributed to standardization and has been praised as the desired goal of Synthetic Biology. In early days of the reincarnation of Synthetic Biology, cells have been depicted as bags of screws, in which just the thread of the screws has to be adjusted to obtain an easy-to-manage system. This primitive view was probably necessary to get the attention of engineers, although it frustrated biologists who abandoned this scheme decades ago. Above the screw, the transistor and the many commercial, well-defined electronic parts were used to symbolize standardization. Plugging genes in a genome, like transistors on a circuit board, is a tempting idea, but unfortunately countermanded by the next level of comparison. At present, many Synthetic Biology talks and some publications compare a cell with a computer (19). Here, a physical layer of, e.g., proteins, is compared to transistors, and biochemical reactions are compared to gates and pathways to modules, such as integrated circuits. The good aspect of this comparison is that biologists and engineers can agree on such an analogy. The bad news is that this comparison leaves little room for off-the-shelf standard parts. Computer chip development is in reach only for a few companies worldwide, and production takes place in foundries costing billions of dollar/euro to set up. At the same time, the typical life span of a computer chip series, which requires highly adapted boards and brides for its life cycle, is about 2 years. Synthetic Biology parts, which become incompatible within the life span of mouse, are unattractive. Within this comparison, Synthetic Biology should position its standards between the off-the-shelf transistor and the modern computer industry.

Even if engineering does not provide a gold standard, Synthetic Biology has much to learn from these disciplines. Foremost, this is the divide and conquers approach to a complex problem by defining abstraction layers, a working set of conditions and specifications for each required part. Most impressive to biologists is the separation of hard- and software in the computer industry. And the soon-to-come programmer of an organism cannot be knowledgeable about all the enzymatic details in a cell.

Another role model of technical engineering is public acceptance. If engineers would develop standards, which would allow cars to drive people twice as fast to work as today, most people would cheer the development. If synthetic biologist would develop strategies that make horses run twice as fast to work, most people would be scared. Thus, the process of setting biological standards has to take into account public opinion and norms, which often entail a love for nature conservatism and to some extent biophobia.

### 4.2. Standardization According to ISO

A well-known organization developing technical standards is the "International Organization for Standardization" (ISO, http://www.iso.org). Although standards published by ISO are widely adopted by the industry because legislative bodies tend to encourage their use, they are not available for many researchers due to the costs for the documents. Some ISO publications provide rules, which could be of interest to scientists. For example, ISO 11737–1:2006 "specifies requirements and provides guidance for the enumeration and microbial characterization of the population of viable micro-organisms on or in a medical device, component, raw material or package." A standard published by the ISO has to pass several hurdles. From the first concept to publication, the following statuses might apply: Preliminary Work Item, New Proposal/New Work Item Proposal, Approved New Work Item, Working Draft, Committee Draft, Final Committee Draft, Draft International Standard, Final Draft International Standard, Proof of a new International Standard, and International Standard. Taken together, these standards are useful for specific tests, but the ISO process of defining standards seems not suited for Synthetic Biology.

### 4.3. Standardization for the Internet

For outsiders, it may seem off topic to discuss standardization of the Internet community. However, within the Synthetic Biology community, the fast-paced development of hardware, software, and the Internet is seen as a paradigm for success. In addition, Randy Rettberg, the prominent figure behind iGEM, has ties to the Internet development. Hence, it is worthwhile to take a look at this example. The Internet is mainly standardized by the "Internet Engineering Task Force" (IETF, http://www.ietf.org), which publishes memorandums named RFC. RFCs cover a wide range of topics comprising methods, behaviors, research, or innovations related to the Internet and its connected systems. The basic definition of the IETF standards process is described in RFC 2026 and its amendments. One section of this RFC is highlighted on the IETF Web site (http://www.ietf.org/rfc/rfc2026.txt) and is given below, as it fits to the interests and problems of Synthetic Biology (see Subheading 5.1):

"In outline, the process of creating an Internet Standard is straightforward: a specification undergoes a period of development and several iterations of review by the Internet community and revision based upon experience, is adopted as a Standard by the appropriate body [...) and is published. In practice, the process is more complicated, due to (i) the difficulty of creating specifications of high technical quality; (ii) the need to consider the interests of all of the affected parties; (iii) the importance of establishing widespread community consensus; and (iv) the difficulty of evaluating the utility of a particular specification for the Internet community [...) The goals of the Internet Standards Process are (i) technical excellence; (ii) prior implementation and testing; (iii) clear, concise, and easily understood documentation;(iv) openness and fairness; and (v) timeliness.

[...) The goal of technical competence, the requirement for prior implementation and testing, and the need to allow all interested parties to comment all require significant time and effort. On the other hand, today's rapid development of networking technology demands timely development of standards. The Internet Standards Process is intended to balance these conflicting goals. The process is believed to be as short and simple as possible without sacrificing technical excellence, thorough testing before adoption of a standard, or openness and fairness."

The IETF RFCs can have increasing degrees of importance. The status can range from "informational," "experimental," "best current practice," and "standards track." Standards track documents are further divided into "proposed standard," "draft standard," and "Internet standard document." Documents, which become obsolete, may get the status "historic."

## 5. Standardization in Synthetic Biology

### *5.1. Overview*

Since Synthetic Biology builds on several life science disciplines and adds its own dimension, it also requires a broad set of standards to enable productive collaboration. Biology alone encompasses diverse samples ranging from small molecules to ecosystems and diverse technologies ranging from biophysics to behavioral analyses. The bold goals of Synthetic Biology aim at providing effective solutions for health, agriculture, renewable resources, and energy based on bioengineering and bioinspired engineering. The current status of Synthetic Biology can still be seen as the enabling technology build-up phase. Most efforts focus on establishing basic genetic engineering in a standardized fashion and demonstrating understanding of biology by bottom-up engineering. In contrast to purely man-made technology, often manifold examples of potential solutions can be found in natural systems; yet harnessing, tweaking, and managing these natural solutions provide a formidable challenge. Next to genetic engineering comes the engineering of other biomolecules, most notably protein engineering. Depending on the set goal, requirements for standardization can vary significantly. For example, the current challenge for biopharmaceuticals often falls within protein engineering while providing future solutions for regenerative medicine requires manipulation of cells and organs. As a consequence of this diversity, Synthetic Biology heavily relies on standards developed in various disciplines. The main task for Synthetic Biology is to clarify the abstraction hierarchies and rules of how to bring together these aspects or to reformat previously developed standards to fit the umbrella idea of building biological systems.

Apart from this, the classical standardization requirements for science apply:

i) Results need to be comparable between samples.
ii) Results need to be comparable between labs.
iii) Experiments need to be repeatable in several labs.
iv) Researchers need to use the same language/naming.
v) Data need to be amenable to electronic data processing.

### 5.2. The Biobricks Foundation, the RFC Process, iGEM, and the Registry of Standard Biological Parts

The most prominent efforts to organize and standardize Synthetic Biology are the formation of the Biobricks Foundation (BBF, http://biobricks.org), the "iGEM" competition (http://www.igem.org), and the "Registry of Standard Biological Parts" (http://partsregistry.org) (see Note 2). This concept evolves around the idea of developing and providing standardized genetics parts named "BioBricks," which can be assembled in a more or less black box fashion to build more and more complex systems. In short:

1. The BioBricks Foundation is the legal owner of the "BioBricks"™ trademark.
2. The student competition iGEM educates future researchers in the use of BioBricks and is at the same time the main provider of BioBricks.
3. The Registry stores the BioBrick DNA and maintains a database with the description of the BioBricks.

There are many positive and interesting aspects of these initiatives, yet here the focus is on the standardization emerging from these efforts. After a few years with no formal process, the BioBricks

Foundation set up an RFC procedure following the style of the IETF RFCs described above (Subheading 4.3). This form of standardization is still in an early stage. Browsing the first set of BBF RFCs, one realizes that the proposed standardizations are mostly linked to issues arising within the iGEM competition. As BBF RFCs are not filtered before publishing, quality and relevance also vary significantly.

### 5.3. The Current Status of BioBrick Cloning

An ongoing discussion within the iGEM community dwells on what DNA assembly methods should or can be used. The current standard is an idempotent cloning strategy named BBF RFC 10 (20) placing each part between a defined prefix and suffix sequence, which each contains two restriction enzymes (EcoRI, XbaI and SpeI, PstI), which are consequently not allowed within the part.

Several modifications of the original standard (RFC 10) exist adapting it to fusion proteins (e.g., RFC 23, RFC 25) or alternative restriction enzymes (e.g., RFC 20, RFC 21). Since the iGEM competition pressures participating teams to adhere to these standards, a lot of effort is invested on making parts compatible with this standard. The seemingly simple issue of how to clone illuminates problems which plague standardization in molecular biology. Choices of RFC 10 are debated, for example, in RFC 12, RFC 20, RFC 21, RFC23, and RFC54. Everybody realizes that automation of a standard cloning method would significantly speed up the process and would have the potential to eliminate human errors. Unfortunately, a "one method fits all," which is inexpensive, sequence independent, and scalable from small parts of a few base pairs to hundreds of kilo bases, remains elusive. Depending on the task, which can be cloning of fusion proteins requiring scar-free cloning, cloning of highly repetitive or self-complementary sequence stalling most known polymerase, or cloning of very large fragments and genome integration, tailored solutions are required, which at present are either performed manually or need individual adaption of a pipetting robot. In some labs, high-content cloning has been established for specific tasks, yet these methods often rely on expensive and/or extensive robots or proprietary vector systems and are thus hard to transfer between labs. In gene synthesis companies, the task of generating genes has reached a higher level of automation and prices dropped significantly to render many material transfer agreements unnecessary. Nonetheless, gene synthesis is not yet a replacement for efficient cloning. In summary, most labs would welcome speedier and more compatible cloning, but the job gets done and thus the activation barrier for a change is very high. Therefore, it requires a catalyst like iGEM to switch to a more common standard, even more so if the standard is far from being perfect and, from an individual standpoint, provides often only minor speed advantages.

*5.3.1. Cloning According to BBF RFC 10 and RFC 25*

The first BioBrick assembly standard was set before the development of the RFC process and is described in a working paper (21), which was then converted to RFC 10 (20). RFC 25 (22) is an extension of RFC 10 enabling the idempotent cloning of fusion proteins while maintaining full compatibility to RFC 10. The following protocol lists the RFC 10 and RFC 25 requirements for cloning BioBricks (Figs. 1 and 2).

RFC 10 Design Requirements

1. The BioBrick DNA sequence must not contain the restriction endonuclease sites: EcoRI GAATTC, XbaI TCTAGA, SpeI ACTAGT, PstI CTGCAG, NotI GCGGCCGC.
2. At the 5′-end, the BioBrick has to be flanked (see Notes 3–5):
   (a) For non protein-coding Biobricks by the prefix 5′-GAATTC GCGGCCGC T TCTAGA G-3′
   (b) For protein-coding Biobricks by the prefix 5′-GAATTC GCGGCCGC T TCTAG-3′; in this case, the start codon must be ATG
3. At the 3′-end, the BioBrick has to be flanked by the suffix 5′-T ACTAGT A GCGGCCG CTGCAG-3′ (see Notes 6 and 7).

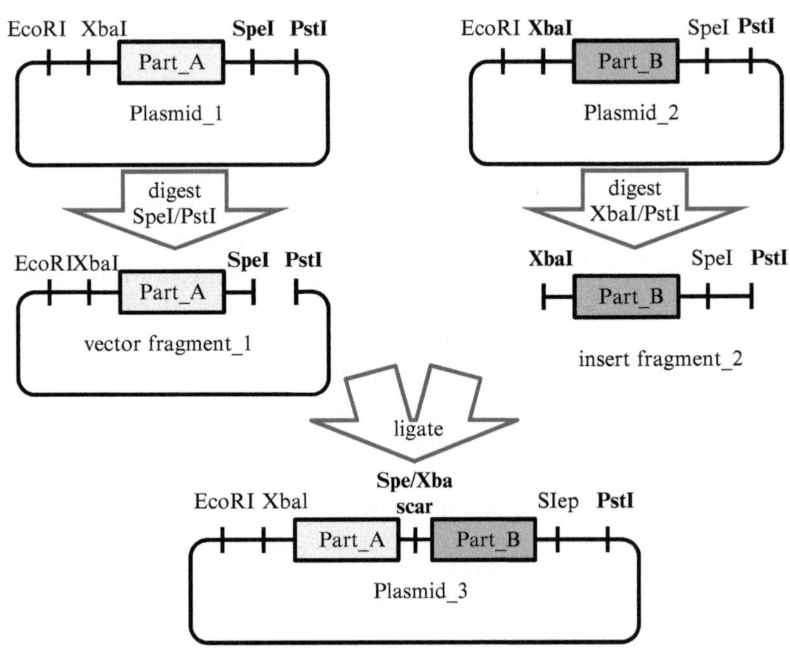


Fig. 1. Scheme of the RFC 10 idempotent cloning strategy. Plasmid_1 containing Part_A is digested with SpeI and PstI. Plasmid_2 containing Part_B is digested with XbaI and PstI. Vector fragment and insert are ligated and yield Plasmid_3 containing the composite part of A and B interspersed by the cloning scar and flanked by the standard prefix and suffix.

```
EcoRI    NotI     XbaI   NgoMIV        AgeI        SpeI     NotI      PstI
  |        |        |         |            |           |        |          |
  GAATTCgcggccgctTCTAGAtgGCCGGCnnnnnnACCGGTtaatACTAGTagcggccgCTGCAG
5'---------+---------+---------+---------+---------+---------+----- 3'
  CTTAAGcgccggcgaAGATCTacCGGCCGnnnnnnTGGCCAattaTGATCAtcgccggcGACGTC
c    I  R  G  R  F  *  M  A  G  ?  ?  T  G  *  Y  *  *  R  P  L  Q
```

Fig. 2. Sequence of prefix and suffix according to RFC 25. The *underlined part* is identical to the RFC10 expression part definition. The additional restriction sites for NgoMIV and AgeI flank the part sequence, which is symbolized with "n." Reading frame c of the shown sequence segment is given. The in-frame restriction sites code for AlaGly and ThrGly. Ligation of the compatible ends of AgeI and NgoMIV generates a scar coding for ThrGly.

4. BioBrick parts must be supplied in plasmids compatible with Registry and assembly procedures and may be pSB-series plasmids or other compliant plasmids.
5. It is recommended that the standard sequencing primers VF2: TGCCACCTGACGTCTAAGAA and VR: ATTACCGCCTTTGAGTGAGC work with the plasmid.
6. For PCR amplification of a gene with simultaneous addition of the prefix and suffix sequences, the following PCR primers are recommended: forward 5′-TCT TCG AAT TCG CGG CCG CTT CTA GAG AG-(18–24 bases of the gene)-3′ and reverse 5′-GTT TCT TCC TGC AGC GGC CGC TAC TAG TA-(18–24 bases of the gene)-3′.

**Cloning According to RFC 10**

The idea of RFC 10 is an idempotent strategy in which always the same restriction sites are used to clone into an existing BioBrick assembly. The scheme is illustrated in Fig. 1. The key feature is that XbaI and SpeI generate compatible ends, which upon ligation generate a cloning "scar" which cannot be cut by either enzyme.

For cloning BioBrick B after BioBrick A in a plasmid, follow this protocol:

1. Digest the plasmid with SpeI and PstI and isolate the vector fragment.
2. Digest the insert with XbaI and PstI, and isolate the insert fragment.
3. Mix insert with vector and ligate.
4. Transform in *E. coli* and plate on agar containing the appropriate antibiotic for the vector plasmid.

For cloning an assembly of BioBrick A-B-C-D, follow this protocol:

1. Assemble A-B and C-D according to the above protocol.
2. Clone C-D behind A-B following the above protocol.

#### RFC 25 Design Requirements and Cloning According to RFC 25

RFC 25 is an extension of RFC 10 utilizing the same idempotent strategy with two additional restriction enzymes located within the RFC 10 restriction sites and in frame with a coding sequence (Fig. 2). The two endonucleases used are NgoMIV and AgeI. The NgoMIV overhang (G'CCGG,C) is compatible with the AgeI overhang (A'CCGG,T) and ligation results in the scar ACCGCC coding for the two amino acids Thr and Gly, which are suited as linkers between protein domains.

The RFC 25 design requirements are the following:

1. Restriction sites for EcoRI, NotI, XbaI, NgoMIV, AgeI, SpeI, NotI, and PstI must not be present in the gene of interest.
2. At the 5′ end of the gene of interest, the following prefix is required: gaattcgcggccgcttctagatggccggc.
3. At the 3′-end of the gene of interest, the following suffix is required: accggttaatactagtagcggccgctgcag.
4. If an N-terminal extension is not intended, e.g., if a protein fold does not permit N-terminal extension, an RFC10 expression part prefix can also be used without the NgoMIV addition.

Cloning of fusion proteins composed of protein domain A followed by domain B according to RFC 25 follows these steps:

1. Digest plasmid containing gene A with AgeI and PstI and isolate the vector fragment.
2. Digest plasmid containing gene B with NgoMIV and PstI and isolate the insert fragment.
3. Ligate the insert and vector fragment
4. Transform in *E. coli* and plate on agar containing the appropriate antibiotic for the vector plasmid.

### 5.4. Description of Parts

A statistic of 2009 published on the Registry Web site lists over 5,000 available BioBricks and states that almost 1,700 plasmids containing parts were sent to each team participating in iGEM. The Registry Web site provides access to the parts by type, e.g., promoter or protein coding; device type, e.g., reporter, inverter; or function, e.g., cell–cell signaling. The DNA sequence of a part and information entered by the users can be downloaded. As most parts are submitted by students under time pressure and up to iGEM 2010 the submission process of the part documentation and annotation was manual, the documentation quality of the parts varies significantly. An example of 116 part submissions from the authors' 2010 iGEM team can be found at http://partsregistry.org/Viral_vectors. The total number of parts in the Registry is impressive for a student competition, and it should not be compared to the number of clones offered by some companies which provide clones of each open reading frame from several organisms. The value of the Registry lies in the fact that over thousands of students get trained in the use each year, the growth with each new student generation, and that over several years documentation and raking of parts evolve. Any lab can contribute to the Registry, although so far only few labs take this opportunity. A new development is the setup of the "Biofab" (http://www.biofab.org), which in the future might contribute parts in a larger scale.

The exact definition of a BioBrick and the handling of variations are not completely settled. At present, only DNA as plasmid is stored in the Registry. The BioBrick DNA sequence is given without the cloning prefix and suffix, since these depend on the cloning strategy chosen by the submitting party. Prefix and suffix are given as additional information. This is a very reasonable decision, but has in specific cases major consequences that are often overlooked, again pointing to problems once it comes to the details of standardization. Different choices of prefix and suffix can add cloning scars to protein-coding BioBricks, which are translated in amino acids. Different terminal ends of a protein can significantly influence translation, degradation (23), and stability (24) of proteins. As an undesired consequence, the description of a protein-coding part cannot be easily abstracted from the cloning strategy. A workaround would be to establish a subclassification at the level of single nucleotides. Together with the Registry, the authors' iGEM 2010 team introduced such versioning for the standard Registry cloning vector, albeit only for mutations within the vector.

Once mutations of parts get versions, a typical biological question arises: Which degree of homology or identity justifies bundling in one part or separation in different parts? There is no objective rule to such a decision, and it is typically made based on the use in the community. For example, the fluorescent proteins GFP, YFP, Venus, CFP, and Cerulean got different names despite the high sequence identity because one important property is different, whereas most enzymes get the same name even at low identity level as long as they catalyze the same reaction. Mutations are inherent to biology and the research on alignments and annotations from protein domains to genomes is dealing with this. As functional tolerance to mutations is not yet fully understood, Synthetic Biology should provide part descriptions at the most detailed sequence level available. The need for an exact and easy accessible sequence for each experiment has resulted in the initiative for the essential information for synthetic DNA sequences (25).

Proposals of how a part description can look like have been published. Canto and colleagues define a standard part as "a genetically encoded object that performs a biological function and that has been engineered to meet specified design or performance requirements" (26). They provide an example of how the function of a device genetically composed of a Tet promoter, a ribosome-binding site, the luxR gene, a terminator, and lux-dependent promoter and transformed into *E. coli* can be characterized. On one page, a prototypical "datasheet," key measurements, and parameters of the input and output characteristics are listed. In addition, the authors provide the general setup of the experiment, such as the host strain, media, and temperature. The visual setup of a facts sheet is good for manually browsing parts; however, in the long run, a more detailed and computer readable format will prevail in case of complex systems. Several shortcomings of an attempt to describe a complex biological experiment with all possible variations can be listed, but the importance of an aggregated dataset should be emphasized and arbitrary but reasonable choices are a hallmark of many biological experiments and accepted as long as they serve the intended use (27).

#### *5.4.1. Requirements for Parts Descriptions*

The description of a BioBrick or any other component for Synthetic Biology is not settled and to some extent always depends on the intended use. As mentioned above, the listing of prefix and suffix sequences which potentially influence function is still debated. The following list, thus, can only serve as guideline:

1. Release of information:
   (a) Information and data should be available in a widely accepted nonproprietary format on an open-access platform.
   (b) To the widest extent possible, information should be made available in human- and machine-readable form.
   (c) Information should be given in a short abstract and a full description.
   (d) Information should be given in a standardized vocabulary if available.
   (e) Information should be linked to the existing knowledge, such as databases.
2. Physical composition:
   (a) History and methods of assembly.
   (b) Complete DNA sequence (should be verified by sequencing).
   (c) Annotation of the DNA sequence to the best-available knowledge, including features of the DNA, mRNA, and protein level, if applicable.
3. Experimental environment:
   (a) Description of the host cell/chassis.
   (b) Detailed description of relevant equipment and materials.
   (c) Detailed description of the experimental parameters.
   (d) Validation and calibration status of equipment.
   (e) Contribution of each involved person.

4. Experimental results:
   (a) Raw data if processing obscures deduction.
   (b) Detailed information of how data are processed.
   (c) Processed and aggregated data.
   (d) Visualization of data.
5. Deduced function and parameters for higher abstraction layers:
   (a) Short description of intended implementation.
   (b) The exact amount of parameters required to implement and control the part.
   (c) Working range of the parameters.
   (d) Estimate of robustness with respect to parameter constellation and time.
6. Security and safety:
   (a) Known and anticipated security issues.
   (b) Known and anticipated safety issues.
7. Legal:
   (a) If available, information on the patent situation.
   (b) Any intellectual property claims made by the authors.
   (c) Conflict of interest by the authors.

## 6. Language and Data Processing Standards in Biological Sciences

A very important aspect for enabling large-scale data integration is the development of a common, standardized language. Data interoperability is achieved using formal representations of facts as well as concepts, linking and processing the data. These ontologies are widespread in biological sciences. As an umbrella organization to coordinate ontologies development, the Open Biological and Biomedical Ontologies foundry (http://www.obofoundry.org/) was formed (28). One of the most prominent ontologies is the gene ontology (GO, http://www.geneontology.org/) (29). This ontology "covers three domains: (1) *cellular component*, the parts of a cell or its extracellular environment; (2) *molecular function*, the elemental activities of a gene product at the molecular level, such as binding or catalysis; and (3) *biological process*, operations or sets of molecular events with a defined beginning and end, pertinent to the functioning of integrated living units: cells, tissues, organs, and organisms."

At present, many ontologies are encoded following the Web Ontology Language (OWL, http://www.w3.org/TR/owl-ref) based on the Resource Description Framework (RDF, http://www.w3.org/TR/rdf-concepts/) and using the Extensible Markup

Language (XML, http://www.w3.org/TR/xml/) serialization. These semantic Web standards have been developed by the World Wide Web Consortium (W3C, http://www.w3.org/) using their standardization process (http://www.w3.org/standards/about.html).

Next to genetics, formal data handling and ontologies are also widespread in systems biology. The System Biology Markup Language (SBML, http://sbml.org) (30) defines an interchange format for computer models of biological processes. The Biopax project (http://www.biopax.org) (31) aims to enable integration, exchange, visualization, and analysis of biological pathway data. An overview of various standards in systems biology is given in (32). The link between systems and Synthetic Biology is evident since both disciplines require an abstracted description of biological processes (33).Currently, there are over 150 software packages supporting SBML, which are also useful for Synthetic Biology (34) and can be used for computational design and engineering of biological parts (34, 35).

Specific for Synthetic Biology is the development of the Synthetic Biology Open Language (SBOL, http://www.sbolstandard.org/). A proposal for the use of RDF in the frame of Synthetic Biology has been published as a BBF RFC 30 (36) and BioBrick information has been made available in SBOL (37).

## 7. Legal Considerations

For international DNA repositories collecting and distributing community-supplied material, the legal framework is of outmost importance (38). The patent situation in biotechnology is hard to oversee. Many researchers, particularly in academic settings, are not sufficiently aware of the patent situation and do not have access to intellectual property specialists. The BioBricks foundation has set up a draft contributor agreement for submitting parts to the Registry (39).

## 8. Notes

1. For certain procedures (for example, sampling in a toxicokinetic study), very exact timing is necessary and the data must demonstrate that the schedule was strictly followed.
2. All three initiatives have mainly been developed by a group of people colocated at the MIT at the time of the conceptual layout.

3. For protein-coding RFC 10 BioBricks, the A nucleotide of the start codon becomes part of the XbaI restriction site.
4. The additional T between NotI and XbaI site is inserted to prevent inadvertent creation of EcoBI or EcoKI methylation sites (21); EcoKI methylase modifies adenine residues in the sequences AAC(N6)GTGC and GCAC(N6)GTT.
5. The additional G after the XbaI site is inserted to prevent creation of a Dam methylation site; Dam methylates the $N^6$-position of the adenine in the sequence GATC. Thus, an XbaI site followed by TC could be blocked.
6. If constructing a primer, this sequence must be reverse complemented.
7. The T preceding the SpeI is a required base.

## References


1. Leduc, S. (1912) La biologie synthétique, étude de biophysique (Poinat, A., Ed.). Paris.
2. Szybalski, W., and Skalka, A. (1978) Nobel prizes and restriction enzymes., *Gene 4*, 181–2.
3. Gaisser, S., Reiss, T., Lunkes, A., Müller, K. M., and Bernauer, H. (2009) Making the most of synthetic biology. Strategies for synthetic biology development in Europe., *EMBO reports 10 Suppl 1*, S5-8.
4. Berman, H. M. (2008) The Protein Data Bank: a historical perspective., *Acta crystallographica. Section A, Foundations of crystallography.* International Union of Crystallography *64*, 88–95.
5. Strasser, B. J. (2008) Genetics. GenBank – Natural history in the 21st Century?, *Science (New York, N.Y.) 322*, 537–8.
6. Quackenbush, J. (2009) Data reporting standards: making the things we use better., *Genome medicine 1*, 111.
7. Kettner, C., Field, D., Sansone, S.-A., Taylor, C., Aerts, J., Binns, N., Blake, A., Britten, C. M., Marco, A. de, Fostel, J., Gaudet, P., González-Beltrán, A., Hardy, N., Hellemans, J., Hermjakob, H., Juty, N., Leebens-Mack, J., Maguire, E., Neumann, S., Orchard, S., Parkinson, H., Piel, W., Ranganathan, S., Rocca-Serra, P., Santarsiero, A., Shotton, D., Sterk, P., Untergasser, A., and Whetzel, P. L. (2010) Meeting Report from the Second "Minimum Information for Biological and Biomedical Investigations" (MIBBI) workshop., *Standards in genomic sciences 3*, 259–66.
8. Brazma, A., Hingamp, P., Quackenbush, J., Sherlock, G., Spellman, P., Stoeckert, C., Aach, J., Ansorge, W., Ball, C. A., Causton, H. C., Gaasterland, T., Glenisson, P., Holstege, F. C., Kim, I. F., Markowitz, V., Matese, J. C., Parkinson, H., Robinson, A., Sarkans, U., Schulze-Kremer, S., Stewart, J., Taylor, R., Vilo, J., and Vingron, M. (2001) Minimum information about a microarray experiment (MIAME)-toward standards for microarray data., *Nature genetics 29*, 365–71.
9. Taylor, C. F., Paton, N. W., Lilley, K. S., Binz, P., Jalain, Jr, R. K. J., R, A., Zhu, W., Apweiler, R., Aebersold, R., Deutsch, E. W., Dunn, M. J., Heck, A. J. R., Leitner, A., Macht, M., Mann, M., Martens, L., Neubert, T. A., Patterson, S. D., Ping, P., Seymour, S. L., Souda, P., Tsugita, A., Vandekerckhove, J., and Vondriska, T. M. (2007) The minimum information about a proteomics experiment (MIAPE), *Nature Biotechnology 25*, 887–893.
10. Vizcaíno, J. A., Martens, L., Hermjakob, H., Julian, R. K., and Paton, N. W. (2007) The PSI formal document process and its implementation on the PSI website., *Proteomics 7*, 2355–7.
11. Gibson, F., Anderson, L., Babnigg, G., Baker, M., Berth, M., Binz, P.-A., Borthwick, A., Cash, P., Day, B. W., Friedman, D. B., Garland, D., Gutstein, H. B., Hoogland, C., Jones, N. A., Khan, A., Klose, J., Lamond, A. I., Lemkin, P. F., Lilley, K. S., Minden, J., Morris, N. J., Paton, N. W., Pisano, M. R., Prime, J. E., Rabilloud, T., Stead, D. A., Taylor, C. F., Voshol, H., Wipat, A., and Jones, A. R. (2008) Guidelines for reporting the use of gel electrophoresis in proteomics., *Nature biotechnology 26*, 863–4.
12. Orchard, S., Salwinski, L., Kerrien, S., Montecchi-Palazzi, L., Oesterheld, M., Stümpflen, V., Ceol, A., Chatr-aryamontri, A., Armstrong, J., Woollard, P., Salama, J. J.,

Moore, S., Wojcik, J., Bader, G. D., Vidal, M., Cusick, M. E., Gerstein, M., Gavin, A.-C., Superti-Furga, G., Greenblatt, J., Bader, J., Uetz, P., Tyers, M., Legrain, P., Fields, S., Mulder, N., Gilson, M., Niepmann, M., Burgoon, L., De Las Rivas, J., Prieto, C., Perreau, V. M., Hogue, C., Mewes, H.-W., Apweiler, R., Xenarios, I., Eisenberg, D., Cesareni, G., and Hermjakob, H. (2007) The minimum information required for reporting a molecular interaction experiment (MIMIx)., *Nature biotechnology 25*, 894–8.

13. Taylor, C. F., Binz, P.-A., Aebersold, R., Affolter, M., Barkovich, R., Deutsch, E. W., Horn, D. M., Hühmer, A., Kussmann, M., Lilley, K., Macht, M., Mann, M., Müller, D., Neubert, T. A., Nickson, J., Patterson, S. D., Raso, R., Resing, K., Seymour, S. L., Tsugita, A., Xenarios, I., Zeng, R., and Julian, R. K. (2008) Guidelines for reporting the use of mass spectrometry in proteomics., *Nature biotechnology 26*, 860–1.
14. Jones, A. R., Carroll, K., Knight, D., MacLellan, K., Domann, P. J., Legido-Quigley, C., Huang, L., Smallshaw, L., Mirzaei, H., Shofstahl, J., and Paton, N. W. (2010) Guidelines for reporting the use of column chromatography in proteomics, *Nature Biotechnology*. Nature Publishing Group *28*, 654–654.
15. Hoogland, C., O Gorman, M., Bogard, P., Gibson, F., Berth, M., Cockell, S. J., Ekefjärd, A., Forsstrom-Olsson, O., Kapferer, A., Nilsson, M., Martínez-Bartolomé, S., Albar, J. P., Echevarría-Zomeño, S., Martínez-Gomariz, M., Joets, J., Binz, P. -A., Taylor, C. F., Dowsey, A., and Jones, A. R. (2010) Guidelines for reporting the use of gel image informatics in proteomics, *Nature Biotechnology*. Nature Publishing Group *28*, 655–656.
16. Domann, P. J., Akashi, S., Barbas, C., Huang, L., Lau, W., Legido-Quigley, C., McClean, S., Neusüß, C., Perrett, D., Quaglia, M., Rapp, E., Smallshaw, L., Smith, N. W., Smyth, W. F., and Taylor, C. F. (2010) Guidelines for reporting the use of capillary electrophoresis in proteomics, *Nature Biotechnology 28*, 654–655.
17. WHO. (2009) HANDBOOK GOOD LABORATORY PRACTICE (GLP) (Kioy, D., Long, D., Bhalla, S., and Seiler, J., Eds.).
18. Schulz, K. F., Altman, D. G., and Moher, D. (2010) CONSORT 2010 statement: updated guidelines for reporting parallel group randomised trials., *BMJ (Clinical research ed.) 340*, c332.
19. Andrianantoandro, E., Basu, S., Karig, D. K., and Weiss, R. (2006) Synthetic biology: new engineering rules for an emerging discipline, *Molecular Systems Biology* 1–14.
20. Knight, T. (2007) BBF RFC 10: Draft Standard for Biobrick Biological Parts, http://dspace.mit.edu/handle/1721.1/45138.
21. Knight, T. (2003) Idempotent Vector Design for Standard Assembly of Biobricks, http://dspace.mit.edu/handle/1721.1/21168 1–11.
22. Müller, K. M., Arndt, K. M., IGEM_Freiburg, 2007, and Grünberg, R. (2008) BBF RFC 25: Fusion Protein (Freiburg) BioBrick assembly standard, http://dspace.mit.edu/handle/1721.1/45140.
23. Hwang, C.-S., Shemorry, A., Auerbach, D., and Varshavsky, A. (2010) The N-end rule pathway is mediated by a complex of the RING-type Ubrl and HECT-type Ufd4 ubiquitin ligases., *Nature cell biology*. Nature Publishing Group *12*, 1177–85.
24. Hecky, J., and Müller, K. M. (2005) Structural perturbation and compensation by directed evolution at physiological temperature leads to thermostabilization of beta-lactamase., *Biochemistry 44*, 12640–54.
25. Peccoud, J., Anderson, J. C., Chandran, D., Densmore, D., Galdzicki, M., Lux, M. W., Rodriguez, C. A., Stan, G.-bart, and Sauro, H. M. (2011) Essential information for synthetic DNA sequences., *Nature biotechnology*. Nature Publishing Group *29*, 22; discussion 22–3.
26. Canton, B., Labno, A., and Endy, D. (2008) Refinement and standardization of synthetic biological parts and devices., *Nature Biotechnology 26*, 787–93.
27. Arkin, A. (2008) Setting the standard in synthetic biology., *Nature biotechnology 26*, 771–4.
28. Smith, B., Ashburner, M., Rosse, C., Bard, J., Bug, W., Ceusters, W., Goldberg, L. J., Eilbeck, K., Ireland, A., Mungall, C. J., Leontis, N., Rocca-Serra, P., Ruttenberg, A., Sansone, S.-A., Scheuermann, R. H., Shah, N., Whetzel, P. L., and Lewis, S. (2007) The OBO Foundry: coordinated evolution of ontologies to support biomedical data integration., *Nature biotechnology 25*, 1251–5.
29. Ashburner, M., Ball, C. A., Blake, J. A., Botstein, D., Butler, H., Cherry, J. M., Davis, A. P., Dolinski, K., Dwight, S. S., Eppig, J. T., Harris, M. A., Hill, D. P., Issel-Tarver, L., Kasarskis, A., Lewis, S., Matese, J. C., Richardson, J. E., Ringwald, M., Rubin, G. M., and Sherlock, G. (2000) Gene ontology: tool for the unification of biology. The Gene Ontology Consortium., *Nature genetics 25*, 25–9.
30. Hucka, M., Finney, A., Sauro, H. M., Bolouri, H., Doyle, J. C., Kitano, H., Arkin, A. P., Bornstein, B. J., Bray, D., Cornish-Bowden, A., Cuellar, A. A., Dronov, S., Gilles, E. D., Ginkel, M., Gor, V., Goryanin, I. I., Hedley, W.

J., Hodgman, T. C., Hofmeyr, J.-H., Hunter, P. J., Juty, N. S., Kasberger, J. L., Kremling, A., Kummer, U., Le Novère, N., Loew, L. M., Lucio, D., Mendes, P., Minch, E., Mjolsness, E. D., Nakayama, Y., Nelson, M. R., Nielsen, P. F., Sakurada, T., Schaff, J. C., Shapiro, B. E., Shimizu, T. S., Spence, H. D., Stelling, J., Takahashi, K., Tomita, M., Wagner, J., and Wang, J. (2003) The systems biology markup language (SBML): a medium for representation and exchange of biochemical network models., *Bioinformatics (Oxford, England) 19*, 524–31.

31. Demir, E., Cary, M. P., Paley, S., Fukuda, K., Lemer, C., Vastrik, I., Wu, G., D Eustachio, P., Schaefer, C., Luciano, J., Schacherer, F., Martinez-Flores, I., Hu, Z., Jimenez-Jacinto, V., Joshi-Tope, G., Kandasamy, K., Lopez-Fuentes, A. C., Mi, H., Pichler, E., Rodchenkov, I., Splendiani, A., Tkachev, S., Zucker, J., Gopinath, G., Rajasimha, H., Ramakrishnan, R., Shah, I., Syed, M., Anwar, N., Babur, Ö., Blinov, M., Brauner, E., Corwin, D., Donaldson, S., Gibbons, F., Goldberg, R., Hornbeck, P., Luna, A., Murray-Rust, P., Neumann, E., Reubenacker, O., Samwald, M., Iersel, M. van, Wimalaratne, S., Allen, K., Braun, B., Whirl-Carrillo, M., Cheung, K.-H., Dahlquist, K., Finney, A., Gillespie, M., Glass, E., Gong, L., Haw, R., Honig, M., Hubaut, O., Kane, D., Krupa, S., Kutmon, M., Leonard, J., Marks, D., Merberg, D., Petri, V., Pico, A., Ravenscroft, D., Ren, L., Shah, N., Sunshine, M., Tang, R., Whaley, R., Letovksy, S., Buetow, K. H., Rzhetsky, A., Schachter, V., Sobral, B. S., Dogrusoz, U., McWeeney, S., Aladjem, M., Birney, E., Collado-Vides, J., Goto, S., Hucka, M., Novère, N. L., Maltsev, N., Pandey, A., Thomas, P., Wingender, E., Karp, P. D., Sander, C., and Bader, G. D. (2010) The BioPAX community standard for pathway data sharing, *Nature Biotechnology 28*, 935–942.

32. Strömbäck, L., Hall, D., and Lambrix, P. (2007) A review of standards for data exchange within systems biology., *Proteomics 7*, 857–67.

33. Church, G. M. (2005) From systems biology to synthetic biology., *Molecular systems biology 1*, 2005.0032.

34. Macdonald, J. T., Barnes, C., Kitney, R. I., Freemont, P. S., and Stan, G.-B. V. (2011) Computational design approaches and tools for synthetic biology., *Integrative biology : quantitative biosciences from nano to macro* 97–108.

35. Matsuoka, Y., Ghosh, S., and Kitano, H. (2009) Consistent design schematics for biological systems: standardization of representation in biological engineering., *Journal of the Royal Society, Interface/the Royal Society 6 Suppl 4*, S393-404.

36. Grünberg, R. (2009) BBF RFC 30: Draft of an RDF-based framework for the exchange and integration of Synthetic Biology data Relation to other BBF RFCs, http://hdl.handle.net/1721.1/45143 1–13.

37. Galdzicki, M., Rodriguez, C., Chandran, D., Sauro, H. M., and Gennari, J. H. (2011) Standard Biological Parts Knowledgebase, *PLoS ONE* (Schönbach, C., Ed.) *6*, e17005.

38. Rai, A., and Boyle, J. (2007) Synthetic biology: caught between property rights, the public domain, and the commons., *PLoS biology 5*, e58.

39. Endy, D., and Grewal, D. (2010) The BioBrick Public Agreement DRAFT Version 1a, http://dspace.mit.edu/handle/1721.1/50999 1–8.

# Chapter 3

## Robust Optimal Design of Synthetic Biological Networks


### Dirk Lebiedz, Marcel Rehberg, and Dominik Skanda



### Abstract

In engineering, the use of mathematical modeling for design purposes has a long history. Long before any technical realization, a system is planned, simulated, and tested extensively on the computer. In biosciences, however, the application of model-based design before going to the wet lab is still rather rare but has particularly high potential in synthetic biology. We demonstrate exemplarily how mathematical modeling and numerical optimization can be used for the design of a circadian rhythm that is supposed to oscillate robustly with respect to uncertainty in system parameters.



**Key words:** Robust optimal design, MinMax optimization, Mathematical modeling, Synthetic networks, Circadian rhythm


## 1. Introduction

With the ongoing advance in synthetic biology, larger and more complex systems can be fabricated with increasing accuracy. However, this complicates predicting and controlling the dynamical behavior of the designed networks (1, 2). As in engineering science, at a certain level of complexity, mathematical modeling has to be employed to understand, simulate, and test different configurations and avoid costly implementations of designs that do not meet designated goals. Any real-life system is subject to random perturbations in its environment, and in particular in biology parameters characterizing design components are often known only up to some uncertainty range. To ensure optimal system performance under perturbations and uncertainty, these issues should already be addressed at early design stages of the system (3). An iterative empirical procedure, namely, implementing a system, observing its behavior, and then based on experience refining and adjusting


Wilfried Weber and Martin Fussenegger (eds.), *Synthetic Gene Networks: Methods and Protocols*, Methods in Molecular Biology, vol. 813, DOI 10.1007/978-1-61779-412-4_3, © Springer Science+Business Media, LLC 2012

parameters, might be too time consuming and costly in particular cases and probably for larger systems.

Oscillating behavior is one the most fascinating and extensively studied dynamical features of biological systems with periods spanning various timescales and ranging from the molecular to the ecologic level (4). Among these rhythms, circadian oscillators attract special attention because of their interesting properties and their role in human health (5). In the past, the underlying mechanisms that lead to circadian oscillations in cells have been unraveled, for example in bacteria (6), insects (7), and mammals (8).

We demonstrate our robust optimal design approach for the example of a differential equation model proposed for the circadian rhythm generation in drosophila (9) which already before served as basis for model-based optimal control studies (10, 11). In the field of synthetic biology, oscillating networks naturally belong to the key research fields, see, e.g., refs. 12, 13, and we refer to 14 for a discussion of the design point of view.

## 2. Methods

### 2.1. Problem Formulation

We denote the state of a system at a time point $t \in [0, T]$ as $x(t) \in \mathbb{R}^{n_x}$ or without the argument time simply as $x$. In our framework, a system is modeled by a set of ordinary differential equations depending on two distinct sets of parameters, $p_d \in \mathbb{R}^{n_{p_d}}$ and $p_r \in \mathbb{R}^{n_{p_r}}$, which we refer to as design and robustification parameters, respectively. Our aim is to determine the set of design parameters $p_d$ and initial values $x_0 = x(0)$ such that the system optimally meets certain design goals allowing the robustification parameters $p_r$ to take their worst values with respect to that design goals. In other words: Given the worst case realization of $p_r$, how do we have to choose $p_d$ and $x_0$ such that the system is optimally close to the desired behavior? Mathematically, this can be stated as a min–max optimization problem with a set of constrains comprising the differential equation model:

$$
\begin{aligned}
&\min_{p_d, x_0} \max_{p_r} \mathcal{J}(x(t; x_0, p_d, p_r)) \\
&\textit{subject to}\, \dot{x} = f(x, p_d, p_r), t \in [0, T], \\
&x(0) = x_0 \in X_0 \in \mathbb{R}^{n_x}, \\
&p_d \in \Omega_d \subset \mathbb{R}^{n_{p_d}}, \\
&p_r \in \Omega_r \subset \mathbb{R}^{n_{p_r}},
\end{aligned}
\tag{1}
$$

where $X_0 = [x_0^{\text{lb}}, x_0^{\text{ub}}]$, $\Omega_d = [p_d^{\text{lb}}, p_d^{\text{ub}}]$, and $\Omega_r = [p_r^{\text{lb}}, p_r^{\text{ub}}]$ are cubes of respective dimension. The notation $x(t; x_0, p_d, p_r)$ indicates that the function $x(t)$ as a solution of an ordinary differential equation implicitly depends on initial values and parameters. The function J is called the objective functional of problem 1. It measures the deviation of a certain configuration or property of the system determined by $p_d$, $x_0$, and $p_r$ from a desired configuration/property in an appropriate mathematical formulation. The minimization procedure is performed over $p_d$ and $x_0$ as we seek to find design parameters and initial values that bring the system as close as possible to the desired configuration. Maximization is performed over $p_r$ to find the values that maximally disturb the design goal.

### 2.2. Solving the Min–Max Problem

In optimization theory, problems like 1 are referred to as semi-infinite min–max problems (SMMPs) (15). The term "semi-infinite" stems from the fact that the design vector $(p_d, x_0) \in \Omega_d \times X_0$ is finite dimensional, but there is an infinite number of functions $J(\ldots, p_r)$ parameterized by all $p_r \in \Omega_r$ (15). Several techniques to solve SMMPs are available. We use the method of outer approximations (15, 16) which has already been successfully applied in the field of optimal experimental design for model discrimination (17). The outer approximation algorithm solves iteratively relaxations (finite counterparts) of the semi-infinite problem refining in each iteration step the approximation until a sufficient consistency with original problem is achieved. The outer approximation algorithm can be formulated as follows.

**Algorithm 1**

Start: Choose initial parameter vector $p_r^1 \in \Omega_r$ define set $\mathrm{P}^1 = \{p_r^1\}$.

*Step 1: Solve current relaxed problem*

$$\left(x_0^k, p_d^k\right) = \arg \min_{\substack{x_0 \in X_0 \\ p_d \in \Omega_d}} \max_{p_r \in \mathrm{P}^k} \mathcal{J}(x(\cdot; x_0, p_d, p_r)). \tag{2}$$

*Step 2: Solve maximization problem*

$$p_r^{k+1} = \arg \max_{p_r \in \Omega_r} \mathcal{J}(x(\cdot; x_0^k, p_d^k, p_r)). \tag{3}$$

*Step 3: Check convergence (user-supplied tolerance ε):*
If

$$\mathcal{J}(x(\cdot; x_0^k, p_d^k, p_r^{k+1})) \leq 1(1-\varepsilon) \max_{p_r \in \mathbb{P}^k} \mathcal{J}(x(\cdot; x_0^k, p_d^k, p_r))$$

stop with optimal design ( $x_0^k, p_d^k$ );
else, set $\mathbb{P}^{k+1} = \mathbb{P}^k \cup \{p_r^{k+1}\}$, go to step 2, and set $k \rightarrow k+1$.

It has been shown by Shimizu et al. (16) that this relaxation strategy stops in a finite number of steps at a solution of problem 1 as also discussed, e.g., by Pronzato and Walter (18). To solve the relaxed optimization problem 2 in step 1 of Algorithm 1, we transform it into the equivalent problem

$$\min_{\substack{\tau \in \mathbb{R} \\ x_0 \in X_0 \\ p_d \in \Omega_d}} \tau$$

$$\text{subject to: } \tau \geq \mathcal{J}(x(\cdot; x_0^k, p_d^k, p_r^i)), i = 1, \ldots, k,$$

which is easily solvable by any standard nonlinear optimization software for inequality-constrained problems. We have implemented this problem using IPOPT, an interior point optimization package by Wächter and Biegler (19, 20), using the linear solver MA27 (21). Derivatives up to second order, which are needed for the calculations of the numerical optimization algorithm in IPOPT, are computed by automatic differentiation using CppAD (22, 23). For the solution of the differential equations within the optimization problem, which are often stiff in biological kinetics, we use a variable-step and variable-order backward differentiation formula (BDF) method based on Nordsiek array polynomial interpolation similar to the EPISODE BDF method by Byrne and Hindmarsh (24) using the step size control strategy of Calvo and Rández (25).

In step 2 of the outer approximation Algorithm 1, whose goal is to find a new robustification parameter $p_r^{k+1} \in \Omega_r$ to augment the list $\mathbb{P}^{k+1} = \mathbb{P}^k \cup \{p_r^{k+1}\}$ of sampling points for the relaxation of the robustifaction space $\Omega_r$, we apply a simple random search approach using a set of randomly generated start values from $\Omega_r$ to start a couple of a local searches for a maximum and use the best result from these searches. For the local search method, we use the same module, which we use for step 1 of Algorithm 1.

## 3. Results for Circadian Rhythm Example

A central goal in designing cellular oscillators is to set their period. Formally, the period of a system is defined as the minimum $\tau > 0$ for which $x(t) = x(t + \tau)$ holds for all $t$. For circadian rhythms, $\tau$ is close to 24 h. The basic feature of circadian oscillators is a negative feedback loop through which certain proteins inhibit their own expression. For the circadian system in *Drosophila*, a ten-component mathematical model has been proposed in 9, an overview is given in Fig. 1, the equations are listed in Note 1, and an example model simulation is shown in Fig. 2. The system models a regulated transcriptional network for the two proteins TIM and PER.

For this model, a sensitivity analysis has been performed by Stelling et al. (26) and they divided the parameters into the three subsets "global," "local," and "mixed" according to their role in the cell and rhythm generation. Global parameters are, for example, reaction rates of unspecific protein degradation that also play a role for

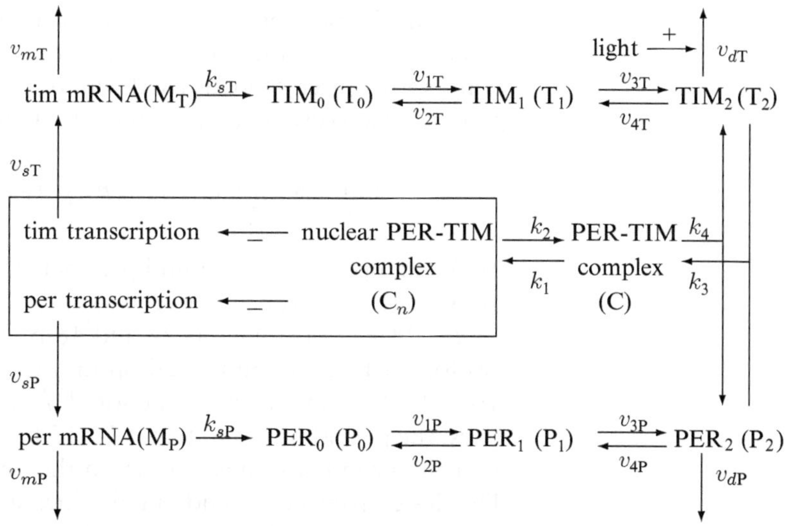


Fig. 1. Structure of circadian model.

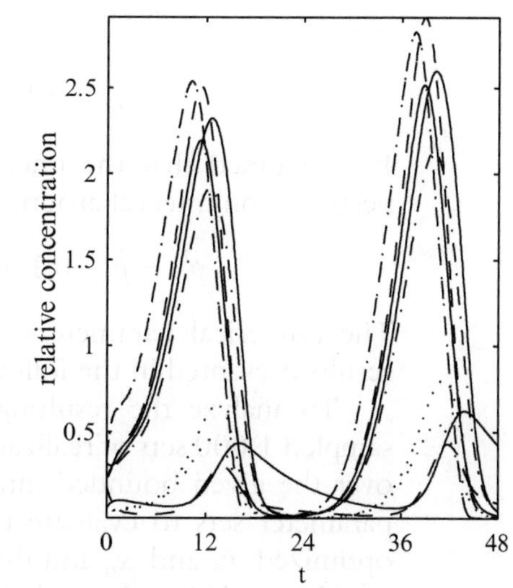


Fig. 2. Example simulation of the model, time course of several species.

other cellular processes. Local parameters, like phosphorylation rates for the circadian clock proteins, are confined to the clock network exclusively. The mixed parameters cannot be easily classified and include, for example, mRNA degradation rates. We sorted all mixed parameters into the global group. The classification is important for us because we adopted it by using the local parameters as design parameters and the global parameters as robustification parameters, reasoning that uncertainties about their true value and random fluctuation between cells and individuals disturb the system. In total, we have 19 design and 19 robustification parameters. An overview over the parameters is given in Notes 3 and 4 respectively. In formulating the optimal design problem, we use the periodicity requirement, i.e., a minimum deviation from a desired $\tau$-period in the objective functional with a (user defined) fixed period $\tau$. To enhance stability of the period, we also evaluate the system at integer multiples of $\tau$:

$$\min_{x_0, P_d} \max_{Pr} \sum_{k=1}^{n} \| x_0 - x(t_k; p_d, p_r) \| \; t_k = \tau, 2\tau, \ldots . \qquad (4)$$

As an example, we aim to find parameters for a period of 24 h. We measure the difference between $x_0$ and $x(t)$ at the two time points $t_k = \tau k = 24\ k$, $k = 1,2$, i.e., at 24 and 48 h. The initial parameter values for starting the numerical optimization Algorithm 1 are taken from the Leloup paper and denoted $\hat{p}_d$ and $\hat{p}_r$. With these parameters, the period of the system is 24.134 h. Additionally, we take $\hat{x}_0$ as arbitrary initial values that lie on the periodic orbit, see Note 2. The design parameters and initial values should be allowed to cover a wide range, and their bounds were set to

$$p_d^{lb} = 0.1 \cdot \hat{p}_d \; and \; p_d^{ub} = 10 \cdot \hat{p}_d$$

and

$$x_0^{lb} = 0.1 \cdot \hat{x}_0 \; and \; x_0^{ub} = 10 \cdot \hat{x}_0 .$$

It is assumed that the uncertain parameters $p_r$ lie within a 10% perturbation interval around the nominal values:

$$p_r^{\mathrm{lb}} = \hat{p}_r - 0.1 \cdot \hat{p}_r \text{ and } p_r^{\mathrm{ub}} = \hat{p}_r + 0.1 . \cdot \hat{p}_r \cdot$$

The numerical parameters (tolerances, etc.) used to obtain the results presented in the following can be found in Note 5.

To analyze the resulting optimal parameter set $(x_0^*, p_d^*)$, we sampled 1,000 sets of realizations of $p_r$ from a uniform distribution over the given bounded intervals of their ranges and used these parameter sets to evaluate the objective functional 3.4 with the optimized $p_d^*$ and $x_0^*$ and the nominal $\hat{p}_d$ (keeping the optimized $x_0^*$). From the resulting simulation output, we computed the values of the objective functional 3.4. A histogram for the results is depicted in Fig. 3. Even on a logarithmic scale, the objective values for the optimized parameters are quite strictly bounded, whereas the objective values for the nominal parameters $\hat{p}_d$ are widely distributed symmetrically which implies much larger variance. Clearly, the optimized parameters lead to smaller objective functional values for all realizations of $p_{r,}$ which shows that with respect to the objective functional the system's performance under random fluctuations is greatly enhanced by the optimally designed $p_d^*$. Another important issue of the optimal design procedure is that we obtain an upper bound for the objective functional value if we assume that the maximization over $p_r$ provides the global maximum.

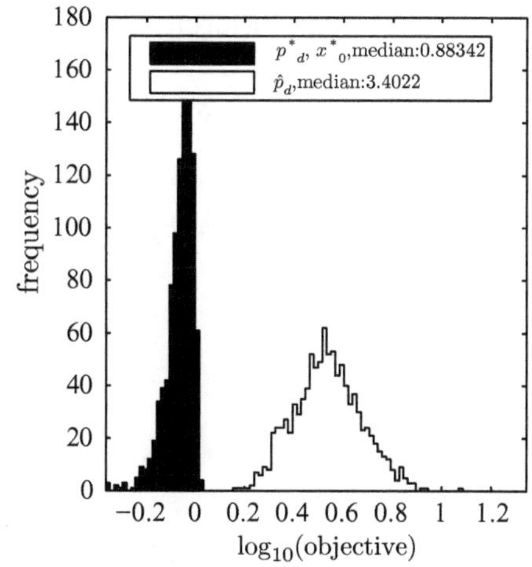


Fig. 3. Histogram of objective values on a logarithmic scale. *White*: initial values, *Black*: optimized parameter values.

## 4. Conclusion

We present a mathematical optimization approach to design synthetic biological systems that takes random perturbations and uncertainties into account already at the early design stage. The procedure incorporates a minimization of some objective function under a worst-case realization of the unknown perturbations. The algorithm may be extended to include optimal control scenarios (27–29) with time-varying parameters into the robust design and even consider dynamic perturbations. The algorithmic approach might prove to be of general use in synthetic biology.

## 5. Notes

1. The circadian rhythm model comprises ten species, and the correspondence between biochemical entities and mathematical symbols is as follows:
   (a) per mRNA: $x_1$, tim mRNA: $x_5$
   (b) PER protein, phosphorylation state 0, 1, 2: $x_2$, $x_3$, $x_4$
   (c) TIM protein, phosphorylation state 0, 1, 2: $x_6$, $x_7$, $x_8$
   (d) PER–TIM complex cytoplasm: $x_9$
   PER–TIM complex nucleus: $x_{10}$.

The differential equations are:

$$\dot{x}_1 = v_{sP}\frac{K_{IP}^n}{K_{IP}^n + x_{10}^n} - v_{mP}\frac{x_1}{K_{mP} + x_1} - k_d x_1 \tag{5a}$$

$$\dot{x}_2 = k_{sP}x_1 - v_{1P}\frac{x_2}{K_{1P} + x_2} + v_{2P}\frac{x_3}{K_{2P} + x_3} - k_d x_2 \tag{5b}$$

$$\begin{aligned}\dot{x}_3 = {} & v_{1P}\frac{x_2}{K_{1P} + x_2} - v_{2P}\frac{x_3}{K_{2P} + x_3} \\ & -v_{3P}\frac{x_3}{K_{3P} + x_3} + v_{4P}\frac{x_4}{K_{4P} + x_4} - k_d x_3\end{aligned} \tag{5c}$$

$$\begin{aligned}\dot{x}_4 = {} & v_{3P}\frac{x_3}{K_{3P} + x_3} - v_{4P}\frac{x_4}{K_{4P} + x_4} - k_3 x_4 x_8 \\ & +k_4 x_9 - v_{dP}\frac{x_4}{K_{dP} + x_4} - k_d x_4\end{aligned} \tag{5d}$$

$$\dot{x}_5 = V_{sT}\frac{K_{IT}^n}{K_{IT}^n + x_{10}^n} - v_{mT}\frac{x_2}{K_{mT} + x_5} - k_d x_5 \tag{5e}$$

$$\dot{x}_6 = k_{sT}x_5 - v_{1T}\frac{x_6}{K_{1T} + x_6} + v_{2T}\frac{x_7}{K_{2T} + x_7} - k_d x_6 \tag{5f}$$

$$\begin{aligned}\dot{x}_7 = {} & v_{1T}\frac{x_6}{K_{1T} + x_6} - v_{2T}\frac{x_7}{K_{2T} + x_7} \\ & -v_{3T}\frac{x_7}{K_{3T} + x_7} + v_{4T}\frac{x_8}{K_{4T} + x_8} - k_d x_7\end{aligned} \tag{5g}$$

$$\dot{x}_8 = \begin{pmatrix} v_{3T}\dfrac{x_7}{K_{3T} + x_7} - v_{4T}\dfrac{x_8}{K_{4T} + x_8} - k_3 x_4 x_8 \\ +k_4 x_9 - v_{dT}\dfrac{x_8}{K_{dT} + x_8} + k_d x_8 \end{pmatrix} \tag{5h}$$

$$\dot{x}_9 = k_3 x_4 x_8 - k_4 x_9 - k_1 x_9 + k_2 x_{10} - k_{dC} x_9 \tag{5i}$$

$$\dot{x}_{10} = k_1 x_9 - k_2 x_{10} - k_{dN} x_{10} \tag{5j}$$

2. Initial values ($x_0$).

$$\begin{aligned} & x_0^{(1)} = 1.5587 && x_0^{(2)} = 0.4474 \\ & x_0^{(3)} = 0.3936 && x_0^{(4)} = 0.2545 \\ & x_0^{(5)} = 1.5587 && x_0^{(6)} = 0.4477 \\ & x_0^{(7)} = 0.3957 && x_0^{(8)} = 0.2728 \\ & x_0^{(9)} = 0.1429 && x_0^{(10)} = 0.5595 \end{aligned} \tag{5k}$$

3. Design parameters (*pd*).

| Parameter | Process | Nominal value $\hat{p}_d$ |
|---|---|---|
| $N$ | Transcriptional repression | 4 |
| $K_{IP}$ | PER autoinhibition | 1 nM |
| $K_{IT}$ | TIM autoinhibition | 1 nM |
| $v_{1P}$ | PER phosphorylation I | 8 nMh$^{-1}$ |
| $v_{1T}$ | TIM phosphorylation I | 8 nMh$^{-1}$ |
| $v_{2P}$ | PER dephosphorylation I | 1 nMh$^{-1}$ |
| $v_{2T}$ | TIM dephosphorylation I | 1 nMh$^{-1}$ |
| $v_{3P}$ | PER phosphorylation II | 8 nMh$^{-1}$ |
| $v_{3T}$ | TIM phosphorylation II | 8 nMh$^{-1}$ |
| $v_{4P}$ | PER dephosphorylation II | 1 nMh$^{-1}$ |
| $v_{4T}$ | TIM dephosphorylation II | 1 nMh$^{-1}$ |
| $K_{1P}$ | PER phosphorylation I | 2 nM |
| $K_{1T}$ | TIM phosphorylation I | 2 nM |
| $K_{2P}$ | PER dephosphorylation I | 2 nM |
| $K_{2T}$ | TIM dephosphorylation I | 2 nM |
| $K_{3P}$ | PER phosphorylation II | 2 nM |
| $K_{3T}$ | TIM phosphorylation II | 2 nM |
| $K_{4P}$ | PER dephosphorylation II | 2 nM |
| $K_{4T}$ | TIM dephosphorylation II | 2 nM |

4. Robustification parameters (*pr*).

| Parameter | Process | Nominal value $\hat{p}_r$ |
|---|---|---|
| $v_{sP}$ | PER transcription | 1 nMh$^{-1}$ |
| $v_{sT}$ | TIM transcription | 1 nMh$^{-1}$ |
| $v_{mP}$ | per mRNA degradation | 0.7 nMh$^{-1}$ |
| $v_{mT}$ | per mRNA degradation | 0.7 nMh$^{-1}$ |
| $v_{dP}$ | PER II degradation | 2 nMh$^{-1}$ |
| $v_{dT}$ | TIM II degradation | 2 nMh$^{-1}$ |
| $K_{mP}$ | PER mRNA degradation | 0.2 nM |
| $K_{mT}$ | TIM mRNA degradation | 0.2 nM |
| $K_{dP}$ | PER II degradation | 0.2 nM |
| $K_{dT}$ | TIM II degradation | 0.2 nM |
| $k_{sP}$ | PER translation | 0.9 h$^{-1}$ |
| $k_{sT}$ | TIM translation | 0.9 h$^{-1}$ |

(continued)

| Parameter | Process | Nominal value $\hat{p}_r$ |
|---|---|---|
| $k_1$ | Transport cytoplasm → nucleus | 0.6 $h^{-1}$ |
| $k_2$ | Transport nucleus → cytoplasm | 0.2 $h^{-1}$ |
| $k_3$ | PER–TIM association | 1.2 $nM^{-1}$ $h^{-1}$ |
| $k_4$ | Complex dissociation | 0.6 $h^{-1}$ |
| $k_d$ | Unspecific degradation | 0.01 $h^{-1}$ |
| $k_{dC}$ | Nucleus-unspecific degradation | 0.01 $h^{-1}$ |
| $k_{dN}$ | Cytoplasm-unspecific degradation | 0.01 $h^{-1}$ |

5. Detailed numerical parameters for Algorithm 1:
   (a) Number of robustification runs (stochastic search): 15.
   (b) Differential equation integration absolute tolerance: $10^{-13}$.
   (c) Differential equation integration relative tolerance: $10^{-11}$.
   (d) IPOPT optimality tolerance: $10^{-5}$.
   (e) Algorithm 1 termination tolerance: $\varepsilon = 10^{-3}$

## References


1. Purnick PEM, Weiss R. The second wave of synthetic biology: from modules to systems. Nat Rev Mol Cell Biol. 2009 Jun;10(6): 410–422. Available from: http://www.dx.doi.org/10.1038/nrm2698.
2. Marchisio MA, Stelling J. Computational design tools for synthetic biology. Curr Opin Biotechnol. 2009 Aug;20(4):479–485. Available from: http://www.dx.doi.org/10.1016/j.copbio.2009.08.007.
3. Mohideen JM, Perkins JD, Pistikopoulos EN. Optimal design of dynamic systems under uncertainty. AIChE J. 1996;42(8):2251–2272. Available from: http://www.dx.doi.org/10.1002/aic.690420814.
4. Kleitman N. Biological rhythms and cycles. Physiol Rev. 1949 Jan;29(1): 1–30.
5. Lévi F, Schibler U. Circadian rhythms: mechanisms and therapeutic implications. Annu Rev Pharmacol Toxicol. 2007;47:593–628.
6. Clodong S, Dühring U, Kronk L, Wilde A, Axmann I, Herzel H, et al. Functioning and robustness of a bacterial circadian clock. Mol Syst Biol. 2007;3:90. Available from: http://www.dx.doi.org/10.1038/msb4100128.
7. Giebultowicz JM. Molecular mechanism and cellular distribution of insect circadian clocks. Annu Rev Entomol. 2000;45:769–793. Available from: http://www.dx.doi.org/10.1146/annurev.ento.45.1.769.
8. Reppert SM, Weaver DR. Molecular analysis of mammalian circadian rhythms. Annu Rev Physiol. 2001;63:647–676.
9. Leloup JC, Goldbeter A. A model for circadian rhythms in Drosophila incorporating the formation of a complex between the PER and TIM proteins. J Biol Rhythms. 1998 Feb; 13(1):70–87.
10. Shaik OS, Sager S, Slaby O, Lebiedz D. Phase tracking and restoration of circadian rhythms by model-based optimal control. IET Syst Biol. 2008 Jan;2(1): 16–23.
11. Rehberg M, Lebiedz D. Phase Tracking of Circadian Rhythm by Model-Based Optimal Control. In: Proceedings of the Third International Conference on Foundations of Systems Biology. in Engineering (FOSBE 2009); 2009.
12. Elowitz MB, Leibler S. A synthetic oscillatory network of transcriptional regulators. Nature. 2000 Jan;403(6767):335–338. Available from: http://www.dx.doi.org/10.1038/35002125.
13. Tigges M, Marquez-Lago TT, Stelling J, Fussenegger M. A tunable synthetic mammalian oscillator. Nature. 2009 Jan;457(7227):309–312. Available from: http://www.dx.doi.org/10.1038/nature07 616.
14. Rand DA, Shulgin BV, Salazar D, Millar AJ. Design principles underlying circadian clocks. J R Soc Interface. 2004 Nov;1(1): 119–130.

Available from: http://www.dx.doi.org/10.1098/rsif.2004.0014.

15. Polak E. Optimization: Algorithms and Consistent Approximations. Springer; 1997.
16. Shimizu K, Aiyoshi E. Necessary conditions for min-max problems and algorithms by a relaxation procedure. IEEE T Automat Contr. 1980;25(1):62–66.
17. Skanda D, Lebiedz D. An optimal experimental design approach to model discrimination in dynamic biochemical systems. Bioinformatics. 2010;26:939–45.
18. Pronzato L, Walter E. Robust experiment design via maximin optimization. Math Biosci. 1988;89(2): 161 – 176. Available from: http://www.sciencedirect.com/science/article/B6VHX-45FKF3H-4W/2/2d1d3988042e1e36ffb9cafde61c1634.
19. Wächter A. An Interior Point Algorithm for Large-Scale Nonlinear Optimization with Applications in Process Engineering. Carnegie Mellon University; 2002.
20. Wächter A, Biegler LT. On the Implementation of a Primal-Dual Interior Point Filter Line Search Algorithm for Large-Scale Nonlinear Programming. Math Program. 2006;106(1): 25–57.
21. HSL. A Collection of Fortran codes for large-scale scientific computation.; 2007. Available from: http://www.hsl.rl.ac.uk.
22. Bell BM, Burke JV. Algorithmic Differentiation of Implicit Functions and Optimal Values. In: Bischof CH, Bücker HM, Hovland PD, Naumann U, Utke J, editors. Advances in Automatic Differentiation. Springer; 2008. p. 67–77.
23. Bell BM. Automatic differentiation software CppAD.; 2010. Available from: http://www.coin-or.org/CppAD/.
24. Byrne GD, Hindmarsh AC. A Polyalgorithm for the Numerical Solution of Ordinary Differential Equations. ACM Transactions on Mathematical Software. 1975;1(1):71–96.
25. Calvo M, Montijano JI, Rández L. On the change of step size in multistep codes. Numer Algorithms. 1993;4:283–304.
26. Stelling J, Gilles ED, Doyle FJ. Robustness properties of circadian clock architectures. Proc Natl Acad Sci USA. 2004 Sep;101(36): 13210–13215.
27. Lebiedz D. Exploiting optimal control for target-oriented manipulation of (bio)chemical systems: a model-based approach to specific modification of self-organized dynamics. Int J Mod Phys B. 2005 August; 19(25): 3763–3798.
28. Lebiedz D, Brandt-Pollmann U. Manipulation of self-aggregation patterns and waves in reaction-diffusion system by optimal boundary control strategies. Phys Rev Lett. 2003 November;91(20):208301.
29. Lebiedz D, Sager S, Bock HG, Lebiedz P. Annihilation of limit cycle oscillations by identification of critical phase resetting stimuli via mixed-integer optimal control methods. Phys Rev Lett. 2005 September;95:108303.

# Chapter 4

# Predicting Synthetic Gene Networks

**Diego di Bernardo, Lucia Marucci, Filippo Menolascina, and Velia Siciliano**

## Abstract

Synthetic biology aims at designing and building new biological functions in living organisms. The complexity of cellular regulation (regulatory, metabolic, and signaling interactions, and their coordinated action) can be tackled via the development of quantitative mathematical models. These models are useful to test biological hypotheses and observations, and to predict the possible behaviors of a synthetic network. Indeed, synthetic biology uses such models to design synthetic networks, prior to their construction in the cell, to perform specific tasks, or to change a biological process in a desired way. The synthetic network is built by assembling biological "parts" taken from different systems; therefore it is fundamental to identify, isolate, and test regulatory motifs which occur frequently in biological pathways. In this chapter, we describe how to model and predict the behavior of synthetic networks in two difference cases: (1) a synthetic network composed of five genes regulating each other through a variety of regulatory interactions in the yeast *Saccharomyces cerevisiae* (2) a synthetic transcriptional positive feedback loop stably integrated in Human Embryonic Kidney 293 cells (HEK293).

**Key words:** Synthetic biology, Mathematical modeling, Positive feedback loop, *S. cerevisiae*, HEK 293, Microfluidics

## 1. Introduction

### *1.1. Synthetic Network in Yeast Saccharomyces cerevisiae*

The synthetic network described herein has been built for benchmarking modeling and reverse-engineering approaches (1, 3). IRMA was designed to be isolated from the cellular environment (Cuccato et al., Heredity 102:527–532, 2009), and to respond to galactose or glucose, which respectively "switch" the network on and off by triggering transcription of its genes. This network (Fig. 1) is very articulated in its interconnections, which include regulator chains, single input motifs, and multiple feedback

Wilfried Weber and Martin Fussenegger (eds.), *Synthetic Gene Networks: Methods and Protocols*, Methods in Molecular Biology, vol. 813, DOI 10.1007/978-1-61779-412-4_4, © Springer Science+Business Media, LLC 2012

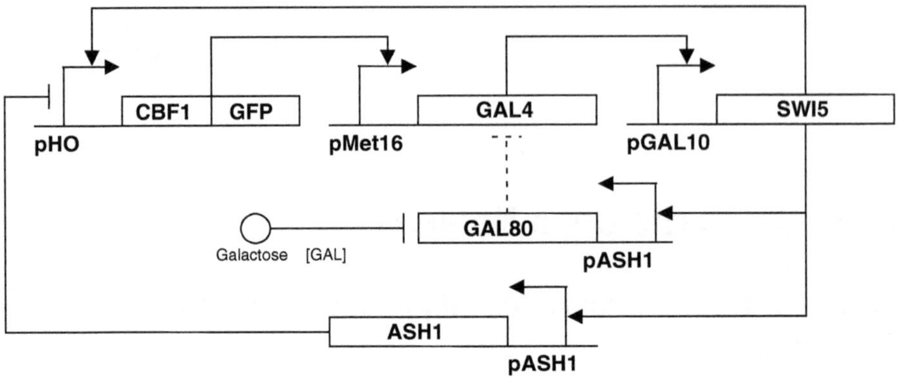


Fig. 1. *Diagram of the synthetic network in yeast. Solid lines model transcriptional interactions while dashed lines are meant to represent protein-protein interactions.*

loops generated by the combination of transcriptional activators and repressors. We selected well-characterized promoter/TF-encoding-genes pairs. We chose nonessential and nonredundant TF-genes that can be knocked out without affecting yeast viability. The following genes have been chosen for IRMA: as activators and repressors encoding genes: *SWI5, ASH1, CBF1, GAL4, and GAL80*; as promoter genes: *HO*, *ASH1*, *MET16*, *GAL10* (Fig. 1).

#### *1.2. Synthetic Transcriptional Positive Feedback Loop in Mammalian Cells*

This synthetic network is shown in Fig. 2. We took advantage of the inducible Tet regulatory system: the expression of Tetracycline-controlled transactivator tTA is self-controlled by a *CMV-TET* promoter responsive to thetTA itself unless Tetracycline, or its analogous Doxycycline, is added to the medium in which cells are grown (4). To follow the protein dynamics of the positive feedback loop, a destabilized yellow variant of the enhanced green fluorescent protein (d2EYFP) (Clontech), with a half-life of approximately 2 h, was expressed together with the tTA transactivator from the same mRNA, via an Intra Ribosomal Entry Sequence (IRES) in between of the transactivator tTA and the d2EYFP (Fig. 2). In order to stably express in HEK293 cells the inducible feedback loop and to better characterize its dynamics overtime, we used a lentiviral vector (5, 6), based on the multisite Gateway technology provided by Invitrogen.

#### *1.3. Derivation of Mathematical Models for Synthetic Network*

When deriving a model from experimental data, three major approaches can be used: white-box, black-box, and gray-box. In white-box modeling, the model and parameter values are entirely derived from first principles, while in black-box, the model is completely derived from input–output data. The third alternative, the so-called gray-box approach (7), combines the two above

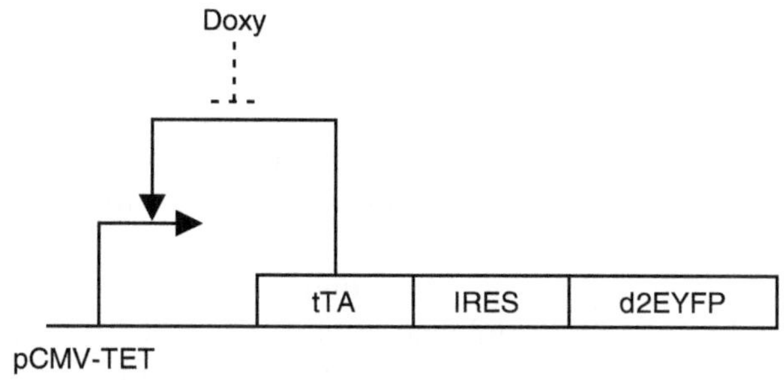


Fig. 2. Design of the positive feedback loop in mammalian cells. The promoter *CMV-TET* consist of seven direct repeats of a 42-bp sequence containing the tet operator sequences (*tetO*), located just upstream of the minimal *CMV* promoter ($P_{minCMV}$). The tetracycline-controlled transactivator *tTA* derives from the addition of the VP16 activation domain to the transcriptional repressor TetR. The d2EYFP is the destabilized *yellow-green* variant of enhanced *green* fluorescent protein with a half-life of approximately 2 h.

approaches. Specifically, first principles are used to partially derive the model structure, while parameters in the model are estimated from experimental data. The approach described in this chapter, both for the yeast and the mammalian synthetic network, is a gray-box one. In this case, modeling entails the following main steps to be executed iteratively: (1) derivation of the model equations and (2) estimation of the model parameters from experimental data and/or literature. Step (1) requires introducing simplifying hypothesis and choosing a proper formal framework. Among the different mathematical formalisms, those based on differential equations are commonly used to describe the average behavior of a population of cells (8). The Differential Equations modeling approach is based on the following biological assumptions: the quantified concentrations are homogeneous in space and they are continuous quantities in time. These assumptions hold true for processes evolving on long time scales in which the number of molecules of the species in the reaction volume is sufficiently large. Step (2) is required to estimate unknown model parameters from the available experimental data. Experimental data are strongly affected by noise. Hence, classical optimization methods, based on gradient descent from an arbitrary initial guess of the parameters' value, can be unfeasible. The above considerations suggest looking at stochastic optimization algorithms, such as Genetic Algorithms (GA) (9), which provide a flexible approach to nonlinear optimization. Their application has been proved to yield good results in the parametrization of synthetic networks (10, 11).

When the network is quite complex in terms of the number of the unknown parameters, there is the need of going through iteration between experiments and modeling, to gather more experimental data if needed. We show an example of such iteration, in the case of the yeast network.

### 1.4. Refining Mathematical Models by Means Of Microfluidics Experiments

Once a model of a synthetic network has been derived, it is possible to refine it by means of several alternative strategies. The primary goal of model refinement is the improvement of model predictions as regards the network dynamics, i.e., how gene and protein expression change in time following a perturbation; in order to accomplish this task, the experimenter may need to stimulate its synthetic network with highly dynamical signals. This is usually done to elicit nonlinear modes or other peculiar characteristics of the network under development. Most of the synthetic networks documented so far (12) use chemical compounds as inducers and thus the previous requirement often translates in the need to quickly change the concentration of these compounds in the media where cells are grown. On the contrary, effective strategies for data acquisition are needed to measure changes in the concentration of the species of interest in live cells. Here, we propose an integrated strategy intended to solve both of these issues at once via "microfluidics" devices and time-lapse fluorescence microscopy. Microfluidics involves the manipulation of very small fluid volumes, enabling creation and control of nanoliter-volume reactors, thus mimicking cellular microenvironments. Microfluidics devices can be conveniently used to finely control the concentration of compounds in the extracellular environment during time lapse microscopy experiments. Data acquired from long term stimulation of cells carrying fluorescent tags and tracked by the microscopy imaging can be conveniently used to improve the quality of the mathematical model of the circuit of interest.

In the following sections, we illustrate how this platform can be adapted in experimental contexts involving both simpler and higher eukaryotic systems, namely mammalian cell lines and *S. cerevisiae.*

## 2. Materials

### 2.1. Yeast Culture, Strains, and Semiquantitative and Quantitative RT-PCR

1. *S. cerevisiae* strains YM4271 background (*MATa ura3-52 his3-Δ200 ade2-101 lys2-801 leu2-3 trp1-901 gal4-Δ542 gal80-Δ538 ade5::hisG*) (3).
2. YEP medium: 10 g/L Bacto yeast extract, 20 g/L Bacto peptone.
3. YEPD medium: YEP containing 2% glucose.
4. YEPGR medium: YEP containing 2% galactose and 2% raffinose.
5. SC medium: 6.7 g/L yeast nitrogen base without amino acids, 1.35 g/L amino acid powder mix.
6. 2 U/ul DNAse I (Roche).
7. Rneasy MiniElute Cleaneup Kit (Quiagen).
8. SuperScript III First-Strand Synthesis System (Invitrogen).
9. Platinum SYBR Green qPCR SuperMix-UDG with ROX (Invitrogen).

10. 7000 ABI Real-Time PCR machine.
11. Applied Biosystems SDS software version 1.2.3 to perform data analyses.

#### 2.2. Microfluidics and Microscopy Apparatus

1. Cole-Parmer nano syringe pump (Cole-Parmer).
2. USB valve control system.
3. Pneumadyne 8-Valve Manifold.
4. Tygon Microbore Tubing I.D.: 0. 25" O.D.: 0.125" (Swagelok).
5. Tygon Microbore Tubing I.D.: 0.020" O.D.: 0.060" (VWR).
6. Double distilled $H_2O$.
7. Sulforhodamine B (Sigma-Aldrich).
8. 60-mL syringe (BD).
9. 10-mL syringes (BD).
10. 22G sterile needles (BD).
11. 2 mL cryovials (BMA).
12. Microfluidic device.
13. Inverted fluorescence microscope with temperature and $CO_2$ conditioning and appropriate fluorescence filters.

#### 2.3. Mammalian Cells Culture and Lentiviral Transduction

1. 293FT cells maintained at 37°C in a 5% $CO_2$-humidified incubator, and cultured in Dulbecco's Modified Eagle's Medium DMEM (GIBCO BRL) supplemented with 10% heat-inactivated fetal bovine serum (FBS) (Invitrogen), 1% L-glutammine, 1% MEM Nonessential Amino Acids, 1% MEM Sodium pyruvate, and 1% antibiotic/antimycotic solution (GIBCO BRL).
2. Hek 293 cells maintained at 37°C in a 5% $CO_2$-humidified incubator, and cultured in Dulbecco's Modified Eagle's Medium DMEM (GIBCO BRL) supplemented with 10% heat-inactivated fetal bovine serum (FBS) (Invitrogen), 1% L-glutamine, and 1% antibiotic/antimycotic solution (GIBCO BRL).
3. Doxycicline (Clontech) dissolved in tissue-culture water to a final concentration of 10 μg/mL, stored in aliquots at −20°C, and then added to tissue culture dishes as required.
4. Polybrene (Invitrogen) dissolved in tissue-culture water to a final concentration of 6 mg/mL, stored in aliquots at −20°C, and then added to tissue culture dishes to a final concentration of 6 μg/mL.
5. Blasticidin (Sigma) dissolved in tissue-culture water to a final concentration of 100 mg/mL stored in aliquots at −20°C, and then added to tissue culture dishes to a final concentration of 3 μg/mL.

### 2.4. PCR and Construction of Gene Circuit

1. Taq Phusion (Fynnzymes) is used to a final concentration of 0.02 U/μL; the HF buffer is added to a final concentration of 1×, primers and dNTPs are supplied to a final concentration of 0.5 μM each and 200 μM each, respectively.
2. Taq DNA polymerase (Invitrogen) is used to a final concentration of 1 U/μL; the PCR buffer is added to a final concentration of 1× $MgCl_2$ 1.5 mM, primers, and dNTPs are supplied to a final concentration of 0.5 μM each and 200 μM each respectively.
3. 5 U/μL of NheI and EcoRV restriction enzymes (Roche).
4. 2 U/μL of T4 DNA Ligase, and T4 DNA Ligase buffer to a final concentration of 1×.
5. LR Clonase II plus enzyme mix (Invitrogen).

### 2.5. Computing System and Algorithms

1. Personal computer equipped with MATLAB Simulink (MathWorks).
2. Image processing algorithm implemented in the control schematic.

## 3. Methods

### 3.1. Construction of S. cerevisiae Strains

To construct the IRMA containing strain, sequential PCR-based genomic integrations were made sequentially. All the integrations were confirmed by PCR.

1. The *2× HA-hphMX4* cassette was amplified by PCR and inserted in front of the stop codon of *ASH1* gene in YM4271 strain resulting in P278 strain.
2. To generate P280 strain *MET16* promoter was amplified from W303 and cloned in YIplac128 between Hind III and Sac I.
3. *GAL4*ORF was cloned between Sac I and Nde I, thus resulting in p*MET16pGAL4.*
4. The *MET16pGAL4-LEU2* cassette was integrated in *SHE2* locus.
5. *CBF1*ORF was amplified from W303 and cloned among Bam HI and Pac I of p*FA6a-GFP (S65T)-kanMX6.*
6. The *CBF1-GFP-kanMX6* cassette was integrated downstream of the *HO* promoter of P280 strain, obtaining P324.
7. *ASH1* promoter was cloned in Pst I and Bam HI of YIplac211, and then *GAL80-3xFLAG* was inserted between Bam HI and Sac I.
8. The *ASH1pGAL80-3XFLAG-URA3* was integrated in *SWI5* locus, thus yielding P326.

9. *ACE2* gene was deleted in the strain P326 by integrating *natMX4cassette* from pAG25.
10. *GAL10pSWI5AAA-MYC9-KlTRP1* was integrated in *CBF1* locus resulting in IRMA containing strain.

Further details can be found in (3).

### 3.2. Quantitative RT-PCR

In order to analyze the dynamic behavior of the network, we analyzed expression profiles of network genes by quantitative real-time RT-PCR following two different perturbation experiments; in the first we shifted cells from glucose (YEPD medium) to galactose (YEPGR medium) ("switch-on" experiments) and from galactose to glucose ("switch-off" experiments). In the second, we overexpressed each of the five network genes in cells that were grown either in glucose or galactose (3).

1. Prepare the total RNA.2. Treat 1 μg of RNA with 2.5 U of DNAseI.
2. Clean up with RNeasy MiniElute Cleaneup Kit (Quiagen).
3. Reverse-transcribe the RNA cleaned using SuperScript III First-Strand Synthesis System.
4. Set up quantitative real-time PCR reactions in duplicates using Platinum SYBR Green qPCR SuperMix-UDG with ROX.
5. Amplify the cDNA thus obtained using a 7000 ABI Real-Time PCR machine.

### 3.3. Construction and Testing of the Synthetic Positive Feedback Loop in Hek293 Cells

We constructed the synthetic positive feedback loop into a lentiviral vector system to allow integration of the circuit in mammalian cells. To this end, we used the ViraPower Promoterless Lentiviral Gateway Expression System (Invitrogen), which takes advantage of the site-specific recombination properties of bacteriophage lambda, making the transfer of single DNA sequences faster than the usual cloning strategies.

1. Cloning of a destabilized yellow-green variant of enhanced green fluorescent protein (d2EYFP) in the p*MAtTA-IRES-EGFP* vector: the *d2EYFP* was amplified from pd2EYFP-1 (Clontech) by PCR using the High-Fidelity DNA Polymerase Taq Phusion, with a forward primer containing a NheI recognition sequence and a reverse primer containing an EcoRV recognition sequence. The PCR product and p*MAtTA-IRES-EGFP* were then digested with NheI-EcoRV restriction enzymes and the *d2EYFP* ligated in place of *EGFP* (ratio ng of vector/ng of d2EYFP = 1/3–1/5).
2. Generating the pENTR vectors: in order to produce a lentiviral vector by using the gateway system we first generated the pENTR vectors containing the genes and the promoters of interest flanked by specific recombination sites. The p*MAtTA-IRES-d2EYFP*

was linearized with the AseI restriction enzyme and recombined with the pDONR221 (invitrogen) following the manufacturer instruction. In this way we generated p*ENTRtTA-IRES-d2EYFP* vector with specific recombination sites. The *CMV-TET* promoter was amplified from pTRE2 (clontech) by PCR. The PCR was performed with the Taq polymerase provided by Invitrogen that adds a single deoxyadenosine (A) to the 3′ ends of PCR products. This allows PCR inserts to ligate efficiently with the pENTR5′-TOPO vector which is supplied linearized with single 3′-deoxythymidine (T) overhangs, obtaining the pENTR5′-TOPO-*CMV-TET* with specific recombination sites. Finally, we performed a recombination reaction between the pENTR*tTA-IRES-d2EYFP*, pENTR5′-TOPO-*CMV-TET*, and the p*Lenti/R4R2/V5-DEST* using the LR clonase enzyme according to manufacturer instructions. The lentivirus was then produced in 293FTcells as described in the instructions provided by Invitrogen.

3. Cell culture for lentiviral transduction: To transduce cells with the virus produced, 500000 HEK293 cells were plated and incubated overnight. The day of transduction the medium was removed and 1 mL of the virus was added to the cells together with polybrene to a final concentration of 6 μg/mL. After an overnight incubation the medium containing the virus was removed and replaced with complete culture medium containing the blasticidin to a final concentration of 3 mg/mL to select for stably transduced cells.
4. To test the dynamics of the autoregolatory loop, we performed two sets of time-series experiments in which stably integrated HEK293 cells were imaged using time-lapse microscopy and fluorescence of d2EYFP was quantified. For both the experimental designs in the first time point, cells were treated with Doxycycline to "switch off" the network, since Doxycycline prevents the tTA transactivator to bind the *CMV-TET* responsive promoter. In the first set of experiments, the dynamics were followed for 37 h at 37°C, while in the second set the temperature was reduced to 32 °C to limit cell motility and thus facilitate image analysis (13).

### 3.4. Mathematical Modeling the Yeast Synthetic Network

#### 3.4.1. Step (i). Derivation of the Model from First Principles

The network of interest it the one showed in Fig. 1. Details about the network construction are reported in Subheadings 2.1 and 3.1. For each species in the network, i.e., each mRNA (capital letters) and correspondent protein concentration (small letters), we wrote one equation, which expresses its change in time as the result of production and degradation:

$$\frac{d[CBF1]}{dt} = \alpha_1 + v_1 H^{+-}([Swi5],[Ash1];k_1,k_2,h_1,h_2) - d_1[CBF1],$$
$$\frac{d[Cbf1]}{dt} = \beta_1[CBF1] - d_2[Cbf1],$$
$$\frac{d[GAL4]}{dt} = \alpha_2 + v_2 H^{+}([Cbf1];k_3,h_3) - d_3[GAL4],$$
$$\frac{d[Gal4]}{dt} = \beta_2[GAL4] - d_3[Gal4],$$
$$\frac{d[SWI5]}{dt} = \alpha_3 + v_3 H^{+}([Gal4^{free}];k_4,h_4) - d_5[SWI5],$$
$$\frac{d[Swi5]}{dt} = \beta_3[SWI5] - d_6[Swi5],$$
$$\frac{d[GAL80]}{dt} = \alpha_4 + v_4 H^{+}([Swi5];k_5,h_5) - d_7[GAL80],$$
$$\frac{d[Gal80]}{dt} = \beta_4[GAL80] - d_8[Gal80],$$
$$\frac{d[ASH1]}{dt} = \alpha_5 + v_5 H^{+}([Swi5];k_6,h_6) - d_9[ASH1],$$
$$\frac{d[Ash1]}{dt} = \beta_5[ASH1] - d_{10}[Ash1].$$

The first two terms on the right-hand side of the mRNA equations represent the production, where $\alpha$ are the basal transcription rates; $v$ are the maximal transcription rates modulated by the Hill functions:

$$H^{+}(y;k,h) = \frac{y^h}{y^h + k^h},$$
$$H^{-}(z;k_1,h_1) = \frac{k_1^{h_1}}{z^{h_1} + k_1^{h_1}},$$
$$H^{+-}(y,z;k,h,k_1,h_1) = H^{+}(y;k,h)H^{-}(z;k_1,h_1).$$

These are used to model transcriptional activation, or repression; $y$ and $z$ represent transcription factor levels, $h$ are the Hill coefficients (pure numbers that refer to the cooperativity of the activation binding reaction) and $k$ are the Michaelis-Menten constants, equal to the amount of transcription factor needed to reach half maximal activation (or repression). For protein equations, the production rates are $\beta$, i.e., the maximal translation rates. Degradations of mRNAs and proteins are represented by $d$, i.e., the degradation constants. The amount of free Gal4 depends on the interactions of the galactose pathway with the network genes. For the units of measurement, please refer to Table 1. Summing up, when writing the above model, we made the following assumptions: (A1) the transcriptional activity of each promoter is leaky ($\alpha$); (A2) the degradation kinetics of both mRNAs and proteins are first-order; (A3) the protein production terms are proportional to the corresponding mRNA concentrations; (A4) the transcriptional activation–repression of each promoter by a transcription factor can be modeled as a Hill function (2).

Note that the concentration of Gal4free is the amount of Gal4 protein that is not involved in the formation of the protein-protein complex with Gal80 and hence activates the *GAL10* promoter driving *SWI5* expression.

**Table 1**
**Parameters of the mathematical model the yeast synthetic network**

| Parameter | Phenomenological model | Refined model | Exp. id. |
|---|---|---|---|
| $k_1$ (a.u.) | 1.884 | 1 | 1 |
| $k_2$ (a.u.) | 30 | 0.035 | 0.035 |
| $k_3$ (a.u.) | 0.229 | 0.037 | 0.037 |
| $k_4$ (a.u.) | 0.216 | 0.09 Glu | 0.09 Glu |
| | | 0.01 Gal | 0.01 Gal |
| $k_5$ (a.u.) | 0.16 | 1.884 | 1.884 |
| $k_6$ (a.u.) | 0.160 | 1.884 | 1.884 |
| $\alpha_1$ (a.u. min.$^{-1}$) | 0 | 0 | – |
| $\alpha_2$ (a.u. min.$^{-1}$) | $1.10 \times 10^{-4}$ | $1.49 \times 10^{-4}$ | – |
| $\alpha_3$ (a.u. min.$^{-1}$) | $3.2 \times 10^{-4}$ | $3 \times 10^{-3}$ | – |
| $\alpha_4$ (a.u. min.$^{-1}$) | 0 | $7.4 \times 10^{-4}$ | – |
| $\alpha_5$ (a.u. min.$^{-1}$) | $7.37 \times 10^{-5}$ | $6.1 \times 10^{-4}$ | – |
| $v_1$ (a.u. min.$^{-1}$) | 0.065 | 0.04 | – |
| $v_2$ (a.u. min.$^{-1}$) | 0.002 | $8.82 \times 10^{-4}$ | – |
| $v_3$ (a.u. min.$^{-1}$) | 0.025 | 0.002 Glu<br>0.020 Gal | $v_{3\,Glu}/v_{3\,Gal}9$ |
| $v_4$ (a.u. min.$^{-1}$) | 0.007 | 0.014 | – |
| $v_5$ (a.u. min.$^{-1}$) | 0.002 | 0.018 | – |
| $d_1$ (min$^{-1}$) | 0.033 | 0.022 | – |
| $d_2$ (min$^{-1}$) | 0.042 | 0.047 | – |
| $d_3$ (min$^{-1}$) | 0.047 | 0.421 | – |
| $d_4$ (min$^{-1}$) | 0.141 | 0.098 | – |

(continued)

**Table 1 (continued)**

| Parameter | Phenomenological model | Refined model | Exp. id. |
|---|---|---|---|
| $d_5$ ($min^{-1}$) | 0.018 | 0.050 | – |
| $h_1$ | 1 | 1 | 1 |
| $h_2$ | 1 | 1 | 1 |
| $h_3$ | 1 | 1 | 1 |
| $h_4$ | 1 | 4 | 4 |
| $h_5$ | 1 | 1 | 1 |
| $h_6$ | 1 | 1 | 1 |
| $h_7$ | 1 | 4 | 4 |
| $\beta_1$ ($min^{-1}$) | 0.223 | 0.201 | – |
| $\beta_2$ ($min^{-1}$) | 0.285 | 0.167 | – |
| $\gamma$ (a.u.) | $10^{-4}$ Glu<br>5.55 Gal | 0.2 Glu<br>0.6 Gal | 0.2 Glu<br>0.6 Gal |
| $\tau$ (min) | 100 | 100 | – |

### *3.5. Preliminary Dataset*

In order to identify model's parameters and validate the model, we collected mRNAs expression levels during a time course experiment, by shifting cells from glucose (YEPD medium) to galactose (YEPGR medium) ("switch-on" experiment) as described in (3) and in Subheadings 2.1 and 3.2. Data are shown in Fig. 3a.

We included as the first point of the time-series the expression level of the network genes after growing cells overnight in glucose, just before shifting them from glucose to galactose (3). The second point, taken after 10 min, is measured just after the shift has occurred. The averaged gene expression profiles (Fig. 3a) show that the standard washing steps, needed to shift cells from glucose medium to the fresh new galactose-containing medium, induce a transient increase in mRNA levels of *GAL4* and *GAL80* (Fig. 3a, gray bars). This effect is not dependent on galactose addition, but uniquely on the washing steps (3), and it is probably due to the transient deprivation of carbon source during washing, which attenuates the degradation levels of *GAL4* and *GAL80* mRNAs (14).

Moreover, the activation of *CBF*1 appears to be delayed with respect to the other Swi5 targets, respectively *GAL80* and *ASH1* (Fig. 3a). Such delay is physically due to the sequential recruitment of chromatin modifying complexes to the *HO* promoter, which follows binding of Swi5 (15, 16).

We then performed four additional experiments, shifting cells from galactose to glucose, thus "switching off" gene expression in the network, as described (3) and in Subheadings 2.1 and 3.2.

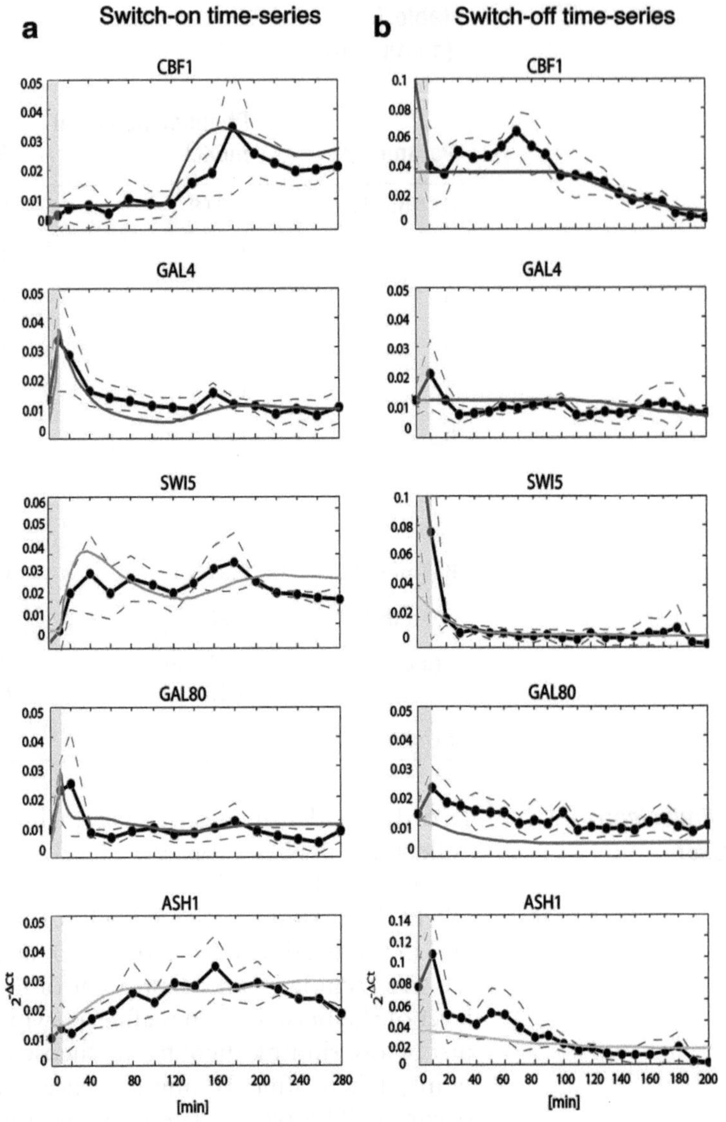


Fig. 3. Identification and validation results of time-series, phenomenological model. *Circles* represent average expression data for each of the IRMA genes at different time points. *Dashed lines* represent standard errors. *Solid lines* represent in silico data. (**a**) Identification results of the phenomenological model on the average 5 h "switch-on" time-series. (**b**) Validation of the phenomenological model on the average 3 h "switch-off" dataset.

The averaged time-series gene expression profiles (Fig. 3b) were used for validating model predictive performance. We refer to this dataset as the "switch-off" dataset.

Finally, we collected another set of experiments to be used to further validate the model's predictive ability. We measured gene expression responses of the five network genes following exogenous overexpression of each of the five genes under the control of a strong constitutive promoter, as described in (3) and in Subheadings 2.1 and 3.2. Such overexpression experiments were performed both in glucose and in galactose. We refer to these two experimental datasets as the "Galactose steady-state" and "Glucose steady-state" (Fig. 4a, c).

*3.5.1. Step (i). Derivation of the Phenomenological Model*

At this stage, we had to properly refine the model both to be able to capture the features highlighted by the gene expression profiles and to reduce the number of parameters to be estimated. First of all, we made the following extra modeling assumptions: (A5) a fix time delay, $\tau$, equal to 100 min, is added in the activation of the *HO* promoter by Swi5; (A6) a transient decrease in the mRNA degradation of *GAL4* and *GAL80* of value $\Delta\beta_1$ and $\Delta\beta_2$ is added for an interval of 10 min to describe the effect of the washing steps.

Due to the lack of protein concentrations measurements, we also assumed that the protein concentrations are monotonically increasing functions of their corresponding mRNA concentrations at any time (A7).

In order to define the active amount of Gal4 (GAL4 free in the above Equations), we needed to describe the effect of the galactose pathway on the network dynamics. In the literature, very detailed models of the galactose pathway have been presented (17). Such paradigms can be simplified in a number of ways, but it often leads to include in the model nonmeasurable complexes concentrations. Thus, we decided to use a phenomenological approach, assuming that (A8) the protein-protein interaction between Gal80 and Gal4 can be modeled as a direct inhibition of *GAL80* on the promoter of *SWI5*, and that the strength of such inhibition depends on the medium (strong inhibition in glucose, weak inhibition in galactose). Actually we assumed that the *GAL10* promoter is activated by *GAL4* and noncompetitively inhibited by *GAL80*. The resulting phenomenological Delay Differential Equations (DDEs) model is:

$$\frac{d[CBF1]}{dt} = \alpha_1 + v_1\left(\frac{[SWI5(t-\tau)]^{h_1}}{K_1^{h_1} + [SWI5(t-\tau)]^{h_1}}\right)\cdot\left(\frac{K_2^{h_2}}{K_2^{h_2} + [ASH1]^{h_2}}\right) - d_1[CBF1],$$

$$\frac{d[GAL4]}{dt} = \alpha_2 + v_2\left(\frac{[CBF1]^{h_3}}{K_3^{h_3} + [CBF1]^{h_3}}\right) - (d_2 - \Delta\beta_1)[GAL4],$$

$$\frac{d[SW15]}{dt} = \alpha_3 + v_3\left(\frac{([GAL4])^{h_4}}{\left(K_4^{h_4} + ([GAL4])^{h_4}\right)\left(1 + \frac{[GAL80]^{h_7}}{\hat{\gamma}^{h_7}}\right)}\right) - d_3[SW15],$$

$$\frac{d[GAL80]}{dt} = \alpha_4 + v_4\left(\frac{[SWI5]^{h_5}}{K_3^{h_3} + [SWI5]^{h_5}}\right) - (d_4 - \Delta\beta_2)[GAL80],$$

$$\frac{d[ASH1]}{dt} = \alpha_5 + v_5\left(\frac{[SWI5]^{h_5}}{K_6^{h_5} + [SWI5]^{h_6}}\right) - d_5[ASH1].$$

We use the symbol ^ to indicate medium-dependent quantities. Thus, we are assuming that the Michaelis-Menten coefficient of the phenomenological description of the inhibition of *GAL80* is dependent on the medium.

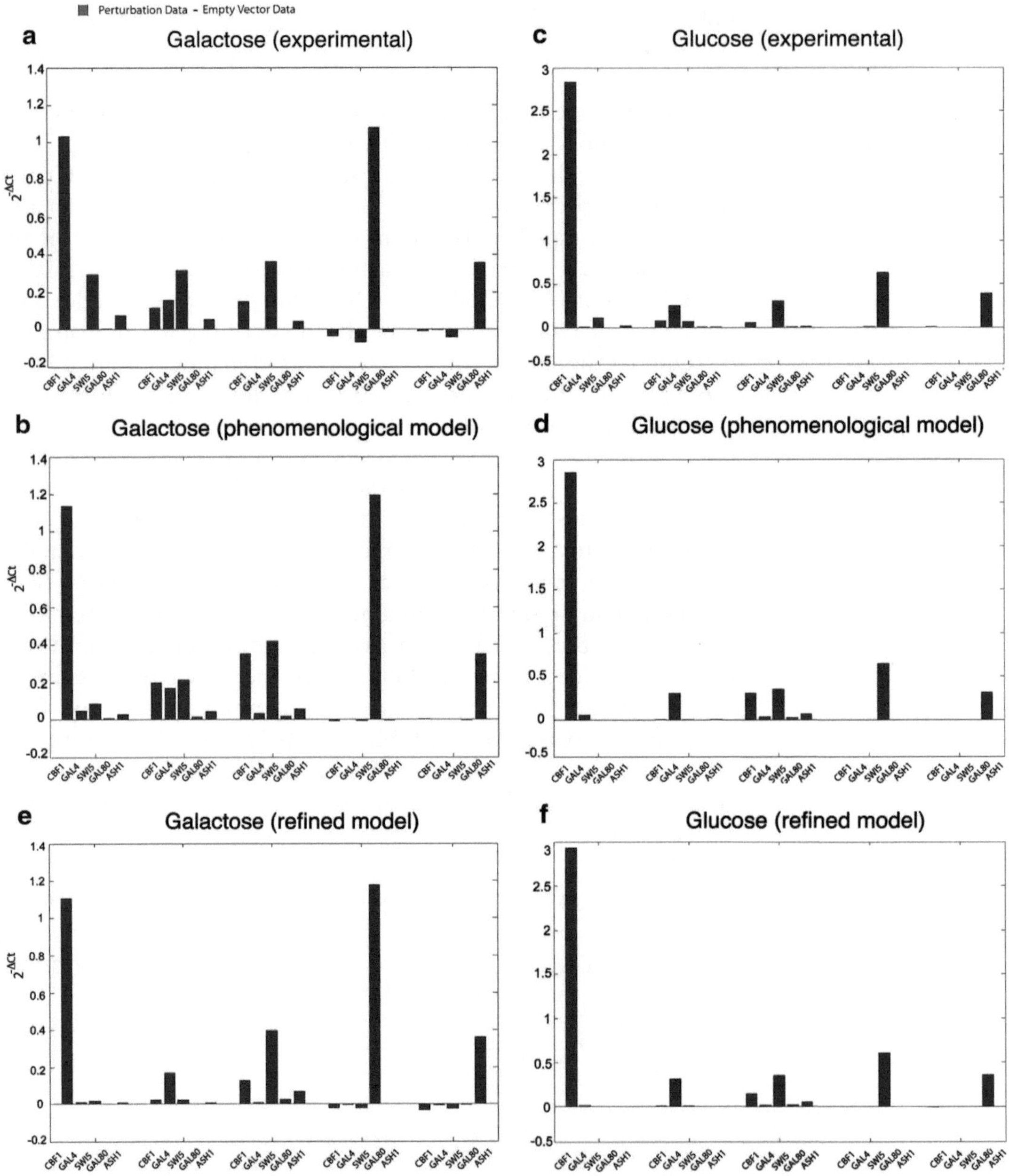


Fig. 4. Experimental and simulated overexpression experiments (**a**, **c**). In vivo expression levels of IRMA genes after overexpression of each gene (perturbed gene; indicated by the *black dots* on the bars) from the constitutive *GPD* promoter (*gray bars*) and after transformation of the empty vector (*white bars*). IRMA cells were transformed with each of the constructs containing one of the five genes or with the empty vector. At least, three difierent colonies were grown in glucose (**b**) and in galactose-rafinose (**a**) up to the steady-state levels of gene expression. Quantitative PCR data are represented (average data from different colonies) (**b**, **d**). In silico expressionlevels of IRMA genes obtained by simulating the overexpression of each gene with the phenomenological model (**e**, **f**). In silico expression levels of IRMA genes obtained by simulating the overexpression of each gene with the refined model.

*3.5.2. Step (II). Identification of Model's Parameters and Model Validation*

For the sake of simplicity, we set all of the Hill coefficients to 1. For the identification of the remaining parameters, we used the "switch-on" dataset, using as initial values the simulated steady-state mRNA levels in glucose. Parameters' identification results, obtained by using the Genetic Algorithm and hand refinement (see Note 6) are shown in Fig. 3a and the inferred parameters in Table 1. The model captures the delay in *CBF1* activation and the small variations of *GAL4* and *GAL80*.

In order to validate the model predictive performance, we used the "Glucose steady-state" and "Galactose steady-state" overexpression experiments, and compared them with their in silico counterparts by simulating the overexpression of each of the five genes (Fig. 4b), (D)). We further validated the predictive performance against the "switch-off" time-series by simulating in silico the "switch-off" experiment (i.e., setting the medium-dependent parameters to their values in glucose and starting the simulation from the steady-state equilibrium in galactose) (Fig. 3b). The phenomenological DDEs model has good descriptive and predictive performance. However, the 24 identified parameter values are likely to be different from their physical values. For example, model parameters (Table 1) indicate that the inhibition of Ash1 on *CBF1* is so weak that can be neglected, even if in the literature it has been reported otherwise (16).

*3.5.3. Additional Experimental Investigation. Promoter Strength Experiments and Reidentification of the Correspondent Parameters (Step (II))*

At this point, we needed to clarify the biological properties of the *HO* promoter by taking direct measurements of the promoters' parameters (3). We thus measured the transcriptional response of the promoters of *GAL10*, *MET16*, *ASH1*, and *HO;* the latter when regulated by both Swi5 and Ash1 (see Note 7). For each promoter, we fitted to data the equation at steady state of the gene whose expression the promoter drives. Of note, it became apparent from the new experimental data and the fitting results that galactose not only weakens the inhibition of Gal80 on the *GAL10* promoter (assumption (A8)), but also allows a faster activation of the *GAL10* promoter. Moreover, in galactose such activation is possible for values of *GAL4* lower than in glucose.

*3.5.4. Step (i). Phenomenological DDEs Refined Model*

In order to capture the behavior observed from the new experiments, we considered two additional parameters in the model to be explicitly dependent on the medium. Thus, we refined the previous model by changing the equation of *SWI5*, which became

$$\frac{\mathrm{d}[SW15]}{\mathrm{d}t} = \alpha_3 + \hat{v}_3\left(\frac{([GAL4])^{h_4}}{\left(\hat{K}_4^{h_4} + ([GAL4])^{h_4}\right)\left(1+\frac{[GAL80]^{h_7}}{\hat{\gamma}^{h_7}}\right)}\right) - \mathrm{d}_3[SW15],$$

where again the symbol ^ indicates parameters dependent on the medium. From the analysis of data, we found that the value assumed by $v_3$ in galactose is 9 times bigger than the one in glucose. Analogously, the value of $k_4$ is 9 times bigger in glucose than in galactose (see Table 1).

*3.5.5. Step (ii). Identification and Validation of the Refined Model*

From the promoter dataset, we estimated 16 parameters, including the medium-dependent ones (Table 1). From such data, we could neither fit degradation constants nor the washing effect parameters. Thus, the remaining 17 parameters were evaluated again from the "switch-on" experiment (Table 1). In simulations, the initial values of mRNA concentrations were set to the steady state values predicted by the model in glucose. The in silico "switch-on" time-series is shown in Fig. 5a. Also in this case, we tested the predictive ability of the model performing in silico the previously described "Glucose steady-state" and "Galactose steady-state" overexpression experiments (Fig. 4e, f) and the "switch-off" time-series (Fig. 5b). Now, the identified parameters confirm that the Ash1 inhibition of the *HO* promoter is indeed strong, as reported in the literature (16).

There are still discrepancies between the in vivo and in silico initial values of *CBF1*, *SWI5*, and *ASH1* in the "switch-off" data-set, and in the predicted steady state of mRNA levels in galactose. We attribute them to the unmodeled effect of protein dynamics, which have been removed from the original model due to the lack of experimental measurements. In particular, we noticed that the Gal4 protein is stable (18), and therefore even a small, or transient, increase in its mRNA level is able to induce the *GAL10* promoter, regulating Swi5 in our network. Since we do not explicitly model protein dynamics, a small increase in *GAL4* mRNA cannot fully activate *GAL10* in the model and does not cause the increase in *SWI5* mRNA seen in vivo. The quality of the fitting and the predictions could be further improved by modeling the proteins levels of all the genes in the network. However, in the actual version of the network, it is not possible to measure protein levels with the exception of only one gene (Cbf1). Thus, the assumption of steady state for protein dynamics is required, not only to simplify the model but also mainly to not introduce the problem of overfitting and nonuniqueness of parameters for proteins.

**3.6. Modeling the Synthetic Transcriptional Positive Feedback Loop in Mammalian Cells**

*3.6.1. Step (i). Derivation of the Mathematical Model*

The network of interest is the one showed in Fig. 2. Details about the network construction are reported in Subheadings 2.3, 2.4 and 3.3. Again, the formal framework we chose is based on Ordinary Differential Equations, since we are measuring the average behavior of a population of cells uniformly infected by the virus.

For each species, i.e., each mRNA (italic capital letters) and correspondent protein concentration (roman small letters), we wrote one equation, which expresses the change in concentration

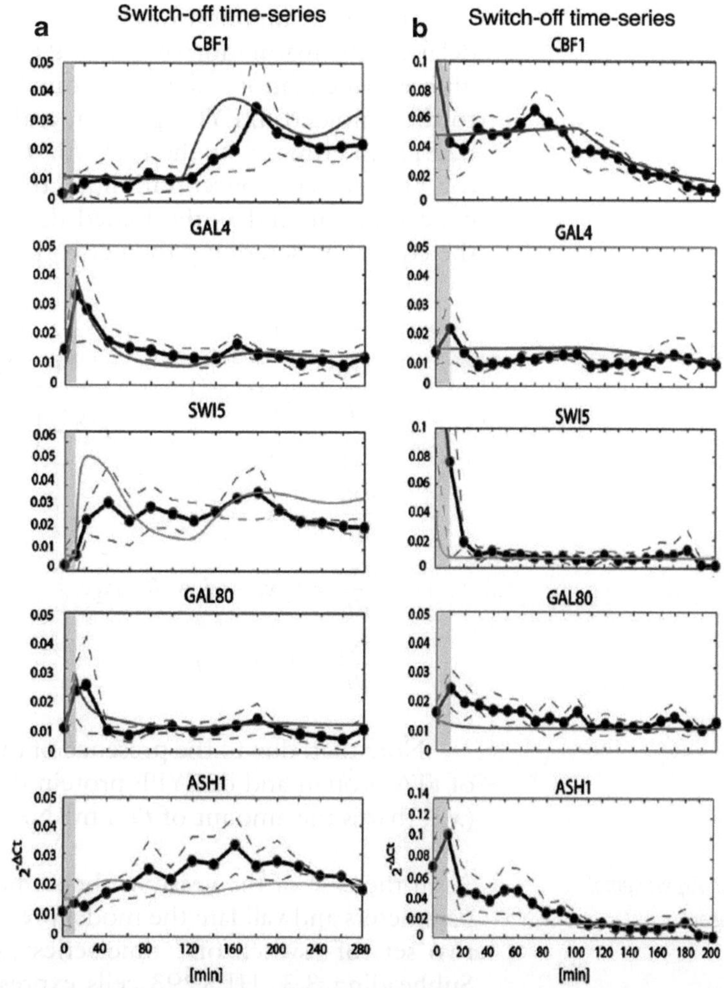


Fig. 5. Identification and validation results of time-series, phenomenological model. *Circles* represent average expression data for each of the IRMA genes at different time points. Dashed lines represent standard errors. *Solid lines* represent in silico data. (**a**) Identification results of the phenomenological model on the average 5 h "switch-on" time-series. (**b**) Validation of the refined model on the average 3 h "switch-off" dataset.

of the species in a given time interval, as the result of a production and a degradation. We assumed:

- Hill functions to model the rate of gene transcription, including basal activity to describe the leakiness of the *CMV-TET* promoter (A1)
- Linear degradation for all genes and proteins (A2)
- Linear dynamics for the translation (A3)
- Michaelis–Menten-like modeling of the effect of the inducer (Doxycycline) (A4)
- Distinct dynamics for the unfolded (inactive) and folded (active) forms of the reporter protein (d2EYFP) (A5)

The last assumption was introduced to take into account time delay for the maturation of d2EYFP protein. Thus, we derived two differential equations as in (11): one for the translation of mRNA to the unfolded d2EYFP protein, and one for the folded protein d2EYFP. Letting $x_1$ be the *tTA/d2EYFP* mRNA concentration, $x_2$ the tTA protein concentration, $x_3$ the unfolded d2EYFP protein concentration, and $x_4$ the folded d2EYFP protein concentration, the equations describing the network become:

$$\frac{dx_1}{dt} = v_1\left(\alpha_1 + (1-\alpha_1)\frac{\left(\frac{\theta}{\theta+D}x_2\right)^{h_1}}{K_1^{h_1} + \left(\frac{\theta}{\theta+D}x_2\right)^{h_1}}\right) - d_1 x_1,$$

$$\frac{dx_2}{dt} = v_2 x_1 - d_2 x_2,$$

$$\frac{dx_3}{dt} = v_2 x_1 - (d_3 + K_f)x_3,$$

$$\frac{dx_4}{dt} = K_f x_3 - d_3 x_4.$$

Note that, due to the presence of the IRES, the concentrations of tTA protein and d2EYPF protein depend on the same variable ($x_1$), that is the amount of *tTA* mRNA.

*3.6.2. Experimental Dataset*

As in the case of the yeast synthetic, network, to identify model's parameters and validate the model predictive ability, we performed two sets of "switch off" time-series experiments as described in Subheading 3.3. HEK293 cells expressing the positive feedback loop, were imaged using time-lapse microscopy and fluorescence of d2EYFP was quantified. At the beginning of the experiment (first time point) cells were treated with Doxycycline to "switch off" the network, since Doxycycline prevents the ttA transactivator to bind the *CMV-TET* responsive promoter. In the first set of experiments the dynamics were followed for 37 h at 37°C, while in the second set the temperature was reduced to 32°C to limit cell motility and thus facilitate image analysis.

*3.6.3. Derivation of Mathematical Models for Synthetic Network*

We first simulated the model using, when available, the parameters values previously estimated in the literature (see ref. 11), and reported in Table 2. However, using such parameters, we were not able to correctly reproduce the experimental "switch-off" dynamic behavior of the postive feedback loop. In particular, using the reference parameters the in-silico simulation shows faster dynamics than the in vivo data. Thus, we used the experimental "switch off" time series to identfy the model parameters using a Genetic Algorithm optimization method, letting the parameters vary in a neighborhood of the literature value. In this case, hand refinement

**Table 2**
**Parameters of the mathematical model of the positive feedback loop in mammalian cells**

| Parameter | Description | Reference value | Estimated value |
|---|---|---|---|
| $K_1$ (nM) | Michaelis-Menten *CMV-TET* promoter | 3 | 3 |
| $\alpha_1$ (nM min$^{-1}$) | Basal activity *CMV-TET* promoter | 0.085 | 0.085 |
| $v_1$ (nM min$^{-1}$) | Maximal Transcription rate *CMV-TET* promoter | 0.055 | 0.35 |
| $v_2$ (min$^{-1}$) | General translation rate | 0.02 | 0.02 |
| $d_1$ (min$^{-1}$) | Degradation rate *tTA* mRNA | 0.017 | 0.017 |
| $d_2$ (min$^{-1}$) | Degradation rate tTA protein | 0.023 | 0.023 |
| $d_3$ (min$^{-1}$) | Degradation rate d2EYFP protein | – | 0.0020 (Dataset 1) 0.0014 (Dataset 2) |
| $h_1$ | Hill coefficient of the *CMV-TET* promoter | 2 | 2 |
| $\theta$ (nM) | Affinity Doxycycline -*CMV-TET* promoter interaction | – | 90 |
| $K_f$ (min$^{-1}$) | Folding rate d2EYFP | 0.015 | 0.015 |

(see Note 6) was not necessary, since the number of parameters to estimate was lower than in the yeast network case, and the search interval was smaller. Of note, during the parameter fitting procedure, we tried to minimize the changes to parameters previously estimated in the literature.

The simulations of the model with the identified parameters (Table 2) are shown in Fig. 6. The model is now able to recapitulate the positive feedback loop dynamics in response to different inducer concentrations and experimental settings (see Subheading 3.3). We estimated a different the degradation rate of the reporter protein (d2EYFP) in the two experimental conditions. As mentioned above, in the first set of experiments (Dataset 1) the cells were kept at 37°C, while, in the second round of experiments (Dataset 2), we used a lower temperature (32°C) to limit cell motility. Figure 6 (A) and (C) show the response of the system using the same amount of inducer (1 μg/mL), with cells at 37°C and 32°C, respectively. The dynamics of the "switch off" are faster if the temperature is higher,

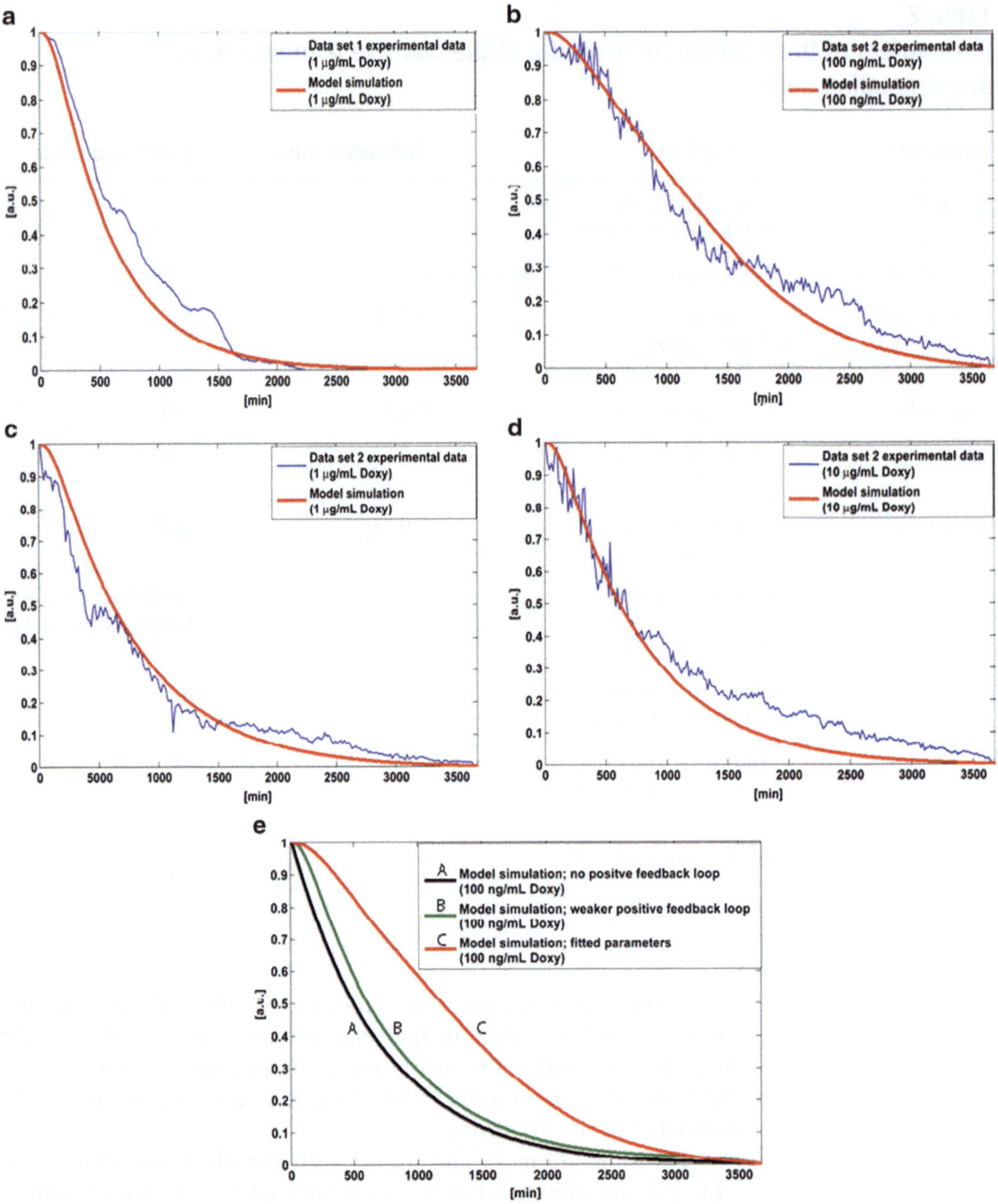


Fig. 6. Dynamical behavior of the positive feedback loop in mammalian cells. In this figure model predictions of the dynamics characterizing the circuit for varying concentrations of Doxycycline (1 μg/mL for (**a**) and (**c**), 100 ng/mL for (**b**) and 10 μg/mL for (**d**)) have been reported. The sample time is equal to 15 min. The cells were treated with the antibiotic at $t=0$ (min). Model predictions are reported in with thick line while experimental results are represented in *blue*. In (**a**), the cells were kept at 37°C and observed up to 37 h. In (**b**–**d**) the cells were kept at 32°C and observed up to 61 h. In (**e**) we report the comparison of the dynamics of the circuit obtained by varying the strength of the positive feedback loop. *C line* = model simulation of the system including the positive feedback loop using the inferred parameter values (Table 2). *B line* = model simulation of the system reducing the strength of the positive feedback loop. *A line* = model simulation of the system removing the positive feedback loop.

as the cells metabolism is faster (13). In the model, we captured this behavior by changing the degradation rate of the reporter protein (parameter $d_3$ in Table 2).

Another parameter we estimated, the maximal transcription rate of the *CMV-TET* promoter (Table 2), was quite different form the reported value in literature. The physical meaning is that the strength of the positive-feedback loop is much stronger than previously estimated, at least in the cell-line we used in this experiment (HEK 293). The presence of the positive feedback loop is the key to understand the dynamics of the network, because it makes it harder for the d2EYFP to be down regulated by Doxycycline. Indeed, in Fig. 6e we analyzed how the presence of the positive feedback loop affects the switch-off dynamics: decreasing its strength (green line) or removing it (black line), the network is switched off much faster. The model was further refined as described in (19).

### 3.7. Time-lapse Experiments Based on Microfluidics Dynamics

In order to carry out experiments with *S. cerevisiae* or mammalian cell lines, appropriate microfluidic device designs must be conceived. The design of such devices should be carefully considered on the basis of the morphological characteristics of the cells used: 10 μm in height channels, for example, may be suited for *S. cerevisiae* but may be not compatible with many mammalian cell types. For this reason the experimenter is highly recommended to check both morphological and physiological properties of the cells to be employed before moving to the device design step. The setup of experiments for *S. cerevisiae* and mammalian cell lines is quite similar and differs only in the preparation of cells and temperature/$CO_2$ conditioning settings. Depending on the motility level of cells you may be interested in lowering it: two effective ways to accomplish this is either (a) by lowering the temperature in the microincubator at 32°C or (b) using drugs such as Cytochalasin D, ladrunculin, Nocodazole, etc. Advantages and disadvantages of both the types of approaches should be carefully evaluated before choosing for one or the other since temperature-based strategy may interfere with cells' metabolism (thus adding uncertainty to the quantification results) while drugs-based strategies (mainly acting on cytoskeleton and microtubules development) may be characterized by unwanted side effects.

#### 3.7.1. Yeast culture Preparation

1. Day 0.
   (a) Fill the 50-mL Falcon tube with 5 mL Synthetic Complete medium.
   (b) Add 500 μL Galactose 20% (w/v).
   (c) Add yeast cells from the solid medium culture.
   (d) Place the Falcon tube in the shaker at 30°C/150–200 rpm.

2. Day 1.
   (a) Measure $O.D._{600}$ and dilute in 5 mL so as to obtain a final $O.D._{600}$ of 0.5.
   (b) Grow for 4–6 h.

This protocol assumes you designed and built a microfluidic device with *n* ports, *w* of which are inlets.

### 3.8. Setting Up the Computing System

This section deals with the specific aspects of the computing system behind the control law. The implementation presented uses MATLAB Simulink but other software packages like National Instruments LabView can be used as well. Here, we provide details of both the steps needed to acquire and process data from the microscope.

1. Program your microscopy control scheme to acquire bright field as well as fluorescent fields at regular time intervals (5 min with exposure times of 200–400 ms have been found to work fine for yeasts).
2. Set up your algorithm so that once images are acquired the image processing routine is launched on them:
   (a) Your image processing algorithm should use bright field image to locate cells.
   (b) Once cells are located, a binary mask image is used to compute a mean fluorescence of cells.
   (c) A mean intensity quantification of the red fluorescence field can be used to match the galactose/raffinose time profile obtained experimentally with the desired one and this information can be used to understand whether clogging or other detrimental phenomena are taking place in the device.

## 4. Notes

1. To obtain a lentiviral vector by recombination of three plasmids, we found that the optimal concentration of each plasmid is: pENTR5′-TOPO promoter 10 ng, pENTR gene 10 ng, p*Lenti/R4R2/V5-DEST* 120 ng, according to the fmoles required for each vector.
2. We suggest to dephosphorylate the p*MAtTA-IRES-d2EYFP* linearized with the AseI restriction enzyme before performing the recombination reaction in pDONR221 to avoid the possibility of forming colonies in the negative control because the plasmid is prone to becoming circular again. We, thus, recommend to add 5 U of the Alkaline Phosphatase, Calf Intestinal (CIP) (New England Biolabs) to the vector linearized

and to leave the reaction at 55°C for 45 min, and then to purify the fragment with the QIA quick PCR purification kit (QIAGEN).

3. It is really necessary to control that all the concentrations of the vectors you find in the ViraPower Promotrless Lentiviral Gateway Expression System are correct.
4. The Doxycycline and Blasticidine working solution can be stored at 4°C up to 1 month.
5. As reported, when you are modeling a synthetic network, to decide what can be simplified, and what needs to be modeled in more details, it is necessary to go through iterative refinement steps both in the model and in the experimental dataset.
6. When the number of unknown parameters is high, as in yeast network case, and the physical feasible range for them is large, it is not easy to identify them. Even though stochastic identification techniques provide a flexible approach to nonlinear optimization, it is good to proceed with hand refinement. Thus, we proceeded in this way:
   (a) Multiple identification of each parameter running GA more than one time letting the parameter vary in a physical feasible range. In this way, we obtained different values for each parameter (even with the same setting, since the algorithm is stochastic) that, however, remain quite close to each other. Thus, this step allows to significantly restricting the search range.
   (b) Hand refinement of the parameters in the narrow range obtained in step (a).
7. Since each experiment is costly and time consuming, the best option in the iterative analysis of a biological network is, at each step, to only perform those experiments that the mathematical modeling deems indispensable. In the case of the yeast network, we could have performed the promoter strength experiments from the beginning, since the Hill functions were almost unchanged during the model refinement, with the exception of the *GAL10* and *HO* promoters modeling. However, the need of performing such experiments arose after the identification of Model C, since we did not trust the identified Hill functions parameters.
8. The quality of the microscopy section is of primary importance in experiments involving *S. cerevisiae* expressing fluorescent reporters due to the thickness of the wall protecting yeasts from the external environment. Yeast experiments of this type are best carried out at magnifications like 40× and 60×. High numerical aperture objectives should be preferred to make the image processing easier. High-quality mercury lamps are

needed to obtain good fluorescent field images; CCD cameras can lower the Signal to Noise Rate (SNR) associated to the images when compared to conventional cameras (20).

9. Attention should be paid to the air-bubbles washing steps: PDMS is porous and the user can take advantage of this property to get rid of the air-bubbles just by injecting $ddH_2O$ at high pressure (just by pressing the plunger of the syringes used to flush the device). Once there are no more bubbles in the device, the experimenter should pay attention to the disconnection of the $ddH_2O$ syringes: the needles should be gently disconnected and pressing the plunger for 10 s before unplugging the related needle can help in generating the pressure needed to prevent air from reentering the port.
10. All media, sugars and dyes to be used should be filtered with 0.22 μm filters so as to avoid the presence of small particles that can accelerate the clogging of the device or, alternatively, mislead the image segmentation algorithm. The preparation of the media in a fume hood does provide a strong advantage in this context.
11. When sugars such as galactose are used as inducers, supplementing with supporting metabolic sources should be considered as an option. Growth of yeast cells in media featuring galactose can be quite slow: having cells slowly moving in a field can be a good advantage from the point of view of the image processing algorithm, unfortunately the metabolic implications of these conditions cannot always be predicted with accuracy thus adding uncertainty to our experiments.
12. In the proposed protocol, we used a red fluorescent dye, namely, Sulforhodamine B, whose excitation and emission spectra largely overlap with red fluorescent protein tag (e.g., Cherry etc.). If the protein to be controlled is tagged with such a fluorescent protein it may help using a different dye (e.g., Atto 655) to track the inducer compound concentration. Nevertheless, it should be noted that choosing such option may imply buying new filters (Atto 655, for example, emits in the far red and requires a specific filter like the Cy5.5-A (Semrock, PN FF685-Di01-25 × 36)).
13. The image processing algorithm is one of the key points of the whole setup: several free software packages have been proposed in the field of cellular microscopy being CellProfiler (21), Cell-ID (22), CellTracker (23). Nevertheless, it should be noted that achieving a good quality of segmentation often may require a long session of parameter fine-tuning and, more often, flexibility of these packages is an added value and not a core characteristic. Experienced users should consider developing their own code on the basis of their specific application and experimental context: testing the dependence of the

segmentation results on the experimental conditions may help in optimizing the code and make it more robust to varying boundary conditions.

## References


1. Cuccato, G., Della Gatta, G., and di Bernardo, D. (2009) Systems and Synthetic biology: tackling genetic networks and complex diseases, *Heredity* **102**, 527–532.
2. Mangan, S., and Alon, U. (2003) Structure and function of the feed-forward loop network motif, *Proc Natl Acad Sci USA* 100, 11980–11985.
3. Cantone, I., Marucci, L., Iorio, F., Ricci, M. A., Belcastro, V., Bansal, M., Santini, S., di Bernardo, M., di Bernardo, D., and Cosma, M. P. (2009) A yeast synthetic network for in vivo assessment of reverse-engineering and modeling approaches, *Cell* 137**,** 172–181.
4. Gossen, M., and Bujard, H. (1992) Tight control of gene expression in mammalian cells by tetracycline-responsive promoters, *Proc Natl Acad Sci USA* 89**,** 5547–5551.
5. Dull, T., Zufferey, R., Kelly, M., Mandel, R. J., Nguyen, M., Trono, D., and Naldini, L. (1998) A third-generation lentivirus vector with a conditional packaging system, *J Virol* 72, 8463–8471.
6. Kafri, T., van Praag, H., Gage, F. H., and Verma, I. M. (2000) Lentiviral vectors: regulated gene expression, *Mol Ther* 1, 516–521.
7. Nelles, O. (2000) *Nonlinear system Identication: From Classical Approaches to Neural*, Springer.
8. de Jong, H. (2002) Modeling and simulation of genetic regulatory systems: a literature review, *J Comput Biol* 9, 67–103.
9. Mitchell, M. (1998) *An Introduction to Genetic Algorithms* The MIT Press.
10. Weber, W., Stelling, J., Rimann, M., Keller, B., Daoud-El Baba, M., Weber, C. C., Aubel, D., and Fussenegger, M. (2007) A synthetic time-delay circuit in mammalian cells and mice, *Proc Natl Acad Sci USA* 104**,** 2643–2648.
11. Tigges, M., Marquez-Lago, T. T., Stelling, J., and Fussenegger, M. (2009) A tunable synthetic mammalian oscillator, *Nature* 457, 309–312.
12. Mukherji, S., and van Oudenaarden, A. (2009) Synthetic biology: understanding biological design from synthetic circuits, *Nat Rev Genet* 10, 859–871.
13. Kaufmann, H., Mazur, X., Fussenegger, M., and Bailey, J. E. (1999) Influence of low temperature on productivity, proteome and protein phosphorylation of CHO cells, *Biotechnol Bioeng* 63, 573–582.
14. Jona, G., Choder, M., and Gileadi, O. (2000) Glucose starvation induces a drastic reduction in the rates of both transcription and degradation of mRNA in yeast, *Biochim Biophys Acta* 1491, 37–48.
15. Bhoite, L. T., Yu, Y., and Stillman, D. J. (2001) The Swi5 activator recruits the Mediator complex to the HO promoter without RNA polymerase II, *Genes Dev* 15, 2457–2469.
16. Cosma, M. P., Tanaka, T., and Nasmyth, K. (1999) Ordered recruitment of transcription and chromatin remodeling factors to a cell cycle- and developmentally regulated promoter, *Cell* 97, 299–311.
17. Bennett, M. R., Pang, W. L., Ostroff, N. A., Baumgartner, B. L., Nayak, S., Tsimring, L. S., and Hasty, J. (2008) Metabolic gene regulation in a dynamically changing environment, *Nature* 454, 1119–1122.
18. M. Muratani, C. K., K.M. Shokat, and W.P. Tansey. (2005) The f box protein dsgl/mdm30 is a transcriptional coactivator that stimulates gal4 turnover and cotranscriptional mrna processing., *Cell* 120, 887–899.
19. Siciliano, V. et al. (2011) Construction and modelling of an inducible positive feedback loop stably integrated in a mammalian cell-line. PLoS Computational Biology, 7(6), p.e1002074.
20. Ferry, M. S., Razinkov, I. A., and Hasty, J. (2011) Microfluidics for synthetic biology: from design to execution. Meth. Enzymol. 497, 295–372.
21. Carpenter, A. E., Jones, T. R., Lamprecht, M. R., Clarke, C., Kang, I. H., Friman, O., Guertin, D. A., Chang, J. H., Lindquist, R. A., Moffat, J., Golland, P., and Sabatini, D. M. (2006) CellProfiler: image analysis software for identifying and quantifying cell phenotypes, *Genome Biol* 7**,** R100.
22. Chernomoretz, A., Bush, A., Yu, R., Gordon, A., and Colman-Lerner, A. (2008) Using Cell-ID 1.4 with R for microscope-based cytometry, *Curr Protoc Mol Biol Chapter 14*, Unit 14 18.
23. Shen, H., Nelson, G., Nelson, D. E., Kennedy, S., Spiller, D. G., Griffiths, T., Paton, N., Oliver, S. G., White, M. R., and Kell, D. B. (2006) Automated tracking of gene expression in individual cells and cell compartments, *J R Soc Interface* 3, 787–794.

# Part II

## Development and Characterization of the Synthetic Biology Toolbox

# Chapter 5

# Reprogramming a GFP Reporter Gene Subjects It to Complex Lentiviral Gene Regulation

**Frank Notka and Ralf Wagner**

## Abstract

Late human immunodeficiency virus (HIV)-derived RNAs encoding relevant therapeutic targets or promising vaccine compounds, such as the HIV-1 group-specific antigen (Gag), are translocated from the nucleus into the cytoplasm via sophisticated export machinery. Relevant steps include the concerted action of several cis-acting RNA elements with the viral Rev-shuttle protein and several cellular components (Ran1/Exportin; Crm1). Based on detailed understanding of the molecular mechanisms guiding this complex process, we used rational codon usage modification to design and reprogram a GFP encoding reporter RNA now exactly mimicking the complex transcriptional and posttranscriptional regulation of late lentiviral mRNAs.

**Key words:** Gene design, Functional reprogramming, Gene synthesis, HIV-1, GFP, Codon usage, Rev, RRE, RNA export

## 1. Introduction

Human immunodeficiency virus-1 (HIV-1), like any other virus, has evolved a complex and highly regulated strategy to organise concerted gene expression. Controlling and timing gene expression are vital to facilitate coordinated dispersal of viral components, i.e. proteins and RNA, which in turn is essential for assembling functional replication competent virus progeny. This complex regulation is necessary because the HIV genome contains only one promoter and hence, after integration of the provirus into the host genome, produces only one primary transcript of 9 kilobases (kb) that not only serves as genomic RNA for progeny virus, but also as the messenger RNA (mRNA) encoding the viral Gag and Gag–Pol proteins. Successful replication and production of next-generation infectious viruses also require the balanced expression of seven additional viral proteins.

Wilfried Weber and Martin Fussenegger (eds.), *Synthetic Gene Networks: Methods and Protocols*,
Methods in Molecular Biology, vol. 813, DOI 10.1007/978-1-61779-412-4_5, © Springer Science+Business Media, LLC 2012

The first step towards providing this proteomic diversity is the generation of more than 40 viral mRNAs by alternative splicing events, basically classified into three species of various lengths (1). The process of RNA splicing is highly orchestrated and requires several sequence motifs within the RNA for recognition by the cellular spliceosome (a 5′ splice donor, a branch point, and a 3′ splice acceptor). HIV-1 uses multiple alternative 5′ and 3′ splice sites to generate the different mRNA species (for a review, see ref. 2). One key mechanism for translating different splice products into controlled gene expression is restricted and organised export of RNA classes from the nucleus into the cytoplasm. This strategy uses viral factors and elements and involves cellular components, such as splice factors, nuclear translocation factors, or compartmentalisation (reviewed in ref. 3).

The central element of this mechanism is the interaction between the viral trans-active Rev protein and a complex 351 nt RNA stem-loop structure, called the Rev-responsive element (RRE), promoting nuclear export of late HIV-1 RNAs (4). The Rev gene product is a shuttle protein, derived from a completely spliced and hence regularly exported RNA species. Binding to the RRE cis-acting target sequence contained in all late, incompletely spliced HIV transcripts directs them into the CRM1-dependent export pathway (5–8). Otherwise, in the absence of Rev protein, late RNA species are retained in the nucleus, trapped in spliceosomes via the strong splice donor site present within the 5′ untranslated region (UTR) of these RNA species.

The association with spliceosomes is necessary to protect the viral pre-mRNAs from nuclear degradation (9). Nuclear degradation of the non-protected, late HI-viral RNA has been associated with diffuse AU-rich RNA sequence elements, the so-called *c*is-acting *r*epressive (CRS) or *in*hibitory (INS) *s*equences (10–15). The repressive function of these sequences could be inactivated by introducing silent mutations, which consequently allowed Gag expression in the absence of Rev (16, 17). Extending this limited mutagenesis to complete codon optimisation showed equivalent results (18–20). These modifications allowed constitutive nuclear export of the corresponding mRNAs even in the absence of any regulatory element required in the natural situation as described above (21). HIV codon usage, relative to human codon usage, is suboptimal for expression in human cells (the natural HIV host), but this apparently contributes only marginally to diminished translational efficiency, compared to the major effect of processing and export of nuclear RNA (22).

Taken together, these findings suggest that Rev dependence of late lentiviral transcripts relies on (1) a functional 5′ splice donor in the UTR, (2) AU-rich codon usage, and (3) the interaction between Rev and RRE for active nuclear export. In order to test whether our understanding of this complex mechanism is valid or at least the basic principle, we wanted to transfer these findings to an unrelated, unrestrictedly expressed common reporter. Choosing the green fluorescent protein (*gfp*) gene, we conferred the same features: (1) a strong 5′ splice donor site, (2) AU-rich codon usage, and (3) RRE to this gene by conventional gene design, optimisation, and gene synthesis. On different experimental levels, our adaptation of the *gfp* reporter gene to HIV codon bias was sufficient to turn this hivGFP RNA into a quasi-lentiviral message (23). Using human cell lines for expression experiments, we demonstrated that hivGFP expression was significantly decreased (as shown by Western blotting), strictly correlating with reduced RNA levels (shown by Northern blotting). The reprogrammed *gfp* gene further followed the rules of late lentiviral gene expression. Accordingly, in the presence of the HIV 5′ major splice donor, hivGFP RNAs were stabilised in the nucleus, and following fusion of the RRE and co-expression of HIV-1 Rev, were efficiently exported to the cytoplasm, as demonstrated in cell culture, co-transfection, and rescue experiments at the RNA level (using Northern blots). This Rev-dependent translocation was specifically inhibited by Leptomycin B, as demonstrated in cell culture inhibition experiments and Western blot analyses, suggesting export via the CRM1-dependent pathway used by late lentiviral transcripts.

## 2. Materials

### 2.1. Gene and Plasmid Synthesis

1. PCR master mix (prepared freshly for 50 reactions): Mix 1 ml $H_2O$, 50 μl dNTPs (100 mM each), 250 μl 10× reaction buffer, 50 μl PAN Polymerase mix (PAN, Aidenbach, Germany) and spin down; store at 2–8°C. The polymerase is stable for more than 12 months if stored at −20°C, but only for a few days at temperatures above −20°C. The dNTP set can be stored for 24 months at −20°C. Avoid multiple freeze/thaw cycles. For long-term storage, aliquoting is recommended.
2. MinElute® PCR Extraction Kit or MinElute® Gel Extraction Kit (Qiagen, Hilden, Germany): Store spin columns at 2–8°C and the remaining kit components at room temperature (RT) for up to 12 months. Buffers PB and ERC contain guanidine hydrochloride and isopropanol and are therefore harmful, irritant, and flammable. Buffer QG contains harmful and irritant guanidine thiocyanate.
3. Elution buffer (EB) (10 mM Tris/HCl, pH 8.5).
4. Restriction enzymes (New England Biolabs (NEB), Ipswich, MA, and Fermentas, Burlington, Canada): Store at −20°C.
5. Tenfold restriction buffer 1–4 (NEB) and Tango-buffer (Fermentas): Store at 2–8°C.

6. TE buffer: 10 mM Tris/HCl, 1 mM ethylenediamine tetraacetic acid (EDTA), pH 8, can be stored at room temperature.
7. 10× loading dye: Mix 250 mg bromophenol blue (Merck, Darmstadt, Germany), 250 mg xylene cyanol (Merck), 58 ml glycerine (86%, Roth, Karlsruhe, Germany) in 42 ml $H_2O$ and aliquot.
8. T4-DNA-Ligase-buffer (NEB): Store at −20°C in aliquots. Thaw and resuspend just before use.
9. T4-DNA-Ligase (NEB): Store at −20°C; be sure to keep on ice while taking a sample.
10. Agarose gel running buffer: TAE buffer (40 mM Tris/HCl, pH 8.0, 20 mM NaAc, 2 mM EDTA).
11. Luria Bertani (LB) medium: In 1 L $H_2O$ mix 10 g Bacto-trypton (Sigma–Aldrich, Deisenhofen, Germany), 5 g yeast extract (Sigma–Aldrich), 10 g NaCl, adjust to pH 7.0 with NaOH, autoclave, and store at 4°C.

### 2.2. Cell Culture

1. Culture medium: Dulbecco's Modified Eagle's Medium (DMEM) (Gibco Life Technologies, Karlsruhe, Germany) supplemented with 10% foetal bovine serum (FBS, PAN).
2. 2.5 M calcium chloride: Dissolve 277.45 g $CaCl_2$ in 1 L $H_2O$. Filter sterilise and autoclave.
3. HEPES-buffered saline (HeBS): Dissolve 8.2 g NaCl, 5.95 g HEPES, 0.105 g $Na_2HPO_4$ in 500 ml $H_2O$, adjust to pH 7.05.
4. Phosphate-buffered saline (PBS): 10 mM $Na_2HPO_4$, 1.8 mM $KH_2PO_4$, 137 mM NaCl, 2.7 mM KCL, adjust to pH 8.0.
5. Teflon® cell scrapers (Fisher Scientific, Pittsburgh, PA).
6. RIPA buffer: Mix 5 ml Tris/HCl (1 M), pH 8.0, 15 ml NaCl (1 M), 1 ml SDS (10%), 1 ml Nonidet P40, 0.5 g Na-deoxycholate and adjust to 100 ml with $H_2O$.
7. Leptomycin B (LMB, Merck).

### 2.3. SDS-Polyacrylamide Gel Electrophoresis

1. Acrylamide/bisacrylamide solution (19:1, Roth): Be careful when handling before polymerisation, wear gloves, and once polymerisation is completed it is no longer toxic.
2. Separating buffer (5×): 1.88 M Tris/HCl, pH 8.8, 0.5% SDS. Store at room temperature.
3. Stacking buffer (5×): 0.625 M Tris/HCl, pH 6.8, 0.5% SDS. Store at room temperature.
4. *N,N,N,N′*-Tetramethyl-ethylenediamine (Roth).
5. Ammonium peroxydisulfate (APS): 10% in $H_2O$ (Roth).
6. SDS-polyacrylamide gel electrophoresis (SDS-PAGE) running buffer: 25 mM Tris/HCl, pH 7.5, 190 mM glycine, 0.1% SDS may be reused two to three times.

7. SDS sample buffer (5×): 125 mM Tris/HCl, pH 6.8, 2% β-mercaptoethanol, 10% glycerine, 1 mM EDTA, 0.005% bromophenol blue (Roth).

### 2.4. Western Blotting

1. Supported nitrocellulose membrane (Schleicher & Schüll, Dassel, Germany).
2. 3MM chromatography paper (Whatman, Maidstone, UK).
3. Transfer buffer: 150 mM glycine, 25 mM Tris/HCl, pH 8.3, 10% methanol. Store in the transfer apparatus at room temperature (with cooling during use, see Note 1).
4. Tris-buffered saline (TBS): Prepare 10× stock with 1.5 M NaCl, 100 mM Tris/HCl, pH 7.5. Dilute 100 ml with 900 ml water for use.
5. TBS with Tween® (TTBS): Prepare 10× stock with 1.5 M NaCl, 100 mM Tris/HCl, pH 7.5, 1% Tween®-20. Dilute 100 ml with 900 ml water for use.
6. Blocking buffer: 5% (w/v) skimmed dry milk in TTBS.
7. Antibody dilution buffer: TTBS supplemented with 2% (w/v) bovine serum albumin (BSA).
8. Primary antibody: Anti-GFP monoclonal antibody (rabbit) (BD Bioscience, Heidelberg, Germany).
9. Secondary antibody: Anti-rabbit IgG conjugated to horseradish peroxidase (HRP) (gout) (Pierce, Rockford, USA).
10. Enhanced chemiluminescent (ECL) reagents SuperSignal® West (Pierce) and Bio-Max ML film (Kodak, Rochester, NY).

### 2.5. Northern Blot

1. Lysis buffer: 50 mM Tris, 140 mM NaCl, 1.5 mM $MgCl_2$, 0.5% NP-40, pH 8.0.
2. RNeasy® Kit from Qiagen.
3. Oligotex® mRNA Kit from Qiagen.
4. 10× E: 0.2 M MOPS, 0.05 M NaAc, and 0.005 M EDTA, adjust to pH 7.0 with NaOH.
5. Sample denaturation mix (100 μl): 64.6 μl formamide, 22.6 μl formaldehyde, and 13 μl 10× E. This mix is not stable; prepare the required amount just before use.
6. Loading dye mix: 50% glycerol, 0.3% xylene cyanol, 0.3% bromophenol blue, and 1 mM EDTA
7. 20× SSC: 3 M NaCl, 0.3 M $Na_3C_6H_5O_7$, adjust to pH 7.0.
8. Methylene blue solution: 0.03% methylene blue (Merck), 0.3 M NaAc.
9. Riboprobe® in Vitro Transcription Systems (Promega, Mannheim, Germany).
10. $^{32}P$-α-CTP (GE Healthcare, Little Chalfont, UK).

11. 20× SSPE: 3 M NaCl, 0.25 M $NaH_2PO_4$, and 0.02 M EDTA, adjust to pH 7.4.
12. 50× Denhardt's: Add to 900 ml distilled $H_2O$ 10 g Ficoll 400, 10 g polyvinylpyrolidone, 10 g BSA and adjust to 1 L with $H_2O$; filter solution prior to storage through a 0.2 μM filter. Store at 4°C.
13. Hybridisation solution: Mix 12.5 ml 20× SSPE, 25 ml formamide, 5 ml 50× Denhartd's, 2.5 ml 10% SDS and add 20 mg t-RNA (Roche).
14. Wash buffer: Add 10 ml of a 10% SDS stock solution and 1 ml 20× SSPE to 900 ml distilled $H_2O$, adjust to 1 L with $H_2O$.

## 3. Methods

### *3.1. Design and Production of Quasi-Lentiviral GFP Reporter Plasmids*

#### *3.1.1. Gene Design and Codon Adaptation*

Generate a synthetic version of a GFP-encoding gene, mirroring the codon bias of HIV-1 gag (Table 1). Starting from the complete sequence of a humanised *gfp* gene according to the GCG database (GENETICS COMPUTER GROUP; GenBank® database: Accession no. M33262), transfer the HIV-1 codon preference as specified in a HIV codon usage table (CUT, http://www.kazusa.or.jp/codon/) to the *gfp* primary gene sequence. Basically, back translate the huGFP amino acid sequence into a gene sequence by exchanging the mammalian-preferred codons with HIV-preferred codons as specified in Table 1 wherever possible. Alternatively, special software tools, e.g. the GeneOptimizer® software, can be used (see Note 2). As a result, the sequence homology between the HIV-adapted synthetic sequence, referred to as hivGFP, and the original codon-optimised ("humanised") variant huGFP should be reduced to 68%, whereas the amino acid composition of the corresponding gene products remains unaffected (Fig. 1).

#### *3.1.2. Gene Synthesis*

1. Sequence breakdown: Divide the hivGFP gene sequence as designed according to Subheading 3.1.1 into overlapping oligonucleotides. Split the sense-strand sequence into sequential L-oligos of 50–60 nucleotides (nt) in length. Make sure that the aligned oligos cover the complete gene sequence without gaps. Split the antisense strand into shorter M-oligos of approx. 40 nt in length partially overlapping the corresponding, complementary L-oligos (see Fig. 2; Note 3).
2. Design of cloning primer: For a second amplification step and in preparation for cloning, design two terminal oligos, primer forward (pf) and primer backward (pb). These terminal oligos should provide a 25-nt overlap with the sequence and an additional 5′ protruding sequence containing restriction enzyme recognition sites, pf-containing sites for *Bam*HI and *Xho*I, and

**Table 1**
**Preferred codon usage of HIV and mammal genes**

| Amino acid | Preferred codon HIV genes | Preferred codon mammal genes |
|---|---|---|
| Ala | GCA | GCC |
| Arg | AGA | AGG |
| Asn | AAT | AAC |
| Asp | GAC | GAC |
| Cys | TGT | TGC |
| End | TAA | TGA |
| Gln | CAA | CAG |
| Glu | GAA | GAG |
| Gly | GGA | GGC |
| His | CAT | CAC |
| Ile | ATA | ATC |
| Leu | TTA | CTG |
| Lys | AAA | AAG |
| Met | ATG | ATG |
| Phe | TTT | TTC |
| Pro | CCA | CCC |
| Ser | AGC | TCC |
| Thr | ACA | ACC |
| Trp | TGG | TGG |
| Tyr | TAT | TAC |
| Val | GTA | GTG |

pb-containing sites for *Eco*RI and *Nco*I for subsequent cloning. The designed oligonucleotides are best produced by conventional oligonucleotide synthesis procedures provided by commercial suppliers (see Note 4). The synthetic hiv–gfp gene exhibiting codon usage of late HIV-1 genes is generated via stepwise PCR amplification, as follows.

3. A first amplification round is called sequential chain reaction (SCR). For each fragment, mix 5 µl of an oligo-pool containing all L- and M-oligos at a final concentration of 15 nM, 18 µl

```
huGFP   1    ATGGTGAGCAAGGGCGAGGAGCTGTTCACCGGGGTGGTGCCCATCCTGGTCGAGCTGGAC
             ||||| ||||| || || ||  | || || || || || || || ||  | ||  | |||
hivGFP  1    ATGGTAAGCAAAGGAGAAGAATTATTTACAGGAGTAGTACCAATATTAGTAGAATTAGAC
              M  V  S  K  G  E  E  L  F  T  G  V  V  P  I  L  V  E  L  D

huGFP   61   GGCGACGTAAACGGCCACAAGTTCAGCGTGTCCGGCGAGGGCGAGGGCGATGCCACCTAC
             || |||||||| || || || || |||||   ||| || || || || || || || || 
hivGFP  61   GGTGACGTAAATGGACATAAATTTAGCGTAAGCGGAGAAGGAGAAGGTGACGCAACATAT
              G  D  V  N  G  H  K  F  S  V  S  G  E  G  E  G  D  A  T  Y

huGFP   121  GGCAAGCTGACCCTGAAGTTCATCTGCACCACCGGCAAGCTGCCCGTGCCCTGGCCCACC
             || ||  | ||  | || || || || || || || || ||  | || || |||||||| || 
hivGFP  121  GGAAAATTAACATTAAAATTTATATGTACAACAGGAAAATTACCAGTACCCTGGCCAACA
              G  K  L  T  L  K  F  I  C  T  T  G  K  L  P  V  P  W  P  T

huGFP   181  CTCGTGACCACCTTCACCTACGGCGTGCAGTGCTTCAGCCGCTACCCCGACCACATGAAG
              | || || || || || || || || || || || ||| | || || ||||| ||||| 
hivGFP  181  TTAGTAACAACATTTACATATGGAGTACAATGTTTTAGCAGATATCCAGACCATATGAAA
              L  V  T  T  F  T  Y  G  V  Q  C  F  S  R  Y  P  D  H  M  K

huGFP   241  CAGCACGACTTCTTCAAGTCCGCCATGCCCGAAGGCTACGTCCAGGAGCGCACCATCTTC
             || || ||||| || ||   ||| ||||| ||||| || || || ||  | || || || 
hivGFP  241  CAACATGACTTTTTTAAAAGCGCAATGCCAGAAGGATATGTACAAGAAAGAACAATATTT
              Q  H  D  F  F  K  S  A  M  P  E  G  Y  V  Q  E  R  T  I  F

huGFP   301  TTCAAGGACGACGGCAACTACAAGACCCGCGCCGAGGTGAAGTTCGAGGGCGACACCCTG
             || || |||||||| || || || ||  | || || || || || || || |||||  | 
hivGFP  301  TTTAAAGACGACGGAAATTATAAAACAAGAGCAGAAGTAAAATTTGAAGGAGACACATTA
              F  K  D  D  G  N  Y  K  T  R  A  E  V  K  F  E  G  D  T  L

huGFP   361  GTGAACCGCATCGAGCTGAAGGGCATCGACTTCAAGGAGGACGGCAACATCCTGGGGCAC
             || ||  | || ||  | || || || ||||| || |||||||| || ||   | || || 
hivGFP  361  GTAAATAGAATAGAATTAAAAGGAATAGACTTTAAAGAGGACGGAAATATATTAGGACAT
              V  N  R  I  E  L  K  G  I  D  F  K  E  D  G  N  I  L  G  H

huGFP   421  AAGCTGGAGTACAACTACAACAGCCACAACGTCTATATCATGGCCGACAAGCAGAAGAAC
             ||  | || || || || || ||||| || || ||||| ||||| ||||| || || || 
hivGFP  421  AAATTAGAATATAATTATAATAGCCATAATGTATATATAATGGCAGACAAACAAAAAAAT
              K  L  E  Y  N  Y  N  S  H  N  V  Y  I  M  A  D  K  Q  K  N

huGFP   481  GGCATCAAGGTGAACTTCAAGATCCGCCACAACATCGAGGACGGCAGCGTGCAGCTCGCC
             || || || || || || || ||  | || || || |||||||| ||||| ||  | || 
hivGFP  481  GGAATAAAAGTAAATTTTAAAATAAGACATAATATAGAGGACGGAAGCGTACAATTAGCA
              G  I  K  V  N  F  K  I  R  H  N  I  E  D  G  S  V  Q  L  A

huGFP   541  GACCACTACCAGCAGAACACCCCCATCGGCGACGGCCCCGTGCTGCTGCCCGACAACCAC
             ||||| || || || || || || || || ||||| || ||  |  | || ||||| || 
hivGFP  541  GACCATTATCAACAAAATACACCAATAGGTGACGGACCAGTATTATTACCAGACAATCAT
              D  H  Y  Q  Q  N  T  P  I  G  D  G  P  V  L  L  P  D  N  H

huGFP   601  TACCTGAGCACCCAGTCCGCCCTGAGCAAAGACCCCAACGAGAAGCGCGATCACATGGTC
             ||  | ||||| ||   |||  | ||||||||||| || || ||  | || || ||||| 
hivGFP  601  TATTTAAGCACACAAAGCGCATTAAGCAAAGACCCAAATGAAAAAAGAGACCATATGGTA
              Y  L  S  T  Q  S  A  L  S  K  D  P  N  E  K  R  D  H  M  V

huGFP   661  CTGCTGGAGTTCGTGACCGCCGCCGGGATCACTCTCGGCATGGACGAGCTGTACAAGTAA
              |  | || || || || || || || || ||  | || ||||||||  | || || |||
hivGFP  661  TTATTAGAATTTGTAACAGCAGCAGGAATAACATTAGGAATGGACGAATTATATAAATAA
              L  L  E  F  V  T  A  A  G  I  T  L  G  M  D  E  L  Y  K  *
```

Fig. 1. Sequence alignment of the humanised (*upper sequence*) and HIV-adapted (*lower sequence*) *gfp* genes with corresponding GFP amino acids indicated below the second base of each triplet. The original AT-content of huGFP (37%) was elevated to 69% in hivGFP, resulting in a final homology of 68% between sequences. Reprinted with permission from Elsevier (23).

$H_2O$, and 27 μl PCR master mix and subject to PCR using the protocol as outlined in Table 2.

4. The SCR product cannot be used for cloning directly, but has to be further amplified using a method called Sequential PCR (SPCR) to introduce the terminal restriction sites. Mix 7 μl of the SCR reaction with each 2 μl of pf and pb (at a concentration

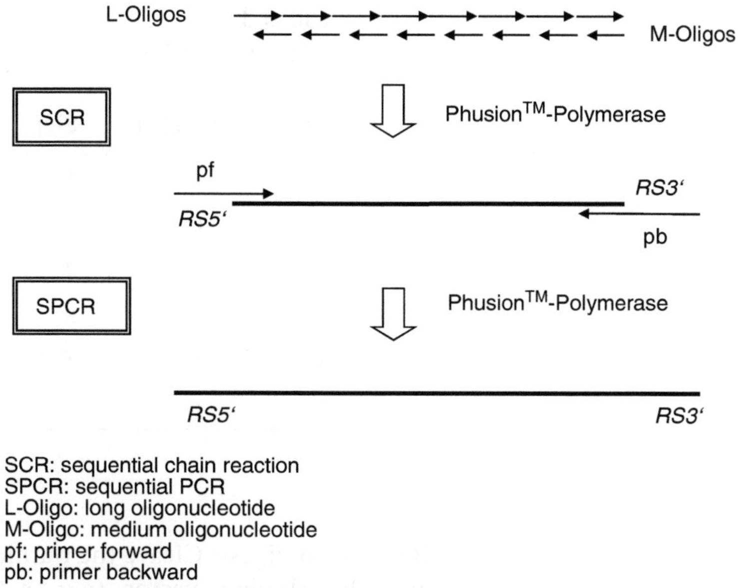


Fig. 2. Schematic overview of a gene synthesis workflow. A set of oligonucleotides synthesised by automated organic chemistry is stepwise assembled, elongated, and amplified into a full-length fragment in two distinct PCR-based process steps.

**Table 2**
**PCR protocol 1**

| Cycle | Number | Step | °C | Time |
|---|---|---|---|---|
| 1 | 1 | 1 | 95 | 04:00 |
| 2 | 30 | 1 | 95 | 00:30 |
| | | 2 | 60→40[a] | 00:30 |
| | | 3 | 72 | 01:00 |
| 3 | 1 | 1 | 72 | 04:00 |
| 4 | 1 | 1 | 4 | ∞ |

[a]Use a touch-down program starting with 60°C and ending with 40°C

of 15 μM each), 27 μl PCR master mix, and 14 μl $H_2O$ and subject to PCR using the protocol as specified in Table 3.

5. Fragment purification: Purify the amplification product using a commercial kit, e.g. the MinElute® PCR Purification Kit (Qiagen), that applies spin columns in a microcentrifuge. Before starting the procedure, it is important to consider and perform the comments and recommendations of the supplier (see Note 5). Elute the DNA with 10 μl buffer EB.

**Table 3**
**PCR protocol 2**

| Cycle | Number | Step | °C | Time |
|---|---|---|---|---|
| 1 | 1 | 1 | 95 | 04:00 |
| 2 | 30 | 1 | 95 | 00:30 |
| | | 2 | 58 | 00:30 |
| | | 3 | 72 | 01:00 |
| 3 | 1 | 1 | 72 | 04:00 |
| 4 | 1 | 1 | 4 | ∞ |

6. Restriction digest: Cleave the purified synthetic gene fragment and the cloning vector (e.g. a pPCR-Script (Stratagene, Heidelberg, Germany) derivative) using *Bam*HI and *Nco*I restriction enzymes. Usually, cleavage is performed in a volume of 30 μl using 1 μg of DNA, 3 μl of tenfold restriction buffer, and 1.5 μl of each restriction enzyme, adjusting the volume with $H_2O$. If using NEB enzymes, use buffer NEB3 and incubate at 37°C for 60 min. When using different restriction enzymes, refer to the buffer usage table provided by each supplier (see Note 6).
7. Fragment purification: Purify both restriction products separately using a commercial kit, e.g. the MinElute® Gel Extraction Kit (Qiagen), a vacuum manifold (e.g. QIAvac 24 Plus or QIAvac 6S, Qiagen) in combination with a vacuum pump (also available from Qiagen), if available, or any other commercially available method, following the instructions of the manufacturer. Mix an aliquot of the purified DNA, usually one-tenth of the reaction volume, diluted in $H_2O$ or TE if necessary, with 1 volume of loading dye per 9 volumes of purified DNA sample for gel analysis and load the mix on a 1% agarose gel (see Note 7). Run the gel at a constant current and voltage (5 V/ $cm^2$, 50–150 V) in TAE buffer.
8. Ligation: Ligate the purified digested vector and synthetic gene in a 1.5 ml reaction tube. Mix 0.5 μl vector preparation and 8 μl of the synthetic gene with 1 μl T4-ligation buffer and 0.5 μl T4-DNA ligase. Be sure to briefly centrifuge the tube to spin down the mix, before incubating the ligation mix for 60 min at room temperature (see Note 8). Then, place the reaction on ice.
9. Transformation: Transform competent *E. coli* (we usually use XL1-gold from Invitrogen by Life Technologies, Carlsbad,

CA, for the high transformation efficiency) with the ligation product. Thaw 100 µl of competent *E. coli* (aliquots stored at −80°C in 0.2 ml reaction tubes) on ice (see Note 9), add the ligation mix to the cells, and incubate for 20 min on ice. To apply a heat shock, incubate the reaction tube for 60 s at 42°C using a water bath or a thermomixer (Thermomixer® comfort, for 1.5 ml reaction tubes, Eppendorf, Hamburg, Germany) (see Note 10). Immediately after the heat shock, put the cells on ice for 5 min. If using ampicillin for selection, plate the cells directly on ampicillin-containing LB plates (100 mg/l) and incubate at 37°C overnight. For other selection agents (e.g. kanamycin or tetracycline), it is essential to incubate the transformed bacteria with LB medium without antibiotics for at least 30 min at 37°C in a shaker in order to provide time for recovery and expression of the respective resistance gene (see Note 11).

#### 3.1.3. Plasmid DNA Construction

In order to mimic the natural shape of unspliced HIV-1 RNAs, equip the GFP variants with the described *cis*-active elements known to contribute to temporally regulated HIV-1 gene expression. Place the 5′ UTR comprising the major splice donor (SD) upstream of the GFP-encoding genes. Insert the RRE downstream of the GFP-encoding sequences to allow Rev binding and Rev-mediated RNA export. Finally, put all *gfp* sequences under the transcriptional control of the CMV immediate-early promoter–enhancer unit provided with the commercially available expression vector pcDNA3.1 (Invitrogen).

1. Design of reference construct: To generate the eukaryotic expression vector pc-huGFP, which serves as a reference expression construct, design a codon-optimised synthetic gene sequence encoding the GFP of *Aequorea victoria* (huGFP; Genbank® accession no. AX511259) according to the described method.
2. Generation of target constructs: To design the required subgenomic HIV-1-derived sequences, refer to the sequence of the provirus HX10 (GenBank® accession no. M15654). Add the UTR located upstream of the gag sequence and the RRE to the huGFP and hivGFP sequences, either the RRE alone or both elements in combination using an appropriate sequence program, e.g. the GeneOptimizer® software. Synthesise the required synthetic genes according to the protocol provided above (see Note 12).
3. Generation of expression constructs: Following synthesis and cloning of the synthetic *gfp* genes, insert all genes into pcDNA3.1 via *Xho*I and *Eco*RI restriction, applying the above-mentioned methods, to obtain the expression plasmids pc-huGFP-RRE, pc-hivGFP-RRE, pc-UTR-huGFP-RRE, and pc-UTR-hivGFP-RRE.

### 3.2. Cell Culture

1. Cultivation of cell lines: Grow the adherent human lung carcinoma cells H1299 in DMEM supplemented with 10% FBS. Cultivate cells in appropriate tissue culture flasks at 37°C in a 5% $CO_2$ incubator.
2. Transfection of cells: For transfection, plate $2.5 \times 10^5$ or $1 \times 10^6$ cells contained in 3 or 13 ml of DMEM on 6-well or 10-cm dishes (Falcon, Heidelberg, Germany), respectively, and incubate for 24 h. One hour prior to transfection, the medium has to be renewed. Using the 6-well plate format, mix 15 μg of DNA in a volume of 270 μl with 30 μl of a 2.5 M $CaCl_2$ solution (sterile filtered) for each well. Carefully pipette this mix dropwise to 3 ml of 2× HeBS while continuously vortexing the HeBS buffer. Incubate the mix for 20 min at room temperature to form a precipitate. Slowly pour the mix into the medium using a 10 ml pipette, dispensing it evenly over the cells. Incubate the cells for 16 h, renew the medium, and harvest cells 48 h later.
3. Breaking open cells and collecting cell proteins: In order to harvest the transfected cells, remove the medium and carefully wash the cells with cold PBS once and scrape the cells in 3 ml PBS quantitatively using a sterile cell scraper. Centrifuge the cells at $8{,}600 \times g$ for 5 min, resuspend the cell pellet in 250 μl RIPA buffer for cell lysis, and incubate for 15 min on ice. For complete lysis, apply a freeze–thaw cycle and remove cell debris by centrifugation at $20{,}000 \times g$ for 5 min. The supernatant containing the soluble proteins can be stored at −20°C.
4. Analytical determination of protein concentration: For subsequent quantitative analyses, it is necessary to apply equal amounts of protein samples. Therefore, determine the total amount of protein using the commercially available Bradford Protein Assay (Bio-Rad) according to the manufacturer's instructions.

### 3.3. SDS-Polyacrylamide Gel Electrophoresis

1. Preparation of glass plates: Clean the glass plates for the gels and the silicone rubber seals with a volatile detergent (e.g. isopropanol) with a fluff-free cleaning cloth before use. The glass plates should be assembled quickly after detergent treatment. For a detailed gel assembly and apparatus description, refer to the manufacturer's instruction, e.g. the maxigel vertical apparatus from Biometra (Göttingen, Germany).
2. Preparation of separation gel: Prepare a 1.0-mm-thick, 12.5% gel by mixing 7.5 ml of acrylamide/bisacrylamide solution, 3.6 ml of separating buffer, 3.3 ml of $H_2O$, 15 μl TEMED, and 90 μl APS. Pour the solution quickly between the assembled glass plates, leaving enough space at the top for a stacking gel, and overlay with water-saturated isobutanol. The gel should be polymerised after about 30 min.

3. Preparation of stacking gel: Pour off the isobutanol and rinse the top of the gel twice with water. Prepare the 5% stacking gel by mixing 0.99 ml of acrylamide/bisacrylamide solution, 1.2 ml of stacking buffer, 2.61 ml of $H_2O$, 6 µl TEMED, and 30 µl APS. Use about 0.5 ml of this to briefly rinse the top of the gel and then pour the stacking gel and insert the comb. The stacking gel should polymerise within 30 min.
4. Gel loading: Remove the comb from the gel, straighten up the lanes, and remove any acrylamide from the wells with a fine pipette tip if required. Add the SDS-PAGE running buffer to the upper and lower chambers of the gel unit. Boil 50 µg of protein for each sample mixed with 1/5 volume SDS sample buffer for 5–10 min and let samples cool down to room temperature. Spin down sample and load mixture in a well. Include one well for a molecular weight marker.
5. Gel run: Complete the assembly of the gel unit and connect to a power supply. The gel can be run (about 5 h) at 20 mA through the stacking gel and 30–40 mA through the separating gel. Run until dye front is 1 cm from the bottom of the gel (the run length can be shorter if desired).

### 3.4. Western Blotting

1. Protein transfer: Transfer the samples separated by SDS-PAGE to a supported nitrocellulose membrane by electrophoresis using a Fast-Blot "semidry" apparatus (Biometra). Prepare a tray containing transfer buffer that is large enough to lay out a transfer cassette with its pieces of foam and two sheets of 3MM paper submerged on one side. Cut a sheet of nitrocellulose just larger than the separating gel and place in a separate tray of distilled water to wet the membrane by capillary action. Submerge the membrane in the transfer buffer on top of the 3MM paper.
2. Disconnect the PAGE unit from the power supply and disassemble. Remove the stacking gel and discard. Mark one corner from the separating gel (with a small cut) to track subsequent orientation. Place the separating gel on top of the nitrocellulose membrane.
3. Place two additional sheets of 3MM paper (soaked with transfer buffer) on top of the gel, ensuring that no bubbles are trapped in the resulting sandwich. Put the second wet foam sheet on top and close the transfer cassette.
4. Place the cassette into the transfer tank; make sure that the nitrocellulose membrane is between the gel and the anode (correct orientation is crucial).
5. Put the lid on the tank and start the transfer at 5 $mA/cm^2$ gel size for 30–60 min.

6. Once transfer is completed, take the cassette out of the tank and carefully disassemble. Leave the gel on top of the nitrocellulose until the orientation has been marked on the membrane.
7. Blocking: Incubate the nitrocellulose in 50 ml blocking buffer for 1 h at room temperature on a rocking platform.
8. Detection: Discard the blocking buffer and rinse the membrane repeatedly with TBS prior to addition of a 1:500 dilution of the primary GFP-specific monoclonal antibody in antibody dilution buffer for 1 h at room temperature on a rocking platform.
9. Remove the primary antibody (see Note 13) and wash the membrane three times for 5 min each with 50 ml TBS.
10. Prepare the secondary antibody freshly for each experiment as 1:2,000-fold dilution in blocking buffer and add to the membrane for 30 min at room temperature on a rocking platform.
11. Discard the secondary antibody and wash the membrane six times for 1 min each with TTBS buffer. During the final wash, warm 2 ml aliquots of each portion of the ECL reagent separately to room temperature. For the remaining steps, change to a darkroom under safe light conditions. Once the final wash is removed from the blot, mix the ECL reagents and add immediately to the blot, ensuring even coverage by rotating the blot.
12. Remove the ECL reagents from the blot, wrap with foil, and place in a Western blot imager ChemiLux Imager 2008 (Intas, Göttingen, Germany) for analysis. An example of the results produced is shown in Fig. 3. Alternatively, the blot can be evaluated using an X-ray film in an appropriate film cassette.

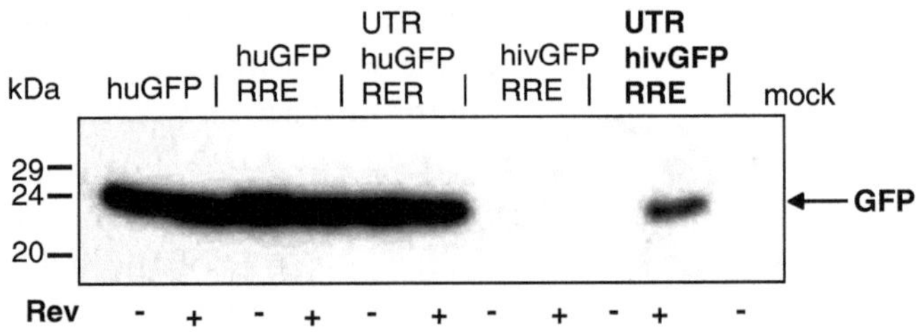


Fig. 3. Analysis of Rev-dependent hivGFP expression. H1299 cells were transiently transfected with 15 μg of the indicated GFP expression plasmids. To determine the Rev dependence of the GFP expression constructs, 15 μg of either pcDNA3.1 (–) or a Rev expression vector (+) were cotransfected as indicated, and cells were harvested 48 h post transfection. For Western blot analysis, 50 μg of cell lysates were separated by 12.5% SDS-PAGE and GFP expression was analysed using a GFP-specific antibody. Positions of molecular weight markers (broad-Marker, Bio-Rad) are indicated on the left. Reprinted with permission from Elsevier (23).

### 3.5. Northern Blot

1. Preparation of RNA: When working with RNA, it is highly recommended to make sure that working solutions are not contaminated with RNases (see Note 14). For preparing nuclear and cytoplasmic RNA species, a commercial kit is recommended (e.g. RNeasy® Kit, Qiagen), following the manufacturer's instructions. After harvesting transfected cells from a petri dish (see 3.2.3), incubate the pelleted cells in 175 μl lysis buffer for 20 min on ice in order to lyse the plasma membranes. Separate nuclei from the cytoplasm by centrifugation (300 × g, 2 min). The upper phase contains the cytoplasmatic fraction; the lower phase contains the nuclei. From the two separated fractions, nuclear and cytoplasmic RNAs can be purified using RNeasy® spin columns. To further separate polyadenylated mRNA from non-mRNAs, use the Oligotex® mRNA Kit from Qiagen following the manufacturer's instructions. For some applications, it is sufficient to continue with total RNA and skip mRNA purification (see Note 15).
2. Separation of RNA: For Northern blot analysis, separate RNA preparations on a 1% agarose gel. Prepare the gel by dissolving 0.5 g agarose in 36 ml boiling water and cool to 60°C. Once cooled to 60°C, place in a fume hood and add 5 ml of 10× E running buffer and 9 ml of formaldehyde. Pour the gel, insert the comb, and allow it to set. Remove the comb, place the gel in the gel tank, and add sufficient 1xE running buffer to cover to a depth of 1 mm.
3. Adjust the volume of each RNA sample to 6 μl with water and add 15 μl of freshly prepared sample denaturation mix. Mix by vortexing, spin down briefly to collect liquid, and incubate for 15 min at 55°C. Cool on ice for 2 min, then add 2 μl loading dye mix, and briefly spin down again.
4. Run the gel in 1×E running buffer at 100 V for 10 min, and then at 65 V for 90 min.
5. RNA transfer: Blot the RNA samples onto a positively charged nylon membrane (Pall, Dreieich, Germany). Place the gel in an RNase-free dish and rinse with sufficient deionised water to cover the gel for 20 min; repeat four times. Fill the glass dish with enough 20× SSC, cut two pieces of Whatman® 1MM paper, place them on the glass plate, and wet with 20× SSC. Place the gel on the filter paper and roll out air bubbles with a glass pipette. Cut four strips of plastic wrap and place over the edges of the gel. Cut a piece of nylon membrane just large enough to cover the gel and pre-incubate in water. Place the wetted membrane on the surface of the gel. Try to avoid air bubbles under the membrane. Flood the surface of the

membrane with 20× SSC. Cut five sheets of Whatman® 3MM paper to the same size as the membrane and place on top of the membrane. Put paper towels on top of the Whatman® 3MM paper to a height of approx. 6 cm, and add a weight to hold the stack in place. Leave overnight.

6. Preparation of the membrane for hybridisation: Remove the paper towels and filter papers and recover the membrane. Mark the position of the wells on the membrane with a pencil and ensure that up–down and back–front orientations are recognisable. The blot can now be stained with methylene blue solution (soft shaking for 45 min) in order to visualise the transferred RNA on a transilluminator. Rinse the membrane in $H_2O$ to remove the staining, mark the prominent 18S and 28S RNA band or the RNA marker if used, and then place it on a sheet of Whatman® 3MM paper and allow to dry. Place the membrane RNA-side down on a UV transilluminator (254-nm wavelength) and irradiate for appropriate length of time (e.g. 1,200 kJ for 1 min).
7. Preparation of RNA probes: The probes are labelled by in vitro run-off transcription of *Xho*I-linearised pCMV-ERR (antisense to the RRE-encoding sequence) or pGEM-β-actin (for a β-actin probe as loading control) using Riboprobe® in Vitro Transcription Systems (Promega). The RNA probe is labelled by including $^{32}$P-α-CTP (10 μCi per reaction) according to the supplier's instructions. Purify the probe using the RNeasy® kit. Collect the purified sample in the bottom of the support tube. For some applications, other types of probes are more appropriate (see Note 16).
8. Pre-hybridisation: Submerge the membrane in hybridisation solution, place it RNA-side up in a hybridisation tube, and add 10 ml hybridisation solution. Place the tube in the hybridisation oven and incubate with rotation for 2 h at 68°C.
9. Hybridisation: Denature the labelled probe by heating in a water bath for 10 min at 100°C, and immediately transfer to ice. Pipette the desired volume of probe into the hybridisation tube and continue to incubate with rotation overnight at 68°C.
10. Visualisation of hybridisation: Wash the membrane three times with wash buffer for 15 min at 68°C. Blot excess liquid, and cover the blot with plastic wrap, if possible shrink wrap. Place the blot in a *Phosphor-Imager* plate and instrument and analyse the results using the *Molecular Analyst Software* (BioRad Laboratories). Alternatively, visualise by exposure to an illumination-sensitive X-ray film in an appropriate cassette for

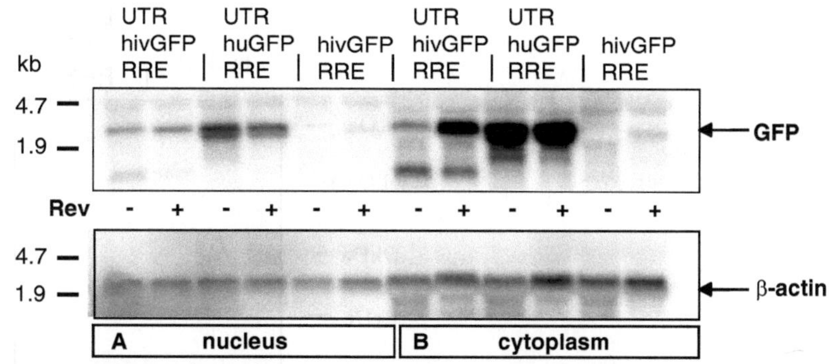


Fig. 4. Cellular distribution of hivGFP and huGFP mRNAs. H1299 cells were transfected with the indicated GFP reporter constructs in the absence (–) or presence (+) of a Rev expression plasmid. Cells were harvested 48 h post transfection and polyadenylated mRNAs isolated from nuclear (**a**) and cytoplasmic (**b**) cell fractions were subjected to Northern blot analysis. GFP-encoding transcripts were detected using a radiolabelled RRE antisense riboprobe (*upper panel*). Beta-actin RNA was used for standardisation (*lower panel*). Reprinted with permission from Elsevier (23).

20–30 min. If necessary, repeat the last step with shorter or longer exposure times, depending on the signal strength. Sometimes, there is background that requires repeating some steps or increasing the stringency of the washing procedure (see Note 17). An example of a Northern blot analysis is shown in Fig. 4.

### 3.6. Analysis of Nuclear Export Pathway Usage

1. Rational: The streptomyces-derived metabolite LMB is a well-known inhibitor of Rev function because it directly interferes with the interaction between Rev's nuclear export signal and CRM1 (5, 24). LMB specifically blocks Rev function in the nanomolar range (2–10 nM) (25, 26), so examining its influence on the quasi-lentiviral reporter gene can help evaluate the nuclear export pathway of the hivGFP mRNA.
2. LMB inhibition: Prior to adding LMB to the cells, evaluate the cytotoxicity of LMB for each cell line separately (see Note 18). Add LMB to cell culture supernatants (transfected/controls) 24 h prior to cell harvesting in increasing amounts, resulting in final concentrations of 0–10 nM. Dilute a 10 mM stock solution in absolute ethanol accordingly (1, 0.5, 0.25 μM) and add 30 μl of the respective solution to 3 ml of cell culture in a 6-well plate (corresponding to a 100-fold dilution). The results of inhibition at 5 nM are shown in Fig. 5.

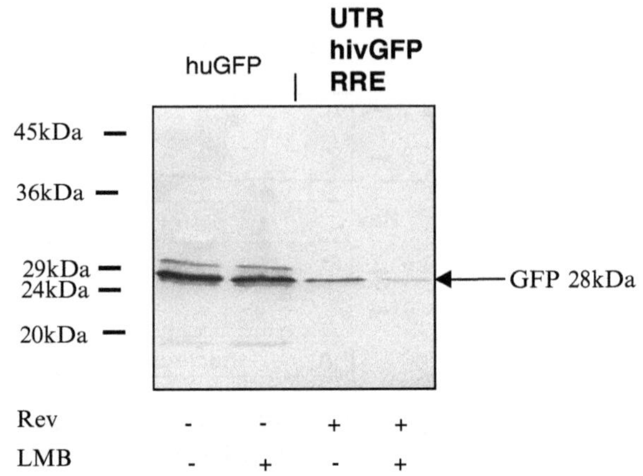


Fig. 5. Effects of LMB on gene expression. To quantify the influence of LMB on GFP expression, H1299 cells transfected with 15 μg of indicated reporter genes were grown for 48 h in the presence or absence of 5 nM LMB, as indicated. 15 μg of either pcDNA3.1 (REV −) or a Rev expression vector (Rev +) were co-transfected as depicted, and cells were harvested 48 h post transfection. For Western blot analysis, 50 μg of cell lysates were separated by 12.5% SDS-PAGE and GFP expression was analysed using a GFP-specific antibody. Positions of molecular weight markers are indicated on the *left*.

## 4. Notes

1. Transfer buffer can be used for up to five transfers within 1 week as long as constant voltage is maintained for each successive run. Cooling using a refrigerated/circulating bath to keep the buffer temperature below room temperature is crucial to prevent heat-induced damage to the gel.
2. This class of sequence manipulation tools usually provides additional options for gene optimisation, such as modifying parameters considered important for a specific application or host organism, or motifs relevant for production, such as restriction recognition sites, in parallel to codon usage.
3. The length of oligos and the required sequence overlap depend on melting temperature and GC content.
4. High oligonucleotide quality is essential to reduce screening work (i.e. identifying a correct clone by sequencing a number of colonies) and avoid correction measures.
5. An essential handling instruction is correct ethanol treatment. A common error is to forget to add ethanol (96–100%) to buffer PE before use, which results in non-binding of DNA to the purification column resin. Analogously, it is essential to remove all residual ethanol from buffer PE before eluting the plasmid DNA. It is, therefore, important to discard the flow-through

before performing the additional centrifugation to remove buffer PE.

6. Since restriction enzymes are stabilised using glycerol, which impedes their activity, take care that the volume of restriction enzyme does not exceed 10% of the total reaction volume. This is particularly important if multiple enzymes are used in one reaction (e.g. double digests). If necessary, increase the overall reaction volume.
7. To prevent the sample rising from the agarose gel slot, mix the solution by pipetting up and down before loading. This stirs up the glycerol contained in the loading dye, so the sample sinks to the bottom of the slot.
8. Ligation of blunt-end DNA fragments requires longer ligation times.
9. Temperature has a significant influence on transformation efficiency. Cells should be thawed on ice and DNA should be added as soon as last traces of ice in the tube disappear.
10. The heat shock step can be performed in a water bath; the recommended time in a water bath is reduced to 30 s, since the water covering the tube instantly heats the solution.
11. Transformation efficiency depends on the outgrowth time and efficiency can be increased by longer incubation (1 h) and shaking/rotating.
12. Assembly of larger genes requires the synthesis of sub-fragments of smaller size. Seamless assembly of these fragments is achieved best by using outside cutter, also called type IIS restriction enzymes, which are characterised by the fact that the recognition site does not overlap with the cleavage site, e.g. *Bbs*I GAAGAC(N)2▼(N)4▲. Thus, one can avoid the need for specific nucleotides at the assembly site, as required with regular restriction enzymes (where the restriction site is usually within the recognition palindromic sequence, e.g. *Bam*HI cutting G▼GATC▲C).
13. The primary antibody can be saved for subsequent experiments by adding 0.02% final concentration sodium azide (usually, dilution from a 10% stock solution; exercise caution since azide is highly toxic) and storage at 4°C.
14. Use RNase-free tips and reagents and wear gloves. Anything that can be autoclaved and does not contain amine groups (such as Tris) can be treated with diethylpyrocarbonate (DEPC, Sigma–Aldrich) and autoclaved. Anything that cannot be autoclaved (some metal salts, etc.) or contains amine groups can be prepared in baked glassware with DEPC-treated water. If using 20× SSC for transfer to a nylon membrane, no special precaution is necessary, since the salt concentration is too high to support RNase activity.

15. The choice of total RNA or mRNA (i.e. poly(A)+RNA) depends largely on the subsequent application for the isolated RNA. Using mRNA usually results in improved sensitivity and possibly more accurate results. If probing for an abundant RNA, total RNA may be sufficient.
16. Basically, there are three types of probe: double-stranded DNA, single-stranded DNA (including oligonucleotides), and single-stranded RNA. Each type can be radiolabelled or non-radiolabelled (e.g. DIG labelled). The best probe in terms of sensitivity is RNA. The design principles do not change significantly: probes must contain antisense sequence to the target RNA, and usually 100–1,000 nt in length. If you want to use an oligonucleotide probe, the antisense sequence should cover at least 15 nt. Especially with oligonucleotide probes, it is important to avoid mismatches to other gene sequences within the genome of the host organism.
17. Excessive background on film: Speckled background indicates that there are unincorporated nucleotides in the probe. Blotchy background indicates non-specific hybridisation, possibly caused by salt still on the blot before hybridisation, but can also indicate insufficient agitation during hybridisation. Be sure to wash the membrane with adequate stringency.
18. Above a certain threshold, LMB is toxic. To clearly discriminate between specific inhibition activity and unspecific cellular toxicity, it is essential to test LMB with each cell type. Use a commercial toxicity test (e.g. XTT-based Cell Proliferation Kit II, Roche) and test concentrations you intend to apply in your experiment. The XTT test is colorimetric, i.e. photometrically determining the modification of XTT by cellular, mitochondrial dehyrogenases. Seed cells in a 24-well plate and cultivate for 24 h. Add increasing amounts of LMB, let the cells sit for another 10–20 h, then add MTT solution, and determine the colour change in a photometer. Calculate the cytotoxic effect in comparison to non-treated cells. Alternatively to the classical MTT assay (27), one can use new non-invasive methods, e.g. the CloneSelect™ Imager system (Genetix, UK) (28).

## References


1. Schwartz S, Felber BK, Benko DM, Fenyo EM, Pavlakis GN (1990) Cloning and functional analysis of multiply spliced mRNA species of human immunodeficiency virus type 1. J. Virol. 64:2519–2529.
2. Tazi J, Bakkour N, Marchand V, Ayadi L, Aboufirassi A, Branlant C (2010) Alternative splicing: regulation of HIV-1 multiplication as a target for therapeutic action, FEBS J. 277(4):867–876.
3. Cullen B (2003) Nuclear mRNA export: insights from virology. Trends Biochem. Sci. 28:419–424.
4. Olsen HS, Nelbock P, Cochrane AW, Rosen CA (1990) Secondary structure is the major determinant for interaction of HIV rev protein with RNA. Science 247:845–848.
5. Fornerod M, Ohno M, Yoshida M, Mattaj IW (1997) CRM1 is an export receptor for leucine-rich nuclear export signals. Cell 90:967–970.

6. Neville M, Stutz F, Lee L, Davis LI, Rosbash M (1997) The importin-ß family member CRM1p bridges the interaction between Rev and the nuclear pore complex during nuclear export. Curr. Biol. 7:767–775.
7. Yi R, Bogerd HP, Cullen BR (2002) Recruitment of the Crm1 nuclear export factor is sufficient to induce cytoplasmic expression of incompletely spliced human immunodeficiency virus mRNAs. J Virol. 76(5):2036–2042.
8. Blissenbach M, Grewe B, Hoffmann B, Brandt S, Ueberla K (2010) Nuclear RNA export and packaging functions of HIV-1 Rev revisited. J Virol. 84(13):6598–6604.
9. Kammler S, Leurs C, Freund M, Krummheuer J, Seidel K, Tange TO, Lund MK, Kjems J, Scheid A, Schaal H (2001) The sequence complementarity between HIV-1 5′ splice site SD4 and U1 snRNA determines the steady-state level of an unstable env pre-mRNA. RNA 7:421–434.
10. Maldarelli F, Martin MA, Strebel K (1991) Identification of posttranscriptionally active inhibitory sequences in human immunodeficiency virus type 1 RNA: novel level of gene regulation. J. Virol. 65:5732–5743.
11. Olsen HS, Cochrane AW, Rosen C (1992) Interaction of cellular factors with intragenic cis-acting repressive sequences within the HIV genome. Virology 191:709–715.
12. Brighty DW, Rosenberg M (1994) A cis-acting repressive sequence that overlaps the Rev-responsive element of human immunodeficiency virus type 1 regulates nuclear retention of env mRNAs independently of known splice signals. Proc. Natl. Acad. Sci. USA. 91: 8314–8318.
13. Cochrane AW, Jones KS, Beidas S, Dillon PJ, Skalka AM, Rosen CA (1991) Identification and characterization of intragenic sequences which repress HIV structural gene expression. J. Virol. 65:5305–5313.
14. Nasioulas G, Zolotukhin AS, Tabernero C, Solomin L, Cunningham CP, Pavlakis GN, Felber BK (1994) Elements distinct from HIV-1 splice sites are responsible for the Rev dependence of env mRNA. J. Virol. 68:2986–2993.
15. Schwartz S, Felber BK, Pavlakis GN (1992) Distinct RNA sequences in the gag region of human immunodeficiency virus type 1 decrease RNA stability and inhibit expression in the absence of Rev protein. J. Virol. 66:150–159.
16. Schneider R, Campbell M, Nasioulas G, Felber BK, Pavlakis GN (1997) Inactivation of the human immunodeficiency virus type 1 inhibitory elements allows Rev-independent expression of Gag and Gag/protease and particle formation. J. Virol. 71:4892–4903.
17. Schwartz S, Campbell M, Nasioulas G, Harrison J, Felber BK, Pavlakis GN (1992) Mutational inactivation of an inhibitory sequence in human immunodeficiency virus type 1 results in Rev-independent gag expression. J. Virol. 66:7176–7182.
18. Haas J, Park EC, Seed B (1996) Codon usage limitation in the expression of HIV-1 envelope glycoprotein. Curr. Biol. 6:315–324.
19. Kotsopoulou E, NarryKim V, Kingsman AJ, Kingsman SM, Mitrophanous KA (2000) A Rev-independent human immunodeficiency virus type 1 (HIV-1)-based vector that exploits a codon-optimized HIV-1 gag-pol gene. J. Virol. 74:4839–4852.
20. Wagner R, Graf M, Bieler K, Wolf H, Grunwald T, Foley P, Ueberla K (2000) Rev-independent expression of synthetic gag-pol genes of human immunodeficiency virus type 1 and simian immunodeficiency virus: implications for the safety of lentiviral vectors. Hum. Gene Ther. 11:2403–2413.
21. Graf M, Bojak A, Deml L, Bieler K, Wolf H, Wagner R (2000) Concerted action of multiple cis-acting sequences is required for Rev dependence of late human immunodeficiency virus type 1 gene. J. Virol. 74: 10822–10826.
22. Ngumbela KC, Ryan KP, Sivamurthy R, Brockman MA, Gandhi RT, Bhardwaj N, Kavanagh DG (2008) Quantitative effect of suboptimal codon usage on translational efficiency of mRNA encoding HIV-1 gag in intact T cells, PLoS One. 3(6): e2356. doi:10.1371/journal.pone.0002356.
23. Graf M, Ludwig C, Kehlenbeck S, Jungert K, Wagner R (2006) A quasi-lentiviral green fluorescent protein reporter exhibits nuclear export features of late human immunodeficiency virus type 1 transcripts. Virology 352:295–305.
24. Kudo N, Wolff B, Sekimoto T, Schreiner EP, Yoneda Y, Yanagida M, Horinouchi S, Yoshida M (1998) Leptomycin B inhibition of signal-mediated nuclear export by direct binding to CRM1. Exp. Cell Res. 242:540–547.
25. Fischer U, Huber J, Boelens WC, Mattaj IW, Luhrmann R (1995) The HIV-1 Rev activation domain is a nuclear export signal that accesses an export pathway used by specific cellular RNAs. Cell 82:475–483.
26. Wolff B, Sanglier JJ, Wang Y (1997). Leptomycin B is an inhibitor of nuclear export: inhibition of nucleo-cytoplasmic translocation of the human immunodeficiency virus type 1

(HIV-1) Rev protein and Rev-dependent mRNA. Chem. Biol. 4:139–147.

27. Mosmann T (1983) Rapid Colorimetric assay for cellular growth and survival – application to proliferation and cyto-toxicity assays. J Immunol Methods, 65(1–2):55–63.

28. Marqués-Gallego P, den Dulk H, Backendorf C, Brouwer J, Reedijk J, Burke JF (2010) Accurate non-invasive image-based cytotoxicity assays for cultured cells. BMC Biotechnology:10:43. http://www.biomedcentral.com/1472-6750/10/43.

# Chapter 6

# A High-Throughput Microfluidic Method for Generating and Characterizing Transcription Factor Mutant Libraries

**Marcel Geertz, Sylvie Rockel, and Sebastian J. Maerkl**

## Abstract

Characterizing libraries of mutant proteins is a challenging task, but can lead to detailed functional insights on a specific protein, and general insights for families of proteins such as transcription factors. Challenges in mutant protein screening consist in synthesizing the necessary expression-ready DNA constructs and transforming them into a suitable host for protein expression. Protein purification and characterization are also non-trivial tasks that are not easily scalable to hundreds or thousands of protein variants. Here we describe a method based on a high-throughput microfluidic platform to screen and characterize the binding profile of hundreds of transcription factor variants. DNA constructs are synthesized by a rapid two-step PCR approach without the need of cloning or transformation steps. All transcription factor mutants are expressed on-chip followed by characterization of their binding specificities against 64 different DNA target sequences. The current microfluidic platform can synthesize and characterize up to 2,400 protein–DNA pairs in parallel. The platform method is also generally applicable, allowing high-throughput functional studies of proteins.

**Key words:** Microarrays, Transcription factor binding sites, High-throughput, Binding affinities, DNA array, Protein array, Surface chemistry, Two-step PCR, Microfluidics

## 1. Introduction

The sequence-specific binding of transcription factors (TF) determines in a large part the connectivity of gene regulatory networks as well as the quantitative level of gene expression. While technological advances increased our knowledge of DNA-binding specificities across a number of TF families (1–7), relatively little is know about the functional consequences of how amino acid substitutions in the DNA-binding domain impact these specificities. The lack of a simple correspondence between amino acid residues and specific DNA bases complicates the identification of such a sequence

Wilfried Weber and Martin Fussenegger (eds.), *Synthetic Gene Networks: Methods and Protocols*,
Methods in Molecular Biology, vol. 813, DOI 10.1007/978-1-61779-412-4_6, © Springer Science+Business Media, LLC 2012

recognition code (8). The more recent development of microfluidic large-scale integrations (MLSI) offers new implementations of experimental approaches to elucidate the sequence–function relationship across a variety of TF families (5).

In this protocol we describe a novel detection method for the characterization of TF-DNA interactions based on the mechanically induced trapping of molecular interactions (MITOMI) (4). MITOMI allows the capture of transient and low-affinity interactions between DNA sequences and TFs at equilibrium, and thus the measurement of absolute binding affinities. In short, MLSI devices are fabricated from polydimethylsiloxane (PDMS) an elastomeric material. These microfluidic chips harbor micron-sized channels with thousands of integrated micromechanical valves to control picoliter-sized reaction chambers. Each reaction chamber is aligned to an array of dsDNA sequences printed onto an epoxy-coated glass slide using standard DNA microarray instruments. Here we modify our MITOMI protocol to map in parallel the contribution of amino acid substitution on TF DNA-binding specificity. By co-spotting PCR products coding for TF variants and cognate target DNA sequences and performing on-chip protein synthesis, MITOMI is capable of screening and characterizing the binding capabilities of several hundreds of TF variants to multiple target DNA sequences.

More generally, the protocol described here can be applied to the on-chip expression and functional characterization of hundreds to thousands of proteins in parallel. The expression templates can be readily derived either from cDNA libraries, generated by PCR, site-direct mutagenesis, or synthesized de novo using gene synthesis approaches.

## 2. Materials

### 2.1. Mask and Wafer Fabrication

#### 2.1.1. Instruments

1. Mask writing on Heidelberg DWL200 laser lithography system (Heidelberg Instruments Mikrotechnik GmbH).
2. Mask and wafer development using DV10 (Süss MicroTec AG).
3. Wafer cleaning with oxygen plasma before processing using Tepla300 (PVA Tepla AG).
4. MA6 Mask Aligner (Süss MicroTec AG) for exposure of wafers.
5. Sawatec LMS200, programmable coater for negative resist and Sawatec HP401Z, programmable hot plate for soft bake (Sawatec AG).
6. Süss RC-8 THP, manual coater and hotplate for positive resist (Süss MicroTec AG).

#### 2.1.2. Materials

Chemicals used in mask and wafer fabrication are from Rockwood Electronic Materials, Gréasque, France, and of Metal-Oxyde-Semiconductor (MOS) quality, unless otherwise stated.

1. Masks: square blank 5″ Nanofilm SLM 5 (Nanofilm).
2. Silicon wafers (diameter: $100 \pm 0.5$ mm, thickness: $525 \pm 25$ μm, conductivity type: P, dopant: Boron, resistivity range: 0.1–100 Ω·cm; Okmetic).
3. Photoresists: AZ9260 positive photoresist (MicroChemicals GmbH), SU-8 negative photoresist GM1060 and GM1040 (Gersteltec).
4. Chrome etching of masks: CR7 consisting of $(NH_4)_2\,Ce(NO_3)_6$; $HClO_4$.
5. Developers: MP 351 for mask and AZ 400K for AZ9260 coated wafers (AZ Electronic Materials); 1-methoxy-2-propyl-acetate (PGMEA) for manual development of SU-8 wafers.
6. Solvents: Remover 1165; 93% (v/v) *N*-methyl-2-pyrrolidone (NMP, Sotrachem Technic), 7% (v/v) PGMEA for masks; isopropyl alcohol (IPA); acetone for wafers.

### 2.2. MITOMI Device Fabrication by Multilayer Soft Lithography

1. Polydimethylsiloxane (PDMS, Dow Corning Sylgard 184) resin: heat-curable silicone elastomer.
2. Trimethylchlorosilane (TMCS, Sigma-Aldrich).
3. Mixing and degassing of PDMS: Thinky Mixer ARE-250 equipped with adaptor for 100 ml disposable PP beakers (C3 Prozess- und Analysentechnik GmbH).
4. Degassing of PDMS control layer: vacuum desiccator (Fisher Scientific AG).
5. Spin coating of PDMS flow layer: programmable spin coater SCS P6700 (Specialty Coating Systems Inc.).
6. Stereomicroscope, SMZ1500 (Nikon AG).
7. Manual hole-punching machine and pin vises, 21 gauge (0.04″ OD) (Technical Innovations, Inc.).

### 2.3. Epoxy Slide Preparation

Chemicals for epoxy coating of glass slides are from Sigma-Aldrich.

1. Standard ($76 \times 26 \times 1$ mm) microscope glass slides (VWR).
2. Milli-Q water.
3. 30% (v/v) Ammonium hydroxide ($NH_4OH$).
4. 30% (v/v) Hydrogen peroxide ($H_2O_2$).
5. Solvents: acetone, toluene, isopropyl alcohol (IPA).
6. 97% (v/v) 3-Glycidoxypropyl-trimethoxymethysilane (3-GPS).
7. Nitrogen gas supply.

### 2.4. DNA Synthesis

#### 2.4.1. Synthesis of Linear Template DNA

All chemicals used for synthesis of DNA are from Sigma-Aldrich. For DNA primer sequences see Table 1.

1. MAX isoform B (NCBI accession #: BC003525) from Open Biosystems (Clone ID: 3607261).

**Table 1**
**Primer sequences used to generate ITT linear templates and target DNA library**

| Name (Comment) | Sequence |
|---|---|
| 5′CompCy5 (Extension primer for target DNA synthesis) | 5′-[Cy5]GTCATACCGCCGGA-3′ |
| Target DNA (Design of target DNA library. Variable binding sites of interest are bracketed by constant linker sequences. 3′ ends consist of reverse complementary sequence of CompCy5 extension primer) | 5′-[5′LINKER]-[BINDING SITE OF INTEREST]-[3'LINKER]-TCCGGCGGTATGAC-3′ |
| Forward-ORF-HIS (This primer primes 3′ of the DNA-binding region of the transcription factor, adding a new binding region variant and a linker region) | 5′-[ggtggagagcgacgct] [GACAAACGGGCTCATCATAATGCACTGGAACGAAAACGTAGG] [gaccacatcaaagacagcttt] -3′<br>5′-[Linker sequence][Basic Region Variants][Gene Specific]-3′ |
| Reverse-ORF (Adds a c-terminal 5×His tag, MAX iso B specific region is in bold, linker is in italics) | 5′-*GTAGCAGCCTGAGTCG*TTATTAGTGGTGGTGGTGGTG **GCTGGCCTCCATCCG**-3′ |
| Forward extension (T7 promoter sequence is indicated in bold) | 5′-gatcttaaggctagagtacTAATACGACTCACTATAGGGAAT ACAAGCTACTTGTTCTTTTTGCActcgagaattcgccacc atgagcgataacgatgacatcgaggtggagagcgacgct-3′ |
| Reverse extension (Poly(A) track is underlined) | 5′-CAAAAAACCCCTCAAGACCCGTTTAGAGGCCCCAAG GGGTTATGCTAG<u>TTTTTTTTTTTTTTTTTTTTTTTTTTTTTT</u> GTAGCAGCCTGAGTCG-3′ |
| Forward final | 5′-GATCTTAAGGCTAGAGTAC-3′ |
| Reverse final | 5′-CAAAAAACCCCTCAAGAC-3′ |

2. DNA primers (Integrated DNA Technologies, IDT).
3. dNTPs (Roche Diagnostics AG).
4. Expand High Fidelity PLUS PCR system (Roche Diagnostics AG).
5. Elution buffer: 10 mM Tris–HCl, pH 8.5.

#### 2.4.2. Synthesis of Target DNA

1. Primers, 5′CompCy5, (Integrated DNA Technologies, IDT).
2. dNTP (Roche Diagnostics AG).
3. Klenow fragment (3′→5′ exo–) (New England Biolabs).
4. Dithiothreitol (DTT).
5. Magnesiumchloride ($MgCl_2$).
6. Buffer: Tris–HCl, pH 7.9.
7. 0.5% (w/v) BSA in $dH_2O$.

### 2.5. Microarraying/Spotting

1. QArray2 microarrayer (Genetix GmbH) equipped with a NanoPrint™ microarray printer printhead and a 946MP3 micro spotting pin (Arrayit Corporation).
2. Bovine serum albumin (BSA, Sigma-Aldrich) re-suspended in DI water to 2 mg/ml.
3. Synthesized target DNA templates (see Subheading 2.4.2).
4. Epoxy-coated microscope glass slides (see Subheading 2.3).

### 2.6. Microfluidic Control Elements

#### 2.6.1. Pressure Regulation and Control

1. Precision pressure regulator, BelloFram Type 10, 2–25 psi; 1/8″ port size (part. no. 960-001-000, Bachofen SA).
2. Bourdon tube pressure gauges, 0–30 psi (0–2.5 bar), G¼ male connection (part. no. NG 63-RD23-B4, Kobold Instruments AG).
3. Custom-designed manual manifolds (rectangular metal casing: 14.5 × 1 × 1″) for control line regulation with 16 detented toggles and barbs for 1/16″ tubing ID; 1/4 NPT connection (Pneumadyne Inc.).
4. Fittings to connect regulators to gauges and to luer manifolds (Serto AG): tee union (SO 03021–8), male adaptor union (SO 01121-8-1/8).
5. Tubing: Tygon ¼″ OD × 1/8″ ID (Fisher Scientific AG).
6. Polycarbonate luer fittings (Fisher Scientific AG): multi-port luer manifolds for flow inlet regulation (e.g., part. no. Cole Parmer 06464–87); male luer to luer connector (part. no. Cole Parmer 06464–90).

#### 2.6.2. Fluidic Connections

Disposable stainless steel dispensing needles to connect to syringe 23 gauge, 1/2″ long, ID 0.33 mm (part. no 560014; I & Peter Gonano)

1. Tubing (Fisher Scientific AG): for liquid/gas; flexible plastic tubing for fluidic connections, Tygon S54HL, ID 0.51 mm.
2. Steel pins for chip-to-tube interface: tube AISI 304 OD/ID × L Ø0.65/0.30 × 8 mm, cut, deburred, passivated (Unimed S.A.).

### 2.7. MITOMI

#### 2.7.1. Surface Chemistry

1. Biotinylated bovine serum albumin (Sigma-Aldrich), reconstituted to 2 mg/ml in DI water (referred to as: BSA-biotin).
2. Neutravidin (Thermo Scientific Pierce), reconstituted to 0.5 mg/ml in PBS (referred to as: NA/PBS).
3. 1:1 solution of biotinylated Penta-histidine antibody (Qiagen AG) in 2% (w/v) BSA.

#### 2.7.2. Transcription Factor Synthesis

1. TNT® T7 coupled wheat germ extract mixture (Promega AG).
2. FluoroTect™ GreenLys tRNA (Promega AG).
3. Linear template DNA coding for transcription factor (see Subheading 2.4.1).

### 2.8. Data Acquisition

1. Modified ArrayWorx scanner (Applied Precision) for detection and softWoRx® software.
2. Axon GenePix (Molecular Devices) for analysis.

## 3. Methods

The entire process involves seven distinct steps that will be described in detail (see Fig. 1). Molds for the two-layer microfluidic device are fabricated on silicon wafers patterned from laser-written chrome masks in a class 100 clean-room environment. A set of control and flow molds is used to produce a double-layer device by multilayer soft lithography (4, 9).

Libraries of linear expression templates, coding for TF variants, and Cy5-labeled target DNA sequences are synthesized and spotted onto epoxy-coated microscope slides. The DNA arrays are aligned to the microfluidic device containing 2,400 unit cells and bonded overnight. Assembled chips are mounted on a microscope stage and connected to a pneumatic setup. Each unit cell of the device can be controlled by three individually addressable micromechanical valves, which allow compartmentalization of each unit cell, control of DNA chamber access and the detection area. A circular button membrane is used to mask a precisely defined area during surface derivatization and for MITOMI.

Upon surface derivatization, the device is loaded with a mixture of wheat germ in vitro transcription/translation (ITT) extract to express TF variants from spotted linear templates. Bound complexes

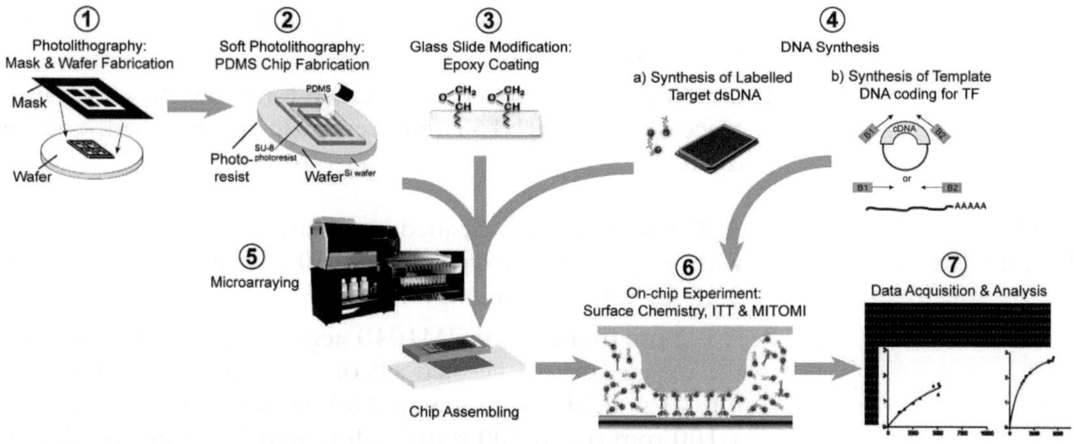


Fig. 1. Workflow of a MITOMI experiment. (**1**) Molds for the two-layer microfluidic device are produced on silicon wafers reproduced from chrome masks. (**2**) A double-layer device is fabricated of PDMS (polydimethylsiloxane) by multilayer soft lithography using a control and flow mold as template. (**3**) Microscope glass slides are surface modified with an epoxysilane coating. (**4a**) Short, fluorescently labeled target DNA sequences are synthesized and (**5**) spotted onto the epoxy-coated glass slides using a microarrayer before aligning and bonding them to microfluidic chip generated in step (**2**). (**6**) After surface derivatization of the glass slide a mixture is loaded containing wheat germ in vitro transcription/translation (ITT) extract and the synthesized linear DNA template (**4b**) coding for the TF of interest. On-chip synthesized TFs are pulled down by immobilized antibody pull-down and can bind spotted target DNA sequences. TF-DNA interactions are captured by a mechanism based on mechanically induced trapping of molecular interactions (MITOMI) and (**7**) binding affinities quantified from detected interactions after scanning.

are trapped at equilibrium using MITOMI. These complexes can subsequently be visualized by scanning the device with a DNA array scanner and relative binding affinities are determined by quantifying the detected signals.

Note for researchers without clean-room and/or PDMS fabrication facilities: the MITOMI devices may also be obtained directly for a nominal fee from the California Institute of Technology (http://kni.caltech.edu/foundry/) and Stanford Microfluidic Foundries (http://www.stanford.edu/group/foundry/index.html).

Up-to-date protocols and microfluidic design files can be found on our laboratory homepage (http://microfluidics.epfl.ch).

### 3.1. Mask and Wafer Fabrication

All processes in this section are performed in a class 100 clean room.

#### 3.1.1. Mask Fabrication

1. The two-layer device is designed in CleWin4 (WieWeb software).
2. Each layer is reproduced as a chrome mask using a Heidelberg DWL200 laser lithography system with a 10-mm writing head and a solid state wavelength stabilized laser diode (max. 110 mW at 405 nm).
3. For the development of masks, first the dispenser arm within the DV10 development chamber is purged for 10 s, then a developer mixture (MP 351:DI water 1:5) is applied twice to

the mask (15 s), agitated for 45 s and drained, before rinsing and drying (50 s).

4. Developed masks are chrome etched for 110 s, rinsed, cleaned twice 15 min in 1,165 remover bath, quick rinsed, and air dried.

*3.1.2. Flow Mold Fabrication*

1. A 3″ silicon wafer is cleaned in a plasma stripper (Tepla 300) with 400 ml/min oxygen gas at 500 W and a frequency of 2.45 GHz for a period of 7 min.
2. A 1–2 μm thin layer of GM1040 negative resist is deposited on the oxygen plasma cleaned silicon wafer by spin coating first for 10 s at 500 rpm (ramp 100 rpm/s), then for 46 s at 1,100 rpm (ramp 100 rpm/s), followed by a short quick spin for 1 s at 2,100 rpm, and finally for 6 s at 1,100 rpm.
3. The pre-coated wafer is baked for 15 min at 65°C and 15 min at 105°C with a ramp of 4°C/min, and allowed to cool down to room temperature.
4. The wafer is exposed for 2 s in flood exposure mode with an alignment gap of 15 μm, using a lamp with a light intensity of 10 mW/cm$^2$ (further settings: WEC type: cont, N2 purge: no, WEC offset: off).
5. The exposed wafer is baked on a hotplate for 35 min at 100°C.
6. Positive resist AZ9260 is spin coated on the pre-coated wafer for 10 s at 800 rpm, followed by 40 s at 1,800 rpm (ramp 1,000 rpm/s) to yield a substrate height of around 14 μm.
7. The coated wafer is then baked on a hotplate for 6 min at 115°C.
8. The soft-baked positive resist is allowed to rehydrate for 1 h.
9. The wafer is exposed on a MA6 mask aligner at two intervals of 18 s with 15 s. Waiting time in photolithography (soft contact) mode at 360 mJ/cm$^2$ with a light intensity of 10 mW/cm$^2$ (broadband spectrum lamp). The alignment gap is set to 15 μm (further settings: WEC type: cont, N2 purge: no, WEC offset: off).
10. Following 1 h relaxation time the wafer is developed in a development chamber (DV10) for 8–12 min based on visual observation after each cycle of the following routine with a total time of 4 min: A development mixture (AZ 400K:DI water 1:4) is dispensed and agitated on the wafer in three cycles, drained, rinsed (total time: 3:15 min), and finally dried.
11. In a final step, channels of the flow mold are annealed and rounded at 160°C for 20 min to create a geometry that allows full valve closure.

*3.1.3. Control Mold Fabrication*

1. Negative photoresist SU-8 GM1060 is spin coated on an oxygen plasma cleaned silicon wafer (see step 1 in Subheading 3.1.2) first for 10 s at 500 rpm (ramp 100 rpm/s), then for 50 s at 1,500 rpm (ramp 100 rpm/s), followed by a short quick spin for 1 s at 2,500 rpm, and finally for 6 s at 1,500 rpm to yield a height of ~14 μm.
2. The coated wafer is baked for 30 min at 130°C and 25 min at 30°C on a hotplate.
3. The wafer is exposed on a MA6 mask aligner at three intervals of 20 s with 15 s (for all other settings see step 9 in Subheading 3.1.2).
4. The exposed wafer is developed manually for 2×5 min in PGMEA, rinsed with IPA, and dried with an air gun.

***3.2. MITOMI Device Fabrication by Multilayer Soft Lithography***

1. The control layer mold is placed in a glass petri dish lined with aluminum foil to facilitate easy removal. Care must be taken that the aluminum foil lining does not contain any holes.
2. To generate a hydrophobic surface, both flow and control mold are exposed to vapor deposits of TMCS for 30 min by placing them into a sealable plastic container with 1 ml TMCS filled into a plastic cap. TMCS treatment is repeated for 10 min each time prior to PDMS chip fabrication.
3. For the control layer, 60 g of a 5:1 Sylgard mixture (50 g Part A : 10 g Part B) is prepared, mixed for 1 min at 2,000 rpm (~400×*g*) and degassed for 2 min at 2,200 rpm (~440×*g*) in a centrifugal mixer.
4. The mixture is poured onto the control layer mold and degassed in a vacuum desiccator for 10 min.
5. For the flow layer, 10.5 g of a 20:1 Sylgard mixture (10 g Part A: 0.5 g Part B) is prepared, mixed for 1 min at 2,000 rpm (~400×*g*) and degassed for 2 min at 2,200 rpm (~440×*g*) in a centrifugal mixer.
6. The mixture is spin coated onto the flow layer with a 15-s ramp and a 35-s spin at 2,200 rpm.
7. After removing the control layer mold from the vacuum chamber any residual surface bubbles are destroyed by blowing on top of the PDMS layer. Any visible particles on top of the control channel grid are carefully removed using a toothpick.
8. Both layers are cured in an oven for 30 min at 80°C.
9. Following polymerization, both molds are taken from the oven and allowed to cool for 5 min.
10. The control layer is then diced with a scalpel and holes (1–8 and B, S, C, O in Fig. 2a) are punched at the control input side using a hole puncher or a 21-gauge luer stub.

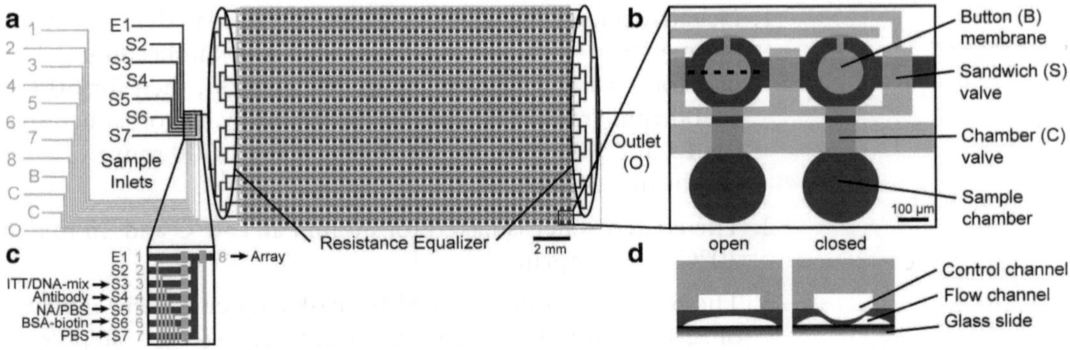


Fig. 2. (**a**) Drawing of a MITOMI device with 768 unit cells in the flow layer (*dark grey*), which are controlled (*light grey*) by 2,388 valves. Resistance equalizers towards solution inlets and outlet ensure equal flow velocities and even derivatization in each row of the channels. (**b**) The magnified view shows two unit cells each controlled by three micromechanical valves: the chamber in each dumbbell-shaped unit cell hosts a different target DNA sequence and is isolated with the chamber valve during surface modification steps for subsequent pull-down to an immobilized antibody in the detection area under the button membrane. For diffusion of samples to the immobilized antibody chamber valves are opened while sandwich valves between individual unit cells are closed in order to prevent cross-contamination between different samples. (**c**) Tygon tubing is loaded with different sample solutions and connected to a metal pin that is plugged into a hole at the end of each channel within the sample inlet tree (also *see* magnified inset of Fig. 3b). Loading of the device with samples via the inlets (S2–S7) is controlled by opening and closing the corresponding control valves. (**d**) A cross-section of one of the unit cells is shown to illustrate the detection mechanism based on mechanically induced trapping of molecular interactions (MITOMI). A thin deflectable membrane can be pushed down by actuating the water-filled control channel and consequently physically trap any material under this area in the flow channel.

11. The channel side of the control layer is thoroughly cleaned with Scotch Magic Tape.
12. The cleaned control layer is then aligned to the flow layer on the stereomicroscope.
13. The device is bonded for 90 min at 80°C in an oven.
14. Bonded devices are removed from the oven and allowed to cool for 5 min.
15. Following the outline of the control layer, each individual device is cut with a scalpel and peeled off the flow layer.
16. Holes are punched for the sample inlet and outlet (S1–S7 and O in Fig. 2a) using a hole puncher.
17. The flow channel side is cleaned thoroughly with tape before aligning the device to a spotted glass slide (see Subheading 3.5).
18. The flow mold is cleaned of any residual polymerized PDMS either by peeling off the thin layer of PDMS using a pair of tweezers or by an additional PDMS layer. For the latter, 11 g of a 10:1 Sylgard mixture (10 g Part A: 1 g Part B) is mixed for 1 min at 2,000 rpm (~400 × *g*), degassed for 2 min at 2,200 rpm (~440 × *g*), poured on the flow mold cured in the oven for 30 min at 80°C, and peeled off after cooling down to room temperature. The control mold is cleaned with a nitrogen air gun of any PDMS debris.

### 3.3. Glass Slide Preparation

#### 3.3.1. Cleaning Procedure

1. All glassware is prepared by rinsing with Milli-Q water.
2. 750 ml Milli-Q water and 150 ml ammonium hydroxide are heated to 80°C in a staining bath.
3. 150 ml hydrogen peroxide is carefully poured to the ammonium solution.
4. Glass slides are added to the staining bath and incubated for 30 min.
5. After removal from the staining bath the glass slides are allowed to cool for 5 min.
6. Glass slides are then rinsed with Milli-Q water in the staining bath.
7. Clean glass slides are dried with nitrogen and stored in a dust-free box.

#### 3.3.2. Epoxysilane Deposition

1. Before epoxysilane deposition all glassware is rinsed with acetone and dried at 80°C.
2. Cleaned glass slides are incubated for 20 min in 891 ml toluene with 9 ml 3-GPS.
3. After rinsing with fresh toluene to remove unbound 3-GPS the glass slides are dried with nitrogen.
4. Glass slides are baked at 120°C for 30 min.
5. Following sonication in toluene for 15 min glass slides are rinsed with fresh IPA.
6. Coated glass slides are dried with nitrogen and stored in a dust-free box under oxygen-free conditions until usage.
7. In case of systematic PDMS chip delamination: Prior to DNA spotting, glass slides are rinsed with toluene and dried with nitrogen.

### 3.4. DNA Synthesis

#### 3.4.1. Synthesis of Linear Template DNA Coding for TF Variants

Generation of linear templates from cDNA (see Note 1) of the TF of interest by a two-step polymerase chain reaction (PCR) method in which the first step amplifies the target sequence and substitutes the DNA-binding domain of the transcription with designed variants. The second step adds required 5′UTR and 3′UTR for efficient ITT.

1. For the first PCR step a mixture of the following components is prepared in a final volume of 50 μl:

   100 nM Forward-ORF primer

   100 nM Reverse-ORF primer

   1 μL Plasmid of MAX iso B cDNA clone

   200 μM of each dNTP of a nucleotide mix

   0.5 units HiFi Plus Polymerase enzyme mixture.
2. After initial denaturation for 7 min at 94°C, the first PCR amplification is performed with 30 cycles as follows:

(a) Denaturation 94°C for 30 s
(b) Annealing 55°C for 60 s
(c) Elongation 72°C for 90 sand finished with a final extension at 72°C for 7 min.
The correct product of this step should be ascertained on a 1% TAE agarose gel.

3. For the second PCR step a mixture of the following components is prepared to yield a final volume of 100 μl:

   5 nM Forward extension

   5 nM Reverse extension

   1 μl PCR product (from previous step)

   200 μM of each dNTP of a nucleotide mix

   1 unit HiFi Plus Polymerase enzyme mixture

4. After initial denaturation for 7 min at 94°C, the second PCR amplification is performed with ten cycles as follows:
   (a) Denaturation 94°C for 30 s
   (b) Annealing 55°C for 60 s
   (c) Elongation 72°C for 90 sand finished with a final extension at 72°C for 7 min

5. To each reaction 2 μl of final primer mix is added (Forward final + Reverse final; each at 5 μM final concentration) and cycling continued for 30 cycles after 4 min at 94°C as follows:
   (a) Denaturation 94°C for 30 s
   (b) Annealing 50°C for 60 s
   (c) Elongation 72°C for 90 sand finished with a final extension at 72°C for 7 min.

6. The final products were then purified on PCR spin columns (Qiagen) with the supplier protocol and eluted in 80 μL of $dH_2O$, pH 8.0–8.5.

*3.4.2. Synthesis of Target DNA*

1. Small Cy5 labeled, dsDNA oligos are synthesized by isothermal primer extension in a reaction of a total volume of 30 μl containing:

   6.7 μM 5′CompCy5

   10 μM Library primer

   5 units Klenow fragment (3′→5′ exo−)

   1 mM of each dNTP from a nucleotide mix

   1 mM Dithiothreitol (DTT)

   50 mM NaCl

   10 mM $MgCl_2$

   10 mM Tris–HCl, pH 7.9.

2. All reactions are incubated at 37°C for 1 h followed by 20 min at 72°C and a final annealing gradient down to 30°C at a rate of 0.1°C/s.
3. After the synthesis, 70 μl of a 0.5% BSA $dH_2O$ solution is added to each reaction.
4. The entire volume is then transferred to a 384-well plate (see Note 2).

### 3.5. Microarraying/Spotting

Spotting target DNA onto epoxy-coated microscope slides is performed by a QArray2 DNA microarrayer.

1. Before each spotting routine an Eppendorf or Falcon tube filled with DI water and the spotting pins is submerged in a sonicator water bath to clean the pins. During the spotting routine a sterilizing loop (1 s DI water, followed by 1 s air drying) between different DNA samples keeps pins clean throughout the spotting procedure.
2. Prior to any template spotting 0.5% BSA $dH_2O$ is spotted to prevent binding of DNA templates to the epoxy surface.
3. Sample plate(s) of ITT template and target DNA are placed in an external source plate stacker before starting the spotting routine (see Note 3).
4. Two rounds of spotting are sequentially performed. In the first round ITT templates, coding for transcription factor variants, are spotted, followed by a second round of spotting of target DNA sequences. Generally three technical repeats of each TF mutant – target DNA combination are spotted.
5. Spotted arrays are aligned manually to a tape-cleaned PDMS device (see Subheading 3.2) on a Nikon SMZ1500 stereoscope and bonded overnight in an incubator at 40°C.
6. DNA arrays can be stored in a sealed box protected from light and dust at room temperature for several weeks.

### 3.6. On-Chip Experiment (Surface Chemistry and MITOMI)

#### 3.6.1. Mounting MITOMI Device to Microfluidic Control Elements

1. The assembled device is mounted on a light microscope and connected to tubing (for details see Fig. 3).
2. Control channels are filled with DI water by actuation with ~5 psi of pneumatic pressure, which forces the air from the dead-end channels into the bulk porous silicone. This procedure eliminates subsequent gas transfer into the flow layer upon valve actuation, as well as prevents evaporation of the liquid contained in the flow layer.
3. Devices are actuated with 15–20 psi in the control lines and between 5 and 8 psi for the flow line.
4. Upon actuation, button membrane and sandwich valves are opened again; chamber valves remain closed during the following

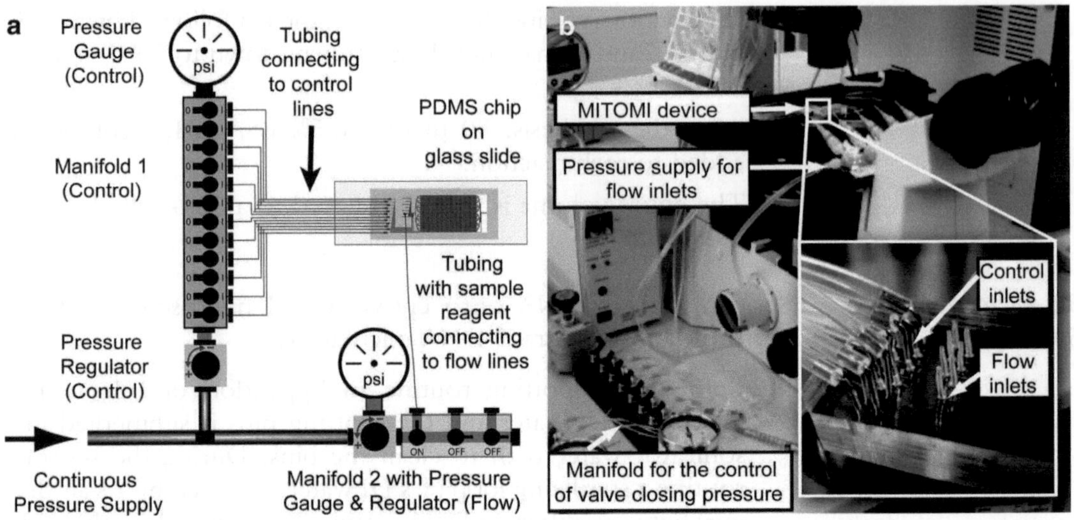


Fig. 3. Setup of the experiment. (**a**) Schematic of fluidic control set-up using regulated pressure and manually controlled manifolds. (**b**) Photograph of an assembled MITOMI device placed on the stage of a microscope where flexible tubing (Tygon) is connected via metal pins to the inlets that actuate the control lines on the device (magnified inset). The tubing is filled with DI water displacing the air in the channels when actuated with pneumatic pressure that is controlled with manifolds. Each reagent for the on-chip experiment is filled into a pin-end flexible tube (Tygon) and connected to the flow inlets. The pressure of the flow can be controlled with a gauge.

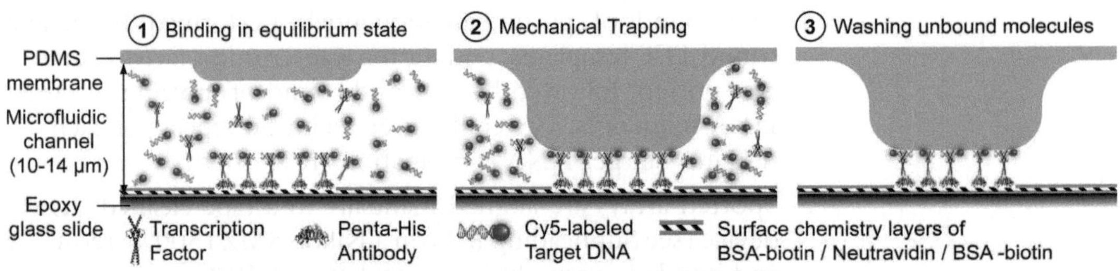


Fig. 4. Schematic of the MITOMI process. The gray structure at the top of each panel represents the deflectable button membrane that can be brought into contact with the glass surface (also *see* Fig. 2d). (**1**) $His_5$-tagged TFs are localized to immobilized penta-His antibody at the epoxy-coated glass slide. Specific binding between Cy5-labeled target DNA and TFs are at steady state when (**2**) the button membrane is actuated and brought into contact with the surface. Any molecules in solution are expelled while surface-bound material is mechanically trapped. (**3**) Unbound material that was not physically protected is washed away and the remaining molecules are quantified.

initial surface derivatization steps to prevent liquid from entering the sample containing chambers.

*3.6.2. Surface Chemistry*

The surface area within the flow channels of the device is modified by depositing layers of BSA-biotin, NA/PBS, and biotinylated antibody onto the epoxy-coated glass slide (see Fig. 4). Using a syringe the different sample solutions are loaded into short pieces (30–40 cm) of clean Tygon tubing that are hooked to a metal pin that is then pushed into the corresponding flow sample inlet holes (see Fig. 2c).

1. Tubing with 30 μl of BSA-biotin is inserted into the sample inlet hole S6. The port on manifold 2 for flow inlet regulation (see Fig. 3) is actuated and valves are opened by switching the corresponding toggles on manifold 1 in the following order: valve 6 controlling sample flow inlet S6, equalizer (1), array inlet (8).
2. Once the air in the channels of the array is displaced with liquid, the outlet valve (O) is opened and the equalizer valve (1) opened.
3. After flowing BSA-biotin for ~20 min the array inlet valve (8) and the sample valve (6) are closed again and the port on manifold 2 turned off.
4. The tubing is disconnected from the port of manifold 2.
5. The process of valve and port opening/closing is performed for the following samples.
6. After BSA-biotin derivatization the surface area is washed for 2–3 min with 5 μl PBS (S7).
7. A 30 μl solution of NA/PBS (S5) is flowed for ~20 min and washed again for 2–3 min with 5 μl PBS.
8. The button membrane (B) is closed and PBS washing continued for 1 min (~2 μl) making sure button is closed.
9. The remaining surface area is passivated with 30 μl BSA-biotin (~20 min) and washed with 5 μl PBS (2–3 min) while button is actuated.
10. 30 μl of a 1:1 solution of biotinylated penta-His antibody in 2% BSA is loaded (S4). To ensure that all channels are saturated with antibody solution the button membrane is opened only after flowing 5 μl.
11. After finishing the antibody deposition (total ~20 min), the surface is washed again with 5 μl PBS for 2–3 min.
12. The surface derivatization procedure is finished with a final 5 μl PBS washing step (see Note 4).

#### *3.6.3. TF Synthesis, DNA Pull-Down and Detection of Interactions On-Chip*

1. 25 μl TNT T7 coupled wheat germ extract is prepared and spiked with 1 μl $tRNA_{Lys}$ – Bodipy – Fl.
2. The mixture is immediately loaded onto the device (S3) and flushed for 10 μl (around 5–7 min) while the button membrane is closed.
3. Chamber valves (C) are opened and the outlet valve (O) is closed to allow for dead-end filling of chambers with wheat germ extract.
4. Chamber valves are closed and the outlet valve opened again and flushing is continued for an additional 10 μl (5–7 min).
5. Sandwich valves (S) that separate each unit cell are closed.

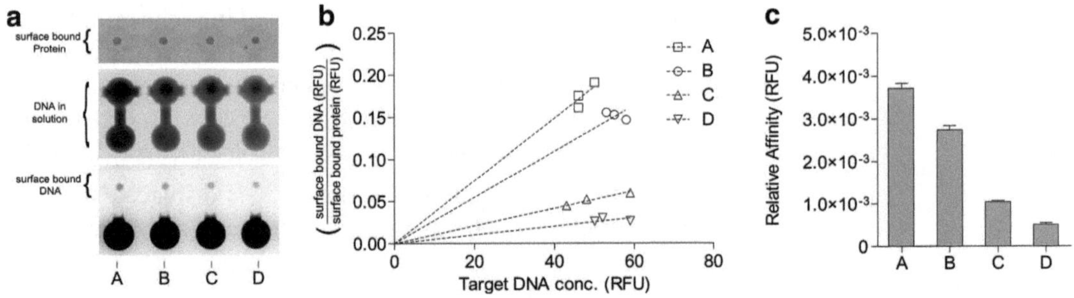


Fig. 5. Steps to analysis of MITOMI experiments. (**a**) Fluorescence scans of subsequent MITOMI steps. Scan of immobilized, fluorescent-labeled protein (*Top*), solubilized target DNA (*Middle*), and surface-bound target DNA after mechanical trapping and washing step (*Bottom*). (**b**) The fraction of surface-bound target DNA is plotted against concentration of target DNA in solution. A linear regression is fitted to each dataset with the origin set to zero. (**c**) Slopes of regressions, as plotted in (**b**), allow for the calculation of relative affinity units (RFU). Care has to be taken to use low DNA concentrations to ascertain that the binding response is in the linear regime.

6. After ensuring that all sandwich valves are closed the chamber valves and button membranes are opened.
7. The device is incubated for 90 min at room temperature or 30°C to allow for protein synthesis and diffusion of the samples to the immobilized antibody under the button membrane.
8. After the incubation period the device is scanned on a modified ArrayWoRx microarray scanner.
9. Button membranes are closed to trap bound samples.
10. Chamber valves are closed, sandwich valves opened and the channels washed with 10 μl PBS (5–7 min).
11. The washed device is scanned once more with closed button membranes to detect the trapped molecules.

### 3.7. Data Acquisition and Analysis

1. For each experiment two images (see Fig. 5a) are analyzed with GenePix3.0 (Molecular Devices): The first image, taken directly after the 60–90-min incubation period before washing, is used to determine the concentration of solution phase or total target DNA concentration (Cy5 channel). The second image, taken after MITOMI and the final PBS wash, is used to determine the concentration of surface-bound protein (FITC channel) as well as surface-bound target DNA (Cy5 channel).

## 4. Notes

1. Linear expression templates can also be synthesized from bacterial cDNA clones after lysing them in 2.5 μl Lyse n'Go buffer (Pierce) at 95°C for 7 min. The lysate serves as template in an

Expand High Fidelity PCR reaction (Roche). The first PCR product is then purified using the Qiaquick 96 PCR purification kit (Qiagen) and eluted in 80 μl of 10 mM Tris–HCl, pH 8.5.

2. The on-chip DNA concentration can be increased by raising the numbers of repetitive stamps of sample DNA per spot during the spotting routine (multiple returns of spotting pin to the same spot).
3. The humidity inside the spotter is set to 65–80% to prevent the samples in the source plate from evaporating during long spotting routines (>2 h).
4. An additional passivation step can be included by flowing 5 μl BSA-biotin for 2–3 min after the antibody immobilization, then closing the button, followed by 2–3 min of 5 μl BSA-biotin. This additional BSA-biotin passivation step was found to reduce background signal.

## References


1. Berger, M. F., and Bulyk, M. L. (2009) Universal protein-binding microarrays for the comprehensive characterization of the DNA-binding specificities of transcription factors, *Nat Protoc 4*, 393–411.
2. Geertz, M., and Maerkl, S. J. (2010) Experimental strategies for studying transcription factor-DNA binding specificities, *Brief Funct Genomics 9*, 362–373.
3. Jolma, A., Kivioja, T., Toivonen, J., Cheng, L., Wei, G., Enge, M., Taipale, M., Vaquerizas, J. M., Yan, J., Sillanpää, M. J., Bonke, M., Palin, K., Talukder, S., Hughes, T. R., Luscombe, N. M., Ukkonen, E., and Taipale, J. (2010) Multiplexed massively parallel SELEX for characterization of human transcription factor binding specificities, *Genome Res 20*, 861–873.
4. Maerkl, S. J., and Quake, S. R. (2007) A systems approach to measuring the binding energy landscapes of transcription factors, *Science 315*, 233–237.
5. Maerkl, S. J., and Quake, S. R. (2009) Experimental determination of the evolvability of a transcription factor, *Proc Natl Acad Sci USA 106*, 18650–18655.
6. Zhao, Y., Granas, D., and Stormo, G. D. (2009) Inferring binding energies from selected binding sites, *PLoS Comput Biol 5*, e1000590.
7. Zykovich, A., Korf, I., and Segal, D. J. (2009) Bind-n-Seq: high-throughput analysis of in vitro protein-DNA interactions using massively parallel sequencing, *Nucleic Acids Research 37*, e151.
8. Matthews, B. W. (1988) Protein-DNA interaction No code for recognition, *Nature 335*, 294–295.
9. Thorsen, T., Maerkl, S. J., and Quake, S. R. (2002) Microfluidic large-scale integration, *Science 298*, 580–584.

# Chapter 7

# Identifying and Optimizing Intracellular Protein–Protein Interactions Using Bacterial Genetic Selection

**Dujduan Waraho and Matthew P. DeLisa**

## Abstract

Protein–protein interactions are crucial for the vast majority of biological processes. To fully understand these processes therefore requires methods for identifying protein interactions within the complex cellular environment. To isolate interacting proteins, we have developed a simple and reliable genetic selection by exploiting the inbuilt "hitchhiker" mechanism of the *Escherichia coli* twin-arginine translocation (Tat) pathway. This method is based on the unique ability of the Tat system to efficiently co-localize noncovalent complexes of two folded polypeptides to the periplasmic space of *E. coli*. The genetic selection is comprised of two engineered fusion proteins: an N-terminal Tat signal peptide fused to the protein of interest, and the known or putative partner protein fused to mature TEM-1 β-lactamase. The efficiency with which co-localized β-lactamase chimeras are exported in the periplasm, and thus confer ampicillin resistance to cells, is directly linked to the relative binding affinity of the protein-ligand system. Thus, ampicillin resistance can be used as a convenient readout for identifying and optimizing protein interactions in *E. coli*. Furthermore, because Tat substrates must be correctly folded for export, our method favors the identification of soluble, non-aggregating, protease-resistant protein pairs. Overall, we anticipate that this new selection tool will be useful for discovering and engineering protein drugs, protein complexes for structural biology, factors that inhibit PPIs, and components for synthetic biology.

**Key words:** Affinity maturation, Antibody engineering, Genetic selection, Intracellular antibodies, Protein–protein interaction, Tat pathway, Bacterial 2-hybrid system

## 1. Introduction

The development of selection and screening methods has been instrumental to our understanding of macromolecular interactions. Examples of such methods for detecting protein–protein interactions (PPIs) include the yeast- and bacterial-based one- and two-hybrid systems, and phage display methods (1–3). These systems are not only useful for identifying interaction partners for specific protein targets but they can also be used in combination with mutagenesis

Wilfried Weber and Martin Fussenegger (eds.), *Synthetic Gene Networks: Methods and Protocols*, Methods in Molecular Biology, vol. 813, DOI 10.1007/978-1-61779-412-4_7, © Springer Science+Business Media, LLC 2012

or randomization strategies to study the details of biologically important interactions or to engineer designer binding proteins. A major advantage of bacterial-based systems compared to their yeast-based counterparts is that they could, in principle, facilitate the rapid analysis of larger libraries due to the higher transformation efficiency and faster growth rate of *E. coli*. Bacterial selection methods could also be superior to phage display because they: (1) permit detection of protein interactions directly in cells, (2) may not require multiple enrichment rounds of affinity purification and amplification; and (3) may be compatible with certain proteins (i.e., large proteins, cytoplasmic proteins) that are not readily displayed on phage particles. Finally, genetic *selections* in which only the desired variants survive are preferable to *screens* in which the specific properties of each variant must be determined individually. Consequently, genetic screens cannot be readily used with libraries larger than ~$10^5$–$10^6$ in size. Here, we describe a novel genetic selection based on the hitchhiker mechanism of the bacterial Tat pathway. The power of this approach is that it directly links the in vivo interaction of two proteins to antibiotic resistance. Importantly, the approach requires no prior structural or functional knowledge and allows selection of PPIs without a need to maintain function of either protein.

A hallmark of the Tat pathway is its remarkable ability to transport folded proteins across tightly sealed, energy-transducing membranes, notably the bacterial cytoplasmic membrane and the chloroplast thylakoid membrane (for recent reviews see refs. 4–8). Substrates of the Tat pathway possess N-terminal signal peptides that contain a highly conserved S/T-R-R-X-F-L-K (where X is any polar amino acid) motif (9–12). In *E. coli*, the Tat system operates alongside the well-characterized Sec pathway and transports two broad classes of protein: globular proteins that fold too rapidly for the Sec system to handle (13) and proteins that are obliged to fold before transport such as those that bind protein subunits (14, 15) or attach redox cofactors (e.g., FeS clusters, molybdopterin centers) (10) in the cytoplasm. The minimal set of components required for Tat translocation consists of the integral membrane proteins TatA, TatB, and TatC (Tha4, Hcf106, and cpTatC in chloroplasts, respectively) (9, 16–18). The TatA and TatB components are single-span integral membrane proteins while TatC has been shown to contain six transmembrane spans (19). Recent cross-linking studies demonstrated that the TatB and TatC proteins form a complex to which substrates initially bind (20). Substrate binding by TatBC appears to trigger the recruitment of TatA oligomers (21), which form a variable diameter ring structure that may serve as a protein-conducting channel or a patch that facilitates translocation by local destabilization of the bilayer (22, 23). However, the subsequent translocation process and the mechanism by which the proton motive force (pmf) energizes vectorial protein transfer across the cytoplasmic membrane are poorly understood.

Studies from our laboratory and elsewhere have firmly established that the Tat pathway preferentially exports folded forms of globular proteins while unfolded proteins or globular proteins that expose hydrophobic patches are not compatible with transport (15, 24–31). These results led us to postulate that a folding quality control mechanism must be an inbuilt feature of the Tat translocon (24); however, the specific details of this quality control process remain a mystery. Nonetheless, the Tat system appears to hold great promise for the expression of secreted proteins and for protein engineering applications. For instance, some recombinant proteins that fail to reach a biologically active form when targeted to the Sec pathway can achieve a functional form by targeting to the Tat pathway, as exemplified by the green fluorescent protein (GFP) (32–35). Moreover, because the Tat machinery does not typically accommodate unfolded proteins, purification from the periplasm can give rise to relatively pure and highly active proteins in one step as highlighted by recent work from our laboratory (26). In addition, we have developed a technique for genetic selection of polypeptides that fold into a stable and soluble conformation (27). The efficiency with which Tat-exported β-lactamase (Bla) chimeras localize in the periplasm, and thus confer ampicillin (Amp) resistance, depends on the folding characteristics of the fusion partner polypeptide. We have extended this strategy to identify protein variants that exhibit higher stability and are less prone to aggregation (25, 28) or small-molecule effectors that alter protein folding and aggregation (36). This method holds promise for the directed evolution of soluble variants of recombinant proteins for biochemical characterization and structural analysis studies.

Along similar lines, we recently developed a reliable genetic selection strategy for isolating interacting proteins using the Tat pathway (15). This method is based on "hitchhiker" export of heterodimeric Tat substrates such as the endogenous *E. coli* hydrogenase-2 (HYD2) complex, comprised of subunits HybO and HybC (14). The large HybC subunit is devoid of any known signal peptide but was reportedly exported to the periplasm via interaction with its partner, HybO, that bears an N-terminal Tat signal peptide (Fig. 1a). Building on this natural example, DeLisa et al. (37) demonstrated that a $F_{AB}$ heavy-chain fused to the Tat-dependent TorA signal peptide (ssTorA) assembled in the cytoplasm with its cognate light chain lacking an export signal, and the heterodimeric $F_{AB}$ complex was exported to the periplasm by the Tat pathway. These findings led us to speculate that the Tat pathway could be exploited to generate a robust, high-throughput technology to identify and isolate intracellular binding proteins of prokaryotic and eukaryotic origin with a built-in quality control feature. To test this notion, we created FLI-TRAP (*F*unctional *L*igand-binding *I*dentification by *T*at-based *R*ecognition of *A*ssociating *P*roteins) (Fig. 1a). In the FLI-TRAP system, two engineered polypeptides are employed: (1) the protein of interest (POI; X in Fig. 1b) is modified with an

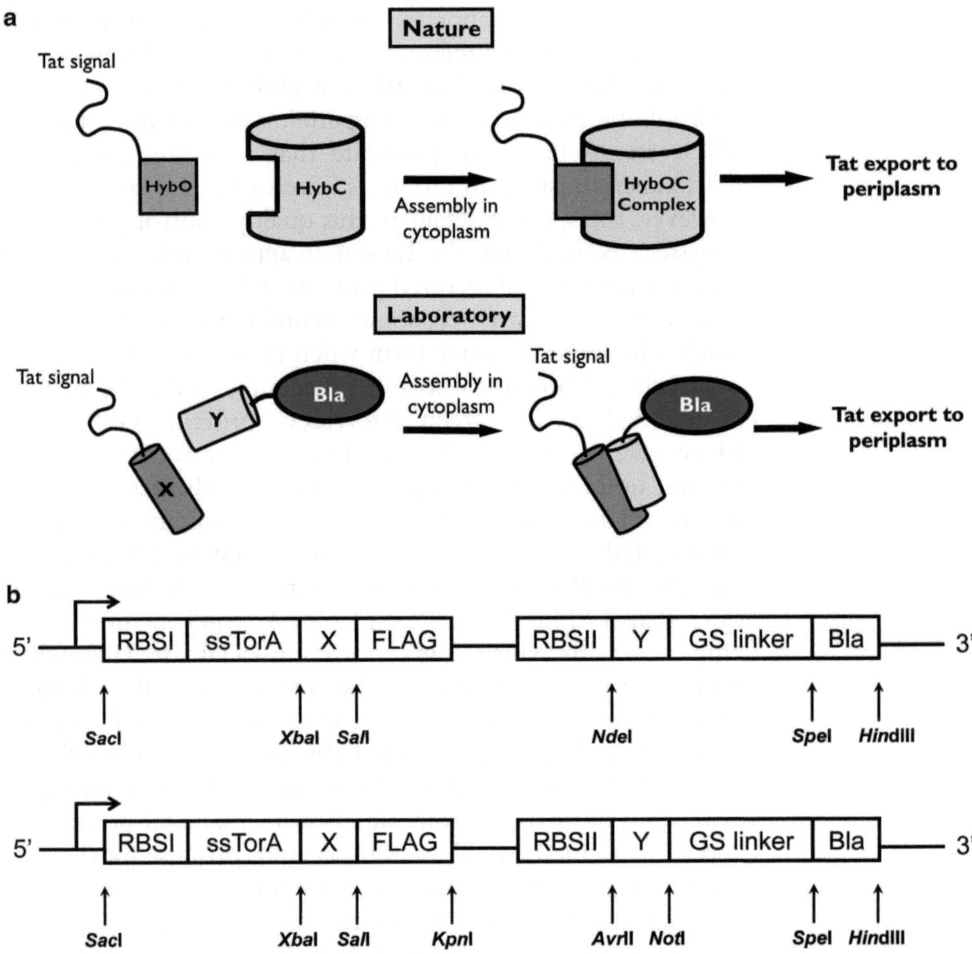


Fig. 1. FLI-TRAP technique for genetic selection of intracellular PPIs. (**a**) Schematic of the FLI-TRAP technique. The method is inspired by the natural hitchhiker export mechanism of the *E. coli* hydrogenase-2 complex. In FLI-TRAP, the hitchhiker mechanism is reconstituted by modifying a POI (X) with an N-terminal Tat-specific signal peptide such as ssTorA from the *E. coli* trimethylamine-*N*-oxide reductase TorA enzyme. The partner protein (Y) is fused to the N-terminus of mature TEM-1 Bla. (**b**) Schematic of two different versions of pDD18-Cm-ssTorA-X-FLAG::Y-Bla showing each element of the bicistronic construction and the restriction sites for cloning (the versions differ in the unique restriction sites flanking each of the genes). The arrow at the 5′ end denotes the L-arabinose-inducible pBAD promoter. The plasmid carries the *cat* gene for chloramphenicol resistance. *RBS* ribosome binding site, *GS* GlySer flexible linker.

N-terminal signal peptide (e.g., ssTorA) for targeting to TatABC and a C-terminal FLAG tag for convenient detection; and (2) the interacting partner (Y) is fused to the N-terminus of mature TEM-1 Bla (Fig. 1b), which acts as a reporter for protein complex transport to the periplasm. The interaction of X and Y produces a physical linkage between ssTorA and Bla in the form of a heterodimeric complex, ssTorA-X::Y-Bla. The subsequent Tat-dependent co-translocation of Bla into the periplasm renders *E. coli* cells resistant to β-lactam antibiotics. By coupling heterodimer formation in the cytoplasm with transport of Bla, simple clonal selection can be used to identify cells manifesting interacting proteins. Bla is an ideal reporter as it allows cells to survive in the presence of β-lactam antibiotics but only if it is transported into the periplasm. Hence, in the absence of an interaction, Bla remains in the cytoplasm and does not confer β-lactam resistance to cells. In addition, Bla is a relatively small (~29 kDa), monomeric protein that can be easily expressed, is not toxic to *E. coli*, and is compatible with the Tat pathway. To date, we have demonstrated FLI-TRAP for diverse interacting protein pairs including basic region-leucine zipper (bZIP) domains of c-Jun and c-Fos proteins and various single chain antibody fragments (scFvs) with their cognate antigens (15). The utility of FLI-TRAP has been further demonstrated through random library selections of amino acid substitutions that restored (1) heterodimerization to a noninteracting FosLZ variant, and (2) antigen binding to a low-affinity scFv antibody. Importantly, the selection procedure bypasses complications associated with purification and immobilization of binding proteins or their ligands as is required for in vitro assays. Moreover, since positive selection requires that the POI and its partner remain correctly folded and tightly associated during membrane transport, FLI-TRAP favors the isolation of high-affinity interactions between soluble protein partners (15). Here we discuss the detailed procedures of FLI-TRAP selection as well as some helpful notes that may assist in library selection when applying this FLI-TRAP strategy to any POI.

## 2. Materials

The FLI-TRAP method converts the Tat pathway's natural mechanism of exporting protein complexes into a versatile in vivo genetic selection technique for intracellular PPIs. To illustrate the approach, we discuss the engineering of a single-chain Fv antibody fragment (scFv) that binds the 47-residue bZIP domain of the yeast transcription factor GCN4 (38). This anti-GCN4 scFv (Ω-graft variant) was previously developed for stable intracellular expression (38) and is a suitable substrate for Tat-mediated export (15). It should be noted, however, that virtually any two candidate proteins or protein domains can be used as the X and Y components in FLI-TRAP. In addition, our system employs a single promoter that directs bicistronic mRNA transcription of the two interacting proteins (i.e., ssTorA-scFv-GCN4-FLAG and GCN4-Bla) (Fig. 1b). Extra copies of the Tat translocon proteins TatA, TatB, and TatC are provided by constitutive expression from a second plasmid to increase transport efficiency. The specific details are presented below.

### *2.1. Determining Selection Conditions*

1. Parental selection plasmid pDD18-Cm-ssTorA-X-FLAG::Y-Bla (Fig. 1b). pDD is derived from pBAD18-Cm (39) and is described in more detail elsewhere (15). The promoter is arabinose inducible. The plasmids containing wild-type anti-GCN4(GLF) or the weakly binding anti-GCN4(GFA) variant, which are used as positive and negative controls, respectively (15), are available on request.
2. pDD322-TatABC. This plasmid encodes TatA, TatB, and TatC expressed from their native promoter and is used to increase Tat transport efficiency and is also available on request.
3. An *E. coli* strain with functioning Tat pathway, such as MC4100, BL21, or TG1.
4. Luria-Bertani (LB) liquid medium (~2 mL for overnight inoculation, 500 μL for resuspension, and 180 μL for each serial dilution, typically six serial dilutions are performed per sample).
5. LB agar medium for spot plating experiments (~12.5 mL for each 100-mm plate).
6. Sterile 100-mm petri dishes.
7. 100 mg/mL carbenicillin (Carb) in sterile water (see Notes 2 and 7).
8. 25 mg/mL chloramphenicol (Cm) in ethanol.
9. 10 mg/mL tetracycline (Tet) in 50% ethanol and 50% sterile water.
10. 20% w/v L-arabinose in sterile water (should be sterile filtered with 0.2 μM syringe filter).
11. 1.7 mL microcentrifuge tubes.
12. Sterile 96-well plate.
13. Multichannel or standard pipettor (with accuracy down to 5 μL).

### *2.2. Library Screening of Intracellular PPIs*

1. Plasmid library under investigation and the parental plasmid.
2. Electrocompetent DH5α (or other high efficiency strain) for plasmid construction in 270 μL aliquots stored at −80°C.
3. Electrocompetent MC4100 (or any strain with functioning Tat pathway) for library screening in 270 μL aliquots stored at −80°C.
4. Electroporation cuvettes with a 2-mm gap (VWR Scientific).
5. Pre-warmed (37°C) SOC liquid medium (3 mL for each transformation plus 1 mL for control).
6. LB liquid medium (250 mL for each transformation, 50 mL for control, 500 μL for resuspension, and 180 μL for each serial dilution).
7. LB agar medium (12.5 mL for each 100-mm plate).

8. Sterile 100-mm petri dishes.
9. 100 mg/mL Carb in sterile water.
10. 25 mg/mL Cm in ethanol.
11. 10 mg/mL Tet in 50% ethanol and 50% sterile water.
12. 20% w/v L-arabinose in sterile water.
13. 1.7 mL microcentrifuge tubes.
14. Sterile 96-well plates.
15. Multichannel or standard pipettor (with accuracy down to 5 μL).

#### 2.3. Verification and Characterization of Isolated Clones

1. LB liquid medium (~2 mL for overnight inoculation, 500 μL for resuspension, and 180 μL for each serial dilution).
2. LB agar medium for spot plating experiments (~12.5 mL for each 100-mm plate).
3. Sterile 100-mm petri dishes.
4. 100 mg/mL Carb in sterile water.
5. 25 mg/mL Cm in ethanol.
6. 10 mg/mL Tet in 50% ethanol and 50% sterile water.
7. 20% w/v L-arabinose in sterile water.
8. 1.7 mL microcentrifuge tubes.
9. Sterile 96-well plates.
10. Multichannel or standard pipettor (with accuracy down to 5 μL).

## 3. Methods

The methods described below are based on the FLI-TRAP technology developed for isolation of intracellular PPIs (15). The natural Tat's pathway unique feature, the hitchhiker mechanism, was converted into a simple genetic selection that can be performed in any laboratory equipped with common microbiology and molecular biology materials. A flow chart of the main steps involved is provided in Fig. 2.

#### 3.1. Determining Selection Conditions

Prior to library selection of improved binding partners, it is important to first identify the selection conditions, i.e. the cell culture dilution factor and the selective Carb concentrations. This section describes the spot plating technique that is used to determine selection conditions. The culture volume used for library screening can be calculated (i.e., scaled up) from spot plating results with the original/parental POI.

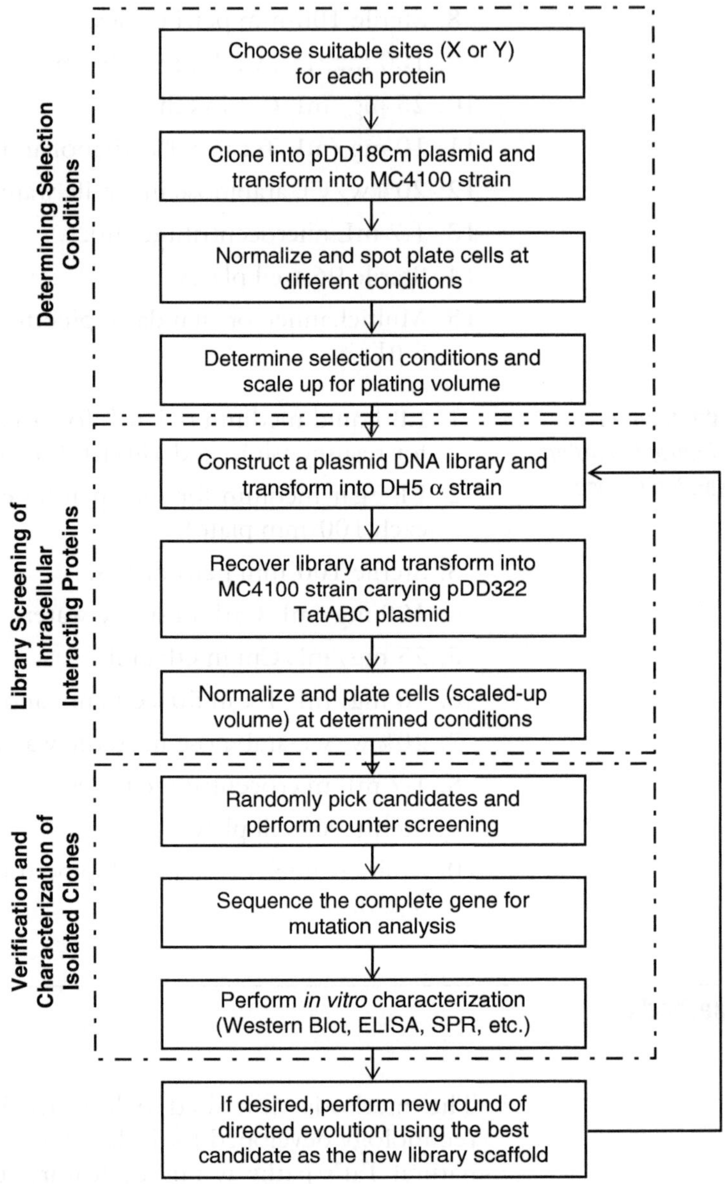


Fig. 2. Flow chart highlighting critical steps in FLI-TRAP method.

1. To generate the bicistronic mRNA transcript of the engineered interacting proteins, the native scFv should be cloned into pDD18-Cm-ssTorA-X-FLAG::Y-Bla at the X position, between the *Xba*I and *Sal*I restriction sites, while its cognate antigen should be cloned at the Y position between *Nde*I and *Spe*I restriction sites (Fig. 1b). It may be useful to clone a non-specific antigen (negative control) to determine the background of growth conferred by non-interacting proteins at each antibiotic concentration. The anti-GCN4(GLF) scFv

and anti-GCN4(GFA) scFv clones may be used as positive and negative growth controls, respectively. The protein being investigated and its partner can be cloned into either X or Y position, thus the suitable position should be determined based on the nature of the POI as described in Note 1.

2. Transform the above plasmids into cells with a functioning Tat pathway such as MC4100. To increase Tat transport efficiency, test plasmids based on pDD18-Cm should be co-transformed with pDD322-TatABC plasmid that expresses the *tatABC* genes from their native constitutive promoter. Plate cells carrying the FLI-TRAP and TatABC plasmids on LB agar plates containing 25 μg/mL Cm and 10 μg/mL Tet followed by incubation at 37°C overnight.
3. Select a single colony from each plate to inoculate 2 mL of Luria-Bertani (LB) medium supplemented with 25 μg/mL Cm and 10 μg/mL Tet. Incubate overnight at 37°C with shaking.
4. Prepare an array of 100-mm LB agar plates containing increasing amounts of Carb ranging from ~12 μg/mL to 1 mg/mL (see Note 2) and increasing amounts of L-arabinose ranging from 0.02 to 1% w/v (see Note 3). As a control, prepare an LB agar plate with only increasing L-arabinose; this will be used to monitor non-selective growth phenotype in the presence of inducer (see Note 4). For example, for each incubation temperature tested, prepare an array of plates each containing 0.2% L-arabinose supplemented with 0, 12, 25, 50, or 100 μg/mL Carb. For another array, prepare plates containing 1% L-arabinose with 0, 12, 25, 50, or 100 μg/mL Carb. To prepare the plates, microwave the LB agar media and leave the bottle in 50°C water bath to cool. Add the appropriate volume of antibiotics and inducer to a sterile 50 mL graduated tube. Add the cooled liquid LB agar medium until the total volume reaches 50 mL. Pour 12.5 mL of the media into a 100-mm petri dish. Let the individual plates solidify. In addition, it is helpful to mark each plate to indicate the top of the plate.
5. The next morning, take an $OD_{600}$ measurement of the overnight culture. Spin down aliquots of each sample in a 1.7 mL microcentrifuge tube and normalize each cell culture in fresh LB media to $OD_{600}=2.5$ (see Note 5).
6. Prepare a dilution series of the cell cultures in fresh LB medium in a sterile 96-well plate. Fill each well with 180 μL of fresh LB medium. Add 20 μL of normalized samples into the first row. Therefore the samples in the first row represent tenfold dilution of overnight samples. Using a multichannel pipettor (see Note 6), pipette up and down to properly suspend cells. Add 20 μL of the first row samples into the second row of the 96-well plate (the samples in the second row represent $10^2$-fold dilution of

the overnight samples or a dilution factor of $10^{-2}$). Using the same tips on the multichannel pipettor, spot plate 5 μl of the first-row samples onto each of the LB agar plates (i.e., all the different Carb and L-arabinose concentrations to be tested). Change tips when spot plating is completed.

7. Repeat step 6. Six consecutive tenfold dilutions (to a dilution factor of $10^{-6}$) are recommended. Since 5 μl of each dilution is needed for each plate be sure that 180 μL is enough for all plates. Only 30 plates (150 μL) are recommended each time as variation of volume is expected.
8. Leave the plates to dry by a flame. The surface droplets should no longer be visible. Incubate the plates upside down in an incubator and observe cell growth between 24 and 48 h (see Note 7). Take pictures of plates at different times post spotting, such as, 24, 40, and 48 h, for comparison later on with isolated clones from the library (refer to Subheading 3.3). The dilution factor and Carb concentration where growth is observed is indicative of the amount of Bla transported by the Tat pathway via the hitchhiker mechanism, and depends on the interacting proteins' solubility and/or affinity.
9. The amount of growth on each plate represents resistance to antibiotic at that selective condition. In addition to Carb concentration, other variables should be tested by spot plating such as incubation period prior to plating, incubation temperature, and induction level (amount of L-arabinose). There is no standard set of conditions that can be applied to all interactions as assay results will vary depending on solubility and affinity of each interacting protein pair. It is recommended to choose several different conditions during library selection (i.e., select on antibiotic concentrations above and below the ideal selection concentration).
10. For increasing the affinity and/or solubility of a target, a library of POIs can be selected using the conditions found above and scaling up. The volume of liquid culture that should be placed on a selection plate can be estimated by scaling up the number of cells in the last spot (most diluted) that grow at each condition. It is generally estimated that there are ~$10^9$ cells/$OD_{600}$ unit in 1 mL of culture. We can use this approximation together with dilution factor to estimate the number of cells that should be placed on the library selection plate. The scaling up can be calculated by assuming that a spot has approximately 6-mm diameter, and a 100-mm petri dish (used as a selection plate) has approximately 85-mm inner diameter. Therefore, 200 times of 5 μL sample is required for a selection plate (see Note 8).

*Example*: If the native scFv co-expressed with its cognate antigen grew slightly at $10^{-3}$ dilution factor and no growth was detected at $10^{-4}$ dilution factor, then $10^{-4}$ dilution factor is used to calculate the dilution used for library screening.

- Hence, 5 μl of $10^{-4}$ dilution in 6-mm spot is scaled up to $5 \times 200 = 1{,}000$ μL of $10^{-4}$ dilution on 100-mm plate. To avoid plating excessive amount of liquid culture, the dilution factor can be adjusted accordingly, i.e. 1,000 μL of $10^{-4}$ dilution is equivalent to 100 μL of $10^{-3}$ dilution of overnight culture; this is a more reasonable volume to apply to the selection plate.
- The approximate library size can then be calculated using the formula below:

Assuming that there are ~$10^9$ cells in 1 mL of overnight culture $OD_{600} = 1$, this scale-up plating allows screening of:

$$OD_{600} \times \frac{10^9 \text{ cells}}{1\text{ml}} \times \frac{1\text{ml}}{1000\mu\text{l}} \times \text{plating volume} (\mu\text{l}) \times \text{plating dilution},$$

where $OD_{600}$ is the normalized $OD_{600}$ of the overnight culture, the plating volume is 100 μL and dilution factor is the plating dilution that is ten times higher than that identified by spot plating. Thus, for this example, the amount of cells plated was:

$$2.5 \times \frac{10^9 \text{ cells}}{1\text{ml}} \times \frac{1\text{ml}}{1{,}000\mu\text{l}} \times 100\mu\text{l} \times 10^{-3} = 2.5 \times 10^5 \text{ cells}.$$

### 3.2. Library Screening of Intracellular PPIs

1. Construct a library of scFv sequences in the X position of the FLI-TRAP system plasmid (Fig. 1b). The method of constructing a library is not covered here. For the anti-GCN4 scFv library, we generated diversity using the GeneMorphII Random Mutagenesis Kit (Stratagene). We recommend electroporating the plasmid library (concentration > 50 ng/mL) into a strain that can be transformed with high efficiency such as DH5α, calculating the library size, recovering the plasmid library by mini- or midi-prep plasmid isolation kit (Qiagen), and finally retransforming the library into MC4100 containing pDD322-TatABC.
2. Prepare DH5α electrocompetent cells. Aliquot 270 μL of cells into sterile 1.7 mL microcentrifuge tubes. Store aliquots at −80°C.
3. Desalt 40 μL aliquots of plasmid DNA library for 20 min by drop dialysis with a 0.025 μm nitrocellulose membrane. Meanwhile, thaw cells on ice. Electroporate each 270 μL aliquot with 30 μL of plasmid DNA library (10% of the total volume) in a 2-mm gap electroporation cuvette. Recover transformed cells by washing three times with 1 mL of pre-warmed SOC and transfer into a 250 mL flask. Repeat until entire plasmid library is transformed. Shake at 37°C for 1 h.

4. To identify how many cells were transformed with plasmid, dilute 20 μL of sample from the flask in 180 μL of SOB in a 1.7 mL microcentrifuge tube. Perform tenfold serial dilution of the sample five more times. Plate 100 μL of a series of tenfold dilutions on LB agar plates supplemented with 25 μg/mL Cm. Scaling up the numbers of colonies on diluted plates allows rough estimation of the library size (total number of transformants in the flask).
5. To recover and enrich only transformed cells prior to selection plating, for each tube of transformed cells, inoculate 250 mL of LB containing 25 μg/mL Cm with the remaining library transformants (after plating in step 4). Shake at 37°C overnight.
6. The next morning, make freezer stock of the library by spinning down 10 mL of cell culture and resuspending in 1 mL of 15% glycerol.
7. Count the colonies on the plate and estimate the library size.

$$\text{Library Size} = \frac{\text{total culture volume (mL)}}{\text{plated volume (mL)}} \times \text{no. of bacterial colony} \times \frac{1}{\text{dilutionfactor}}$$

For example, if 100 colonies resulted from plating 100 mL of a $10^{-5}$ dilution of a 1 mL culture, it can be estimated that:

$$\text{Library Size} = \frac{1}{0.1} \times 100 \times \frac{1}{10^{-5}} = 10^{8} \ \text{library members.}$$

8. Mini-prep several aliquots of overnight culture to recover the plasmid DNA library. The volume of overnight culture that should be mini-prepped depends on library size. For example, if the library size is $10^8$, assuming there are $10^9$ cells in 1 mL $OD_{600}$, 1 mL of culture should represent tenfold coverage of the library (based on genetic diversity). However, typically, only a very small fraction of cells is transformed with plasmid so transformation efficiency of electrocompetent cells should be taken into account. For example, if the electrocompetent cells have transformation efficiency of $10^8$ cfu/μg, 1 μg of plasmid DNA library is required to cover the library size. At least tenfold coverage of library size should be mini-prepped.
9. For library selection of intracellular PPIs, a strain that has a functioning Tat pathway (e.g., MC4100) already transformed with pDD322-TatABC should be used for transformation. Repeat steps 2–6 with this strain. Note that the antibiotics used for these steps are 25 μg/mL Cm and 10 μg/mL Tet. In parallel, transform the parental plasmid used for library construction as a control (this control can be neglected if using naïve library for selection) (see Note 9). For the control plasmid, in

step 3, desalt 10 μL of plasmid DNA and electroporate 5 μL of the parental plasmid into a 45 μL aliquot of cells in a 1-mm gap electroporation cuvette. Recover transformed cells by washing once with 1 mL of pre-warmed SOB. In step 5, use 50 mL cultures.

10. Count the numbers of cells on serial dilution plates and estimate the total number of transformed clones. Multiple-fold coverage of library size is recommended (the number of transformed cells is higher than the number of members of the constructed library). If the number of transformed clones is less than the library size, transformation should be repeated to ensure adequate library coverage.
11. Prepare selection plates with desired Carb and L-arabinose concentrations (refer to step 9 of Subheading 3.1).
12. For each selection condition (refer to dilution factor and volume of liquid culture as identified in step 10 of Subheading 3.1), calculate total amount of cells to be plated (volume in each plate × number of plates). This value should *at least* equal the library size. Multiple-fold coverage is again recommended.
13. Take $OD_{600}$ measurements of overnight library cultures and perform dilution.
14. Plate the appropriate culture volume onto each selection plate. This should be done for the library and the control (parental plasmid) culture.
15. Incubate plates at selected temperature for 24–48 h as identified (see Note 7).
16. Monitor growth after 12–48 h incubation period. Mark clones that grow much faster than expected. From our experience, clones that grow after only 12 h usually represent false positives (see Note 10). After growth selection is complete, individual colonies representing positive clones can be randomly chosen; these should be inoculated in fresh LB and grown at 37°C overnight for verification and characterization. This may be done in test tubes or a 96-well plate depending on the number of positive clones. A few expected false positives may be selected for comparison. If there are more clones than can be characterized at once, keep plates at 4°C for further characterization.

### 3.3. Verification and Characterization of Isolated Clones

1. To confirm that the isolated clones are genuine, they should be serially diluted and spot plated (repeat steps 5–8 of Subheading 3.1) at the same selection conditions used for isolation. The parental plasmid control should be included for each condition tested (see Note 11). An example is shown in Fig. 3a. The second plate has the same Carb concentration (500 μg/mL) as the selection plate while the third plate has almost twice as much Carb (900 μg/mL).

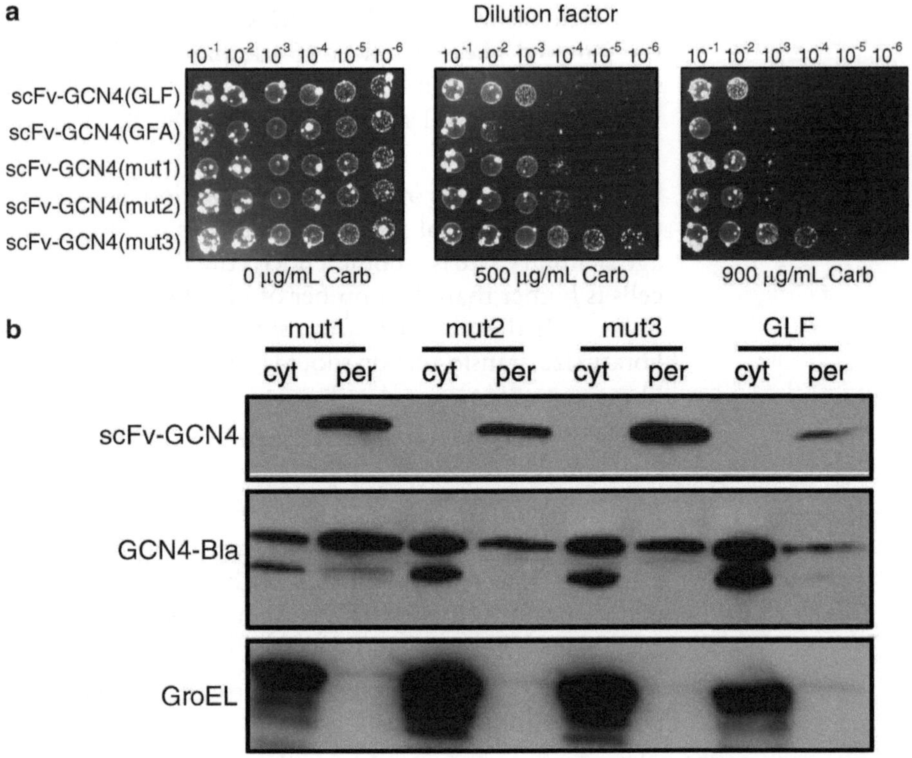


Fig. 3. Isolation of improved scFvs using FLI-TRAP. (**a**) Phenotypic characterization of parental scFv-GCN4(GLF) and three positive clones (mut1, mut2 and mut3) isolated following one round of mutagenesis and selection with FLI-TRAP. Clones were isolated from an error-prone library of scFv-GCN4(GLF), which was screened on either 500 or 600 μg/mL Carb. Following selection, plasmids isolated from positive clones were back-transformed into MC4100 cells carrying pDD322-TatABC and the transformed cells were spot plated at the dilutions and Carb concentrations indicated. When co-expressed with the GCN4-Bla antigen, all three clones grew better than parental scFv-GCN4(GLF) on 500 and 900 μg/mL Carb, with clone mut3 conferring the greatest enhancement in growth. The weakly binding scFv-GCN4(GFA) was also spot plated and served as a negative control. (**b**) Western blot analysis of cytoplasmic and periplasmic fractions generated from MC4100 cells carrying pDD322-TatABC and co-expressing GCN4-Bla with either ssTorA-scFv-GCN4(GLF)-FLAG or the clones mut1, mut2 or mut3. Detection of scFv constructs was with anti-FLAG antibodies and detection of GCN4-Bla was with anti-Bla antibodies. GroEL was detected using anti-GroEL antibodies and served as a subcellular fractionation marker.

2. Mini-prep and sequence plasmids to identify mutations in positive clones. Sequencing the entire mutated gene will confirm presence of the desired insert and absence of frameshifts or other abnormalities.
3. The phenotype of the positive clones should be confirmed. It is important to confirm that antibiotic resistance arises from the plasmid-expressed clones and not from mutations that may spontaneously arise in the *E. coli* host cells. Thus, plasmids from positive clones should be mini-prepped, back-transformed into fresh *E. coli* cells (e.g., MC4100), and spot plated to confirm the phenotype is plasmid associated. True positives should show the same antibiotic resistance as the cells directly isolated from the library.

4. An assay to characterize isolated clones should be performed. For example, Western blotting of the cytoplasmic and periplasmic fractions (see Note 12) of each sample will confirm hitchhiker transport of the interacting proteins as well as visualization of the relative amount of scFv or antigen-Bla in the cytoplasm and periplasm (Fig. 3b). ELISA and Biacore analysis can be used to quantitatively compare the binding activity and affinity, respectively, of each clone. The protocols for Western blot, ELISA and Biacore analysis are not covered here.
5. To improve solubility and/or affinity of a candidate, it is possible to perform multiple rounds of mutagenesis and selection on LB agar plates that contain progressively higher amounts of Carb.

## 4. Notes

1. Examples of considerations are as follows. Some proteins may have limitation regarding fusion to their N- or C-termini so they should be positioned accordingly. For ssTorA fusions, the POI is fused to the C-terminus of the signal peptide. On the other hand, for Bla fusion, the protein ligand is fused to the N-terminus of mature Bla. If the ligand is a small peptide, fusion to globular Bla may act as a scaffold thus providing stability to the fused peptide. It should also be pointed out that it is possible during library selection to identify an scFv (or other binding protein) that has affinity towards Bla and not the target ligand (more profound when using naïve library). Thus, ELISA or other binding assay (using an unfused version of the ligand) should be used to confirm that the isolated binding protein is specific for the target and not the Bla reporter.
2. Some protein pairs require longer incubation time (up to 48–60 h). We recommend Carb instead of Amp because it is more stable over longer incubation periods. To achieve better reproducibility, freeze-thawing of Carb should be minimized. We recommend that smaller aliquots of Carb stock solution (100 mg/mL) should be stored at −20°C until needed and should be thawed on ice only once. Making fresh batch of Carb each time may result in variations due to human error. For initial test, we recommend using 12, 25, 50, and 100 μg/mL Carb. Carb concentration can be varied later depending on the initial test result. We find that adding only Carb to the test plate is sufficient to maintain both plasmids in the cells as both are required in order to confer resistance to Carb. The addition of Cm and Tet may increase pressure on cells and slow the growth rate, resulting in longer incubation periods. Plates can be prepared on the day spot plating will be performed (preferable) or prepared the night before and stored overnight at 4°C.

3. Using the pBAD expression vector, L-arabinose induction level can be varied between 0.02 and 0.2% (w/v) (39). In our experience, cells can be induced with up to 1% L-arabinose. For scFv-antigen experiments, 1% L-arabinose is necessary for the positive control (GLF) to grow distinctively from the negative control (GFA) (15), while for Jun-Fos leucine zipper experiments, 0.2% L-arabinose is sufficient. We recommend testing each new POI with a range of L-arabinose concentrations. The stock solution of 20% L-arabinose should be stored at 4°C.
4. We found that >1.2% L-arabinose can be toxic to certain strains (e.g., MC4100). This can be circumvented by isolating L-arabinose-resistant derivatives of the host strain, which is described elsewhere (34). Otherwise, L-arabinose levels must be maintained at a sub-lethal concentration during all selection experiments.
5. Cell normalization can be performed by calculating the volume of cell culture needed to achieve $OD_{600}$ of 2.5 when resuspended in 500 μL of LB. First measure the $OD_{600}$ of the overnight culture. Then spin down cells, carefully remove supernatant as much as possible using pipettor. Resuspend the pellet in 500 μL of fresh LB medium. Use the following formula to calculate cell culture volume required:

$$\text{overnight cell culture volume} = \text{resuspension volume} \times \frac{\text{desired } OD_{600}}{OD_{600} \text{ of overnight culture}}$$

For example, if $OD_{600}$ of a sample is 3, then 417 μL (500 μL × 2.5/3) of cells need to be spun down.
6. A standard pipettor can be used if a multichannel pipettor is not available. However, when using a multichannel pipettor, glance at the sample level in all tips to make sure they are approximately the same level to keep constant volume.
7. Incubation period and temperature may vary depending on the nature of each interacting protein pair. We recommend different incubation periods ranging between 12 and 48 h and temperatures of 25°C, 30°C, and 37°C. Growth phenotype should be monitored and pictures of the plates should be taken periodically such as 12, 24, 40, and 48 h after incubation. If selection is performed within 12 h of incubation, Amp may be used as an alternative to Carb.
8. An example of scale-up calculation is given below.
   - The scale-up plating volume will be $X$ times the spot plating volume identified earlier:

$$X = \frac{A_p}{A_s} = \frac{d_p^2/4}{d_s^2/4} = \frac{85\text{mm} \times 85\text{mm}}{6\text{mm} \times 6\text{mm}} \approx 200,$$

where $X$ is the library of POIs to spot plating ratio, $Ap$ and $As$ are the surface area of the plate and spot, respectively, $dp$ and $ds$ are the diameter of the plate and spot, respectively. This formula can be used when a different plate size is used (such as 150-mm plates).

9. The parental plasmid control is used to verify the suitability of selection conditions because cells containing parental plasmid should not grow under these selection conditions.
10. In theory, FLI-TRAP should only select for soluble proteins that interact with one another. Unfortunately, due to high selection pressure, it is possible that false positives may still arise. We recommend avoiding clones that grow much faster than the parental POI. From experience, we find that clones exhibiting incremental improvements in solubility and/or binding are less likely to be false positives. However, some false positives may be selected to confirm whether the far superior growth phenotype is an artifact or real.
11. If many candidates arise, cells may be grown in 96-well plates and only spot plated at dilution factors of $10^{-4}$–$10^{-6}$ instead of $10^{-1}$–$10^{-6}$.
12. Subcellular fractionation methods can be performed as follows. Grow cells overnight in 2 mL LB with appropriate antibiotics. The following day, subculture cells in 10–50 mL of fresh LB (depending on the amount of cells needed to give detectable protein expression levels) containing appropriate antibiotics and shake at 37°C until $OD_{600}$ reaches 0.5–0.7. Induce with appropriate amount of L-arabinose and incubate for 4–6 h at the selection temperature of 30°C or 37°C or overnight at room temperature. Measure $OD_{600}$ for normalization. Calculate volume of cells required using the following formula:

$$\text{required sample volume} = \text{resuspension volume} \times \frac{\text{final (desired) } OD_{600}}{OD_{600} \text{ of sample}}$$

Keep final $OD_{600}$ constant to ease direct comparison of samples from each experiment. Typically, final equivalent $OD_{600}$ values of 50–100 after resuspension are preferable. For example, if $OD_{600}$ of a sample is 0.9 and cells will be resuspended in 250 μL of 5 mM $MgSO_4$ with the desired $OD_{600}$ of 75, 20.8 mL of sample (0.25 mL×75/0.9) will be spun down. Note that required volume should be calculated prior to harvesting the cells. If the sample volume is not sufficient, resuspension volume may be varied but $OD_{600}$ should be kept constant among samples.

Harvest cells by transferring appropriate volume of each cell culture into centrifugation tube and pellet by centrifugation at 1,900 × *g*, 4°C for 10 min. Resuspend pellets in 1 mL of cell fractionation buffer (for 1 L of fractionation buffer, add 3 mL of 1 M Tric–HCl (pH 8), 20 g of sucrose, and 200 μL of 0.5 M $Na_2EDTA$) and transfer liquid to 1.5 mL microcentrifuge tube. Incubate at 25°C for 10 min. Centrifuge cells at 9,300 × *g* for 2 min. Discard the supernatant. Resuspend the pellet in determined volume (used for normalization calculation) of ice-cold 5 mM $MgSO_4$ stock solution. Incubate cells on ice for 10 min. Centrifuge cells at 15,700 × *g* for 2 min. Collect supernatant as the *periplasmic fraction*. Resuspend the pellet in 500 μL PBS to rinse. Centrifuge cells at 15,700 × *g* for 2 min. Discard the supernatant. Resuspend the pellet in the same volume of PBS. Sonicate for 30 s. Centrifuge at 15,700 × *g* for 10 min. Collect the supernatant as the *cytoplasmic fraction*.

## References


1. Smith, G. P., and Petrenko, V. A. (1997) Phage Display, *Chem Rev 97*, 391–410.
2. Hu, J. C., Kornacker, M. G., and Hochschild, A. (2000) *Escherichia coli* one- and two-hybrid systems for the analysis and identification of protein-protein interactions, *Methods 20*, 80–94.
3. Phizicky, E. M., and Fields, S. (1995) Protein-protein interactions: methods for detection and analysis, *Microbiol Rev 59*, 94–123.
4. Robinson, C., and Bolhuis, A. (2004) Tat-dependent protein targeting in prokaryotes and chloroplasts, *Biochim Biophys Acta 1694*, 135–147.
5. Robinson, C., and Bolhuis, A. (2001) Protein targeting by the twin-arginine translocation pathway, *Nat Rev Mol Cell Biol 2*, 350–356.
6. Muller, M., and Klosgen, R. B. (2005) The Tat pathway in bacteria and chloroplasts (review), *Mol Membr Biol 22*, 113–121.
7. Muller, M. (2005) Twin-arginine-specific protein export in *Escherichia coli*, *Res Microbiol 156*, 131–136.
8. Lee, P. A., Tullman-Ercek, D., and Georgiou, G. (2006) The bacterial twin-arginine translocation pathway, *Annu Rev Microbiol 60*, 373–395.
9. Weiner, J. H., Bilous, P. T., Shaw, G. M., Lubitz, S. P., Frost, L., Thomas, G. H., Cole, J. A., and Turner, R. J. (1998) A novel and ubiquitous system for membrane targeting and secretion of cofactor-containing proteins, *Cell 93*, 93–101.
10. Berks, B. C. (1996) A common export pathway for proteins binding complex redox cofactors? *Mol Microbiol 22*, 393–404.
11. Berks, B. C., Sargent, F., De Leeuw, E., Hinsley, A. P., Stanley, N. R., Jack, R. L., Buchanan, G., and Palmer, T. (2000) A novel protein transport system involved in the biogenesis of bacterial electron transfer chains, *Biochim Biophys Acta 1459*, 325–330.
12. Berks, B. C., Sargent, F., and Palmer, T. (2000) The Tat protein export pathway, *Mol Microbiol 35*, 260–274.
13. Ribnicky, B., Van Blarcom, T., and Georgiou, G. (2007) A scFv antibody mutant isolated in a genetic screen for improved export via the twin arginine transporter pathway exhibits faster folding, *J Mol Biol 369*, 631–639.
14. Rodrigue, A., Chanal, A., Beck, K., Muller, M., and Wu, L. F. (1999) Co-translocation of a periplasmic enzyme complex by a hitchhiker mechanism through the bacterial Tat pathway, *J Biol Chem 274*, 13223–13228.
15. Waraho, D., and DeLisa, M. P. (2009) Versatile selection technology for intracellular protein-protein interactions mediated by a unique bacterial hitchhiker transport mechanism, *Proc Natl Acad Sci USA 106*, 3692–3697.
16. Bogsch, E. G., Sargent, F., Stanley, N. R., Berks, B. C., Robinson, C., and Palmer, T. (1998) An essential component of a novel bacterial protein export system with homologues in plastids and mitochondria, *J Biol Chem 273*, 18003–18006.
17. Sargent, F., Bogsch, E. G., Stanley, N. R., Wexler, M., Robinson, C., Berks, B. C., and Palmer, T. (1998) Overlapping functions of components of a bacterial Sec-independent protein export pathway, *EMBO J 17*, 3640–3650.
18. Settles, A. M., Yonetani, A., Baron, A., Bush, D. R., Cline, K., and Martienssen, R. (1997) Sec-independent protein translocation by the maize Hcf106 protein, *Science 278*, 1467–1470.

19. Behrendt, J., Standar, K., Lindenstrauss, U., and Bruser, T. (2004) Topological studies on the twin-arginine translocase component TatC, *FEMS Microbiol Lett* ***234***, 303–308.
20. Alami, M., Luke, I., Deitermann, S., Eisner, G., Koch, H. G., Brunner, J., and Muller, M. (2003) Differential interactions between a twin-arginine signal peptide and its translocase in *Escherichia coli*, *Mol Cell* ***12***, 937–946.
21. Mori, H., and Cline, K. (2002) A twin arginine signal peptide and the pH gradient trigger reversible assembly of the thylakoid (Delta] pH/Tat translocase, *J Cell Biol* ***157***, 205–210.
22. Cline, K., and McCaffery, M. (2007) Evidence for a dynamic and transient pathway through the Tat protein transport machinery, *EMBO J* ***26***, 3039–3049.
23. Gohlke, U., Pullan, L., McDevitt, C. A., Porcelli, I., de Leeuw, E., Palmer, T., Saibil, H. R., and Berks, B. C. (2005) The TatA component of the twin-arginine protein transport system forms channel complexes of variable diameter, *Proc Natl Acad Sci USA* ***102***, 10482–10486.
24. DeLisa, M. P., Tullman, D., and Georgiou, G. (2003) Folding quality control in the export of proteins by the bacterial twin-arginine translocation pathway, *Proc Natl Acad Sci USA* ***100***, 6115–6120.
25. Fisher, A. C., and DeLisa, M. P. (2009) Efficient isolation of soluble intracellular single-chain antibodies using the twin-arginine translocation machinery, *J Mol Biol* ***385***, 299–311.
26. Fisher, A. C., Kim, J.-Y., Perez-Rodriguez, R., Tullman-Ercek, D., Fish, W., Henderson, L. A., and DeLisa, M. P. (2008) Exploration of twin-arginine translocation for the expression and purification of correctly folded proteins in *Escherichia coli*, *Microbial Biotechnol* ***1***, 403–415.
27. Fisher, A. C., Kim, W., and DeLisa, M. P. (2006) Genetic selection for protein solubility enabled by the folding quality control feature of the twin-arginine translocation pathway, *Protein Sci* ***15***, 449–458.
28. Gushchina, L. V., Gabdulkhakov, A. G., and Filimonov, V. V. (2009) Design and structural and thermodynamic studies of a chimeric protein derived from spectrin SH3-domain, *Mol Biol (Mosk)* ***43***, 483–491.
29. Richter, S., and Bruser, T. (2005) Targeting of unfolded PhoA to the Tat translocon of *Escherichia coli*, *J Biol Chem* ***280***, 42723–42730.
30. Sanders, C., Wethkamp, N., and Lill, H. (2001) Transport of cytochrome c derivatives by the bacterial Tat protein translocation system, *Mol Microbiol* ***41***, 241–246.
31. Bruser, T., Yano, T., Brune, D. C., and Daldal, F. (2003) Membrane targeting of a folded and cofactor-containing protein, *Eur J Biochem* ***270***, 1211–1221.
32. Feilmeier, B. J., Iseminger, G., Schroeder, D., Webber, H., and Phillips, G. J. (2000) Green fluorescent protein functions as a reporter for protein localization in *Escherichia coli*, *J Bacteriol* ***182***, 4068–4076.
33. Mansell, T. J., Linderman, S. W., Fisher, A. C., and Delisa, M. P. (2010) A rapid protein folding assay for the bacterial periplasm, *Protein Sci* ***19***, 1079–1090.
34. Santini, C. L., Bernadac, A., Zhang, M., Chanal, A., Ize, B., Blanco, C., and Wu, L. F. (2001) Translocation of jellyfish green fluorescent protein via the Tat system of *Escherichia coli* and change of its periplasmic localization in response to osmotic up-shock, *J Biol Chem* ***276***, 8159–8164.
35. Thomas, J. D., Daniel, R. A., Errington, J., and Robinson, C. (2001) Export of active green fluorescent protein to the periplasm by the twin-arginine translocase (Tat) pathway in *Escherichia coli*, *Mol Microbiol* ***39***, 47–53.
36. Lee, L. L., Ha, H., Chang, Y. T., and DeLisa, M. P. (2009) Discovery of amyloid-beta aggregation inhibitors using an engineered assay for intracellular protein folding and solubility, *Protein Sci* ***18***, 277–286.
37. DeLisa, M. P., Tullman, D., and Georgiou, G. (2003) Folding quality control in the export of proteins by the bacterial twin-arginine translocation pathway, *Proc Natl Acad Sci USA* ***100***, 6115–6120.
38. der Maur, A. A., Zahnd, C., Fischer, F., Spinelli, S., Honegger, A., Cambillau, C., Escher, D., Pluckthun, A., and Barberis, A. (2002) Direct *in vivo* screening of intrabody libraries constructed on a highly stable single-chain framework, *J Biol Chem* ***277***, 45075–45085.
39. Guzman, L. M., Belin, D., Carson, M. J., and Beckwith, J. (1995) Tight regulation, modulation, and high-level expression by vectors containing the arabinose PBAD promoter, *J Bacteriol* ***177***, 4121–4130.

# Chapter 8

# Zinc-Finger Nucleases-Based Genome Engineering to Generate Isogenic Human Cell Lines

Anne-Kathrin Dreyer and Toni Cathomen


## Abstract

Customized zinc-finger nucleases (ZFNs) have developed into a promising technology to precisely alter mammalian genomes for biomedical research, biotechnology, or human gene therapy. In the context of synthetic biology, the targeted integration of a transgene or reporter cassette into a "neutral site" of the human genome, such as the *AAVS1* locus, permits the generation of isogenic human cell lines with two major advantages over standard genetic manipulation techniques: minimal integration site-dependent effects on the transgene and, vice versa, no functional perturbation of the host-cell transcriptome. Here we describe in detail how ZFNs can be employed to target integration of a transgene cassette into the *AAVS1* locus and how to characterize the targeted cells by PCR-based genotyping.


**Key words:** *AAVS1* site, Homologous recombination, Gene targeting, Safe harbor, Zinc-finger nuclease

## 1. Introduction

Methods of introducing precise and stable alterations in complex genomes hold great potential for the study of gene function and open a broad range of possibilities in basic research, biotechnology, and human gene therapy (1). One favored technology is known as gene targeting and it relies on the homologous recombination (HR) pathway. Since gene targeting frequencies in mammalian cells are rather low ($10^{-6}$) (2), methods of enhancing HR-based gene targeting are particularly promising in order to avoid screening of large amounts of clones to identify cells with the correct genome modification. In this respect, zinc-finger nucleases (ZFNs) have proven useful for targeted genome engineering in a variety of cell types and whole organisms (3). ZFNs enhance gene targeting by

Wilfried Weber and Martin Fussenegger (eds.), *Synthetic Gene Networks: Methods and Protocols*,
Methods in Molecular Biology, vol. 813, DOI 10.1007/978-1-61779-412-4_8, © Springer Science+Business Media, LLC 2012

their ability to create a site-specific DNA double strand break (DSB) within a complex genome. The DSB activates the cellular DNA damage response, including HR-based repair pathways, and, as a consequence, enhances gene targeting frequencies some 1,000-fold (4).

A ZFN is a heterodimeric designer nuclease (5, 6). Each ZFN subunit consists of a tandem array of linked zinc-finger motifs (Fig. 1a), which confers specific binding to the DNA target sequence, and the catalytic domain of the endonuclease *Fok*I (7). Binding of two juxtaposed ZFN subunits to their target site leads to the dimerization and activation of the endonuclease domain (8). The resulting DSB can be repaired by the cell by one of the two major DNA repair mechanisms: HR-based homology-directed repair (HDR) or non-homologous end joining (NHEJ). In the presence of high concentrations of an extrachromosomal donor DNA that harbors sufficient sequence homology to the respective target site, HDR is the preferred repair pathway (9), in which course the DNA sequence between the two homology arms is copied into the genomic locus (Fig. 1b, c). Depending on the

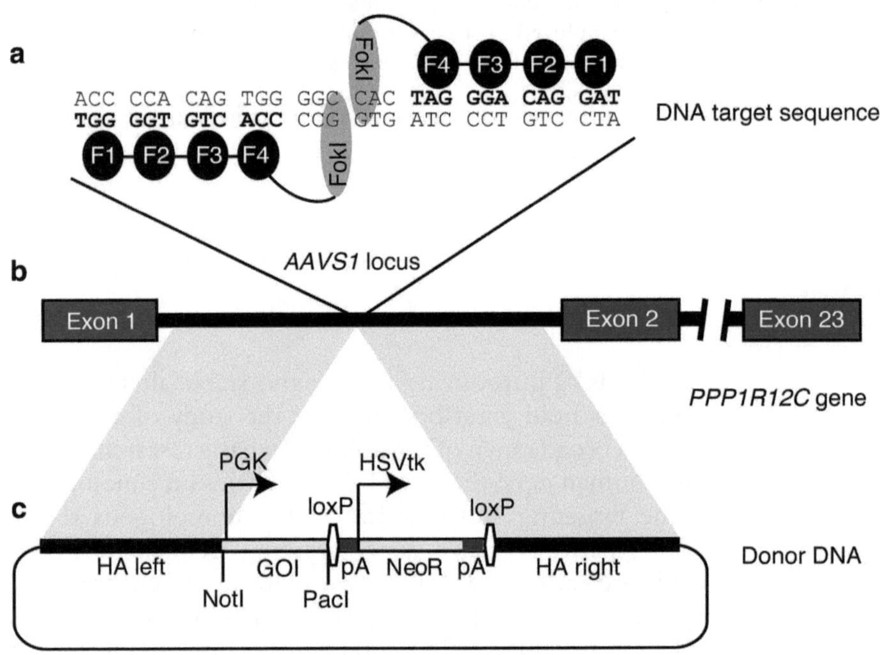


Fig. 1. ZFN-mediated gene targeting at the *AAVS1* site. (**a**) *AAVS1*-specific ZFN pair bound to the 24-bp long target sequence. The ZFN target subsites (*bold*) are separated by a 6-bp spacer sequence. Each ZFN subunit contains an array of four zinc-fingers (F1–F4), which is fused to the catalytic domain of the *Fok*I endonuclease via a short linker. (**b**) *AAVS1* locus. The *AAVS1* site is located in intron 1 of the *PPP1R12C* locus on chromosome 19. (**c**) Donor DNA. The donor plasmid contains ~750-bp long homology arms (HA *left* and HA *right*) that flank a gene of interest (GOI; in our case an *EGFP* gene) under control of the PGK promoter followed by an HSVtk promoter driven neomycin resistance gene (*Neo*$^R$). The GOI can be exchanged with any desired transgene via the unique restriction sites *Not*I and *Pac*I. Upon optional expression of Cre recombinase, the sequence between the two *loxP* sites can be excised, thereby removing the *Neo*$^R$ selection cassette and both poly adenylation (pA) signals.

specific needs, the introduced sequence can encode recombinase recognition sites (such as *FRT* or *loxP*) for further genetic modifications, a complete transgene expression cassette, or a reporter cassette. In the latter case, the reporter can be designed (i) to determine the effects of *trans*-regulatory factors (e.g. variants of a transcription factor) on the cells' physiology, (ii) to assess the function of *cis*-regulatory elements, (iii) to read out the response to intrinsic or extrinsic factors in high-throughput functional genomics and proteomics approaches, or (iv) to add a functional tag to an endogenous locus.

For all of these applications, the generation of isogenic cell lines, that is cells in which the engineered alteration in the genome is the only genetic difference, is essential to avoid epigenetic interference of the readout. Consequently, the locus for integration of the reporter gene has to be carefully chosen and such a "neutral site" should fulfill at least two minimal criteria: Native insulator regions should protect the reporter from integration site-related influences, hence avoiding the issues of variegation or silencing of reporter gene expression. On the other hand, the insulators should prevent the reporter gene promoter from affecting endogenous promoters and so perturbing the cell's transcriptome. The *AAVS1* site, which is located in intron 1 of the human *PPP1R12C* locus (Fig. 1b) on chromosome 19, has been a prime candidate to fulfill these criteria since its initial description as the preferred integration site of adeno-associated virus type 2 (10), an apathogenic human DNA virus, and the discovery of adjacent insulator regions (11). In the interim, several reports have verified sustained expression from transgene cassettes integrated into *AAVS1* in several human cell lines, including pluripotent stem cells and their differentiated daughter cells (12–14). Moreover, a ZFN pair recognizing a 24-bp target site within the *AAVS1* locus (Fig. 1a) has been described and shown to promote robust targeted integration of transgenes into this "neutral site" (13, 14).

As an exemplary approach, we describe in detail how the ZFN technology can be employed to achieve the targeted integration of an EGFP marker cassette into the *AAVS1* locus in the myelogenous human cell line K562 and explain how cells can be genotyped to verify targeted integration.

## 2. Materials

### 2.1. Generation of ZFN and Donor Plasmids

1. A ZFN pair targeting the *AAVS1* locus has been described in detail (13, 14) and can either be generated by gene synthesis (see Note 1) or purchased from Sigma-Aldrich (CompoZr Targeted Integration Kit-AAVS1). The nucleotide and amino acid sequences of the zinc-finger arrays of the left and right *AAVS1*-specific ZFN subunits are shown in Fig. 2.

***AAVS1*-specific zinc-finger array: right**

```
BglII                                                   XbaI
AGA TCT ACC ATG GGC TAC CCC TAC GAC GTG CCC GAC TAC GCC tct aga CCC AAG AAA AAG
           >M   G   Y   P   Y   D   V   P   D   Y   A   S   R   P   K   K   K
 CGG AAA GTG GGC ATC CAC GCT AGC CCT GCC GCC ATG GCT GAG AGG CCC TTC CAG TGC CGG
>R   K   V   G   I   H   A   S   P   A   A   M   A   E   R   P   F   Q   C   R
 ATC TGC ATG CGG AAC TTC AGC caa tcc tct aac CTG gca cgc CAC ATC CGG ACC CAC ACC
>I   C   M   R   N   F   S   Q   S   S   N   L   A   R   H   I   R   T   H   T
 GGC GAG AAG CCC TTC GCC TGC GAC ATC TGC GGC CGG AAG TTC GCC cgt acg gac tac CTC
>G   E   K   P   F   A   C   D   I   C   G   R   K   F   A   R   T   D   Y   L
 gtg gac CAC ACC AAG ATC CAC ACA GGC GAA AAG CCT TTC CAA TGC AGA ATA TGT ATG AGA
>V   D   H   T   K   I   H   T   G   E   K   P   F   Q   C   R   I   C   M   R
 AAC TTC AGC tac aac acc cac CTG acg agg CAC ATC AGA ACA CAC ACA GGG GAG AAA CCG
>N   F   S   Y   N   T   H   L   T   R   H   I   R   T   H   T   G   E   K   P
 TTT GCC TGC GAT ATA TGT GGG AGA AAG TTT GCC cag ggg tac aac CTC gcc ggg CAC ACA
>F   A   C   D   I   C   G   R   K   F   A   Q   G   Y   N   L   A   G   H   T
         Bsu36I     BamHI
AAG ATC CAC CTG AGG GGA TCC
>K   I   H   L   R
```

***AAVS1*-specific zinc-finger array: left**

```
BglII                                                   XbaI
 AGA TCT ACC ATG GGC TAC CCC TAC GAC GTG CCC GAC TAC GCC tct aga CCC AAG AAA AAG
           >M   G   Y   P   Y   D   V   P   D   Y   A   S   R   P   K   K   K
 CGG AAA GTG GGC ATC CAC GCT AGC CCT GCC GCC ATG GCT GAG AGG CCC TTC CAG TGC CGG
>R   K   V   G   I   H   A   S   P   A   A   M   A   E   R   P   F   Q   C   R
 ATC TGC ATG CGG AAC TTC AGC tat aac tgg cat CTG caa aga CAC ATC CGG ACC CAC ACC
>I   C   M   R   N   F   S   Y   N   W   H   L   Q   R   H   I   R   T   H   T
 GGC GAG AAG CCC TTC GCC TGC GAC ATC TGC GGC CGG AAG TTC GCC aga agc gat cac CTC
>G   E   K   P   F   A   C   D   I   C   G   R   K   F   A   R   S   D   H   L
 aca acg CAC ACC AAG ATC CAC ACA GGC GAA AAG CCT TTC CAA TGC AGA ATA TGT ATG AGA
>T   T   H   T   K   I   H   T   G   E   K   P   F   Q   C   R   I   C   M   R
 AAC TTC AGC cac aat tac gct cgc gat tgc CAC ATC AGA ACA CAC ACA GGG GAG AAA CCG
>N   F   S   H   N   Y   A   R   D   C   H   I   R   T   H   T   G   E   K   P
 TTT GCC TGC GAT ATA TGT GGG AGA AAG TTT GCC cag aat agt acg aga att ggt CAC ACA
>F   A   C   D   I   C   G   R   K   F   A   Q   N   S   T   R   I   G   H   T
          Bsu36I     BamHI
 AAG ATC CAC CTG AGG GGA TCC
>K   I   H   L   R
```

Fig. 2. Amino acid and nucleotide sequence of *AAVS1*-specific zinc-finger arrays. The recognition helices that specify DNA binding are highlighted in *gray*.

2. ZFN expression vectors containing *Xba*I and *Bam*HI restriction sites for cloning of the zinc-finger arrays to the N-terminus of obligate heterodimeric *Fok*I domains (15, 16). Such plasmids can be obtained from Addgene or from the authors upon request (see Note 2).
3. TAE running buffer; 40 mM Tris–HCl pH 8.0, 20 mM acetic acid, 1 mM ethylenediaminetetraacetic acid (EDTA).
4. 10 mg/ml ethidium bromide.
5. 6× loading dye; 20% (w/v) Ficoll 400, 0.25% (w/v) bromophenol blue, 0.25% (w/v) xylene cyanol FF, dissolved in water.
6. Agarose gel electrophoresis apparatus.
7. DNA extraction kit, such as QIAGEN Gel Extraction Kit (Hilden, Germany).

8. T4 DNA ligase (400,000 U/ml) supplied with 10× ligase buffer, and restriction endonucleases *Bam*HI and *Xba*I from New England BioLabs (Ipswich, MA, USA).
9. XL1-Blue cells (Stratagene, La Jolla, CA, USA).
10. Donor vector containing your sequence of interest (here EGFP cassette) in combination with a selection marker (here neomycin resistance gene), flanked by homology arms of ~750 bp (Fig. 1c). The described donor plasmid can be obtained from the authors upon request (see Note 3).
11. EndoFree Plasmid Maxi Kit (QIAGEN, Hilden, Germany).

### 2.2. Targeted Integration into AAVS1

1. K562 cells (ATCC CCL-243).
2. Endotoxin-free plasmid maxi preparations of donor plasmid and ZFN expression vectors.
3. Amaxa Cell Line Nucleofection Kit V (Lonza, Basel, Switzerland).
4. Amaxa Nucleofector (Lonza, Basel, Switzerland).
5. RPMI medium supplemented with 10% fetal bovine serum and penicillin/streptomycin (100× solution).
6. Phosphate buffered saline.
7. G418.
8. Propidium iodide staining solution; 5 μg/ml propidium iodide, 5% (v/v) FBS in PBS.
9. QIAamp DNA Blood Mini Kit (QIAGEN, Hilden, Germany).
10. Flow cytometer, such as FacsCalibur (BD Bioscience, Heidelberg, Germany), and software for analysis, such as FlowJo 7.2.5 from Tree Star, Inc. (Ashland, OR, USA).
11. Tissue culture plates 6-well format for suspension cells.

### 2.3. Genotyping

1. Primers for detection of gene targeting event: forward primer GT-F 5′-CCAGCTCCCATAGCTCAGTCTG and reverse primer GT-R 5′-ATACTTTCTCGGCAGGAGCA. Primers for the detection of donor DNA: forward primer DD-F 5′-CCCCTTACCTCTCTAGTCTGTGC and reverse primer DD-R 5′-ATACTTTCTCGGCAGGAGCA. Oligonucleotides (High Purity Salt Free Purification) were ordered from Eurofins MWG Operon (Ebersberg, Germany).
2. Phire® Hot Start II DNA Polymerase supplied with 5× Phire Buffer (Finnzymes, Espoo, Finland).
3. dNTPs (40 mM).
4. Round bottom tissue culture plates 96-well format for suspension cells.

## 3. Methods

***3.1. Generation of ZFN and Donor Plasmids***

1. Clone the left and right *AAVS1*-specific zinc-finger arrays into the mammalian ZFN expression vector using the *Xba*I and *Bam*HI sites (Fig. 2), using standard molecular biology techniques.
2. Perform a test digest of mini prep DNA using *Xba*I and *Bam*HI. Analyze digest on a 1.5% agarose gel containing 1 μg/ml ethidium bromide and verify the presence of the 393-bp fragment (see Note 4).
3. Prepare an endotoxin-free plasmid maxi prep of the ZFN expression plasmid using the QIAGEN EndoFree Plasmid Kit. Elute in 90 μl of EndoFree TE buffer provided in the kit (see Note 5).
4. The donor DNA can be designed in various ways but has to fulfill certain requirements: The homology arms (stretches of sequence homology on either side of the ZFN cleavage site in the *AAVS1* locus) should be ~750-bp long. Any kind of transgene or reporter cassette can be cloned between the two homology arms. Here, the sequence flanked by the two homology arms comprises a PGK-driven EGFP cassette in combination with an HSVtk-driven neomycin-resistance cassette (Fig. 1a) (see Note 6). The PGK-driven EGFP cassette can be exchanged via the unique restriction sites *Not*I and *Pac*I with any desired transgene or reporter cassette (see Note 7).
5. Prepare an endotoxin-free plasmid maxi prep of the donor plasmid using the QIAGEN EndoFree Plasmid Kit. Elute in 90 μl of EndoFree TE buffer provided in the kit.

***3.2. ZFN-Mediated Integration into AAVS1***

1. For ZFN-mediated gene targeting in a certain cell type, the optimal amount and ratio of ZFN expression plasmid to donor DNA should be determined. Also, keep in mind that optimal amount and ratio may change when using a different transfection protocol. Here, we nucleofected human K562 cells with 5 μg of DNA using three ZFN:donor ratios (1:3, 1:1 and 3:1).
2. Culture K562 cells in RPMI medium supplemented with penicillin/streptomycin and 10% FBS in a humidified incubator at 37°C in 5% $CO_2$.
3. Mix the plasmid DNA in a microfuge tube in the respective ratios. The 5 μg of plasmid DNA should be dissolved in a total volume of 1–5 μl of EndoFree TE. Do not exceed the volume of 5 μl. For a 1:3 ratio, mix 600 ng of each ZFN expression plasmid (1.2 μg in total) with 3.8 μg of donor DNA; for the 1:1 ratio 1.25 μg of each ZFN expression plasmid (2.5 μg in total) with 2.5 μg of donor DNA; for the 3:1 ratio 1.9 μg of

each ZFN expression plasmid (3.8 μg in total) with and 1.2 μg of donor plasmid.

4. For each single nucleofection, spin $1 \times 10^6$ K562 cells for 5 min at $300 \times g$ and take off the supernatant. Do not forget a control sample that is mixed with the nucleofector solution but is not pulsed (see Note 8).
5. Prepare a 12-well plate with 1.5 ml of RPMI medium supplemented with penicillin/streptomycin and 10% FBS and pre-equilibrate the medium by placing it in the incubator at 37°C.
6. For each sample, combine 82 μl of Nucleofector Solution V with 18 μl of the supplement solution and mix by pipetting up and down several times.
7. Resuspend the cell pellet in 100 μl of the mixed Nucleofector Solution by carefully pipetting up and down. Do not leave the cells in this solution for longer than 15 min, as this reduces viability.
8. Combine the cell suspension with the 5 μg of DNA (step 3) and mix by pipetting up and down. Be careful not to produce air bubbles.
9. Transfer the cell/DNA suspension into the cuvette. After having checked for the absence of air bubbles close the cuvette with the cap.
10. Select Nucleofector Program T-003, insert the cuvette into the cuvette holder and press the X-button to apply the electric pulse to the cells. Successful termination is indicated by "OK" on the display (see Note 9).
11. Immediately after the electric pulse, add 500 μl of the pre-equilibrated RPMI medium (containing penicillin/streptomycin and 10% FBS) and gently transfer the cells into the prepared 12-well plate containing the 1.5 ml medium (see Note 10).
12. Incubate the transfected K562 cells in a humidified incubator at 37°C in 5% $CO_2$. The next day, add 1 ml of fresh medium to the cells.
13. At day 2 after nucleofection, one third of the cells is harvested for flow cytometric analysis (steps 14–16) and one third will be used for extraction of genomic DNA (steps 17–19). The remaining cells are put under selective pressure for 7 days by adding 800 μg/ml of G418 to the culture medium (steps 20–22).
14. For flow cytometry, spin down one third of the cells of each sample at $300 \times g$ in a table top centrifuge for 5 min at RT. Carefully aspirate the supernatant without disturbing the cell pellet and resuspend the cells in 1 ml of propidium iodide (PI) staining solution (see Note 11). Incubate for 5 min at RT, then pellet cells at $300 \times g$ for 5 min at RT, aspirate supernatant and resuspend your pellet in flow buffer (PBS with 10% FBS).

15. Set up your flow cytometer: For a FacsCalibur (BD Bioscience), we typically use the following settings for Voltage and AmpGain: FSC (E00/1.00), SSC (350/1.00), FL1 (600/1.00), and FL2 (550/1.00). Set compensation between FL1 and FL2 as "FL2 – 18% FL1" (FSC, forward scatter; SSC, side scatter; FL1, green channel; FL2, propidium iodide channel). Settings can slightly change, depending on your flow cytometer, the age of the laser, and the cell batch.
16. Count 20,000 cells and evaluate the percentage of EGFP-positive as well as PI-positive cells (see Note 12).
17. For genomic DNA extraction, spin down one third of the cells at $300 \times g$ in a table top centrifuge for 5 min at RT. Carefully aspirate the supernatant without disturbing the cell pellet and resuspend the cells in 1 ml of PBS. Spin down cells again at $300 \times g$ for 5 min at RT, aspirate supernatant and resuspend cells in 200 μl of PBS.
18. Extract genomic DNA according to the QIAamp DNA Blood Mini Kit protocol for DNA Purification from Blood or Body Fluids. Elute the DNA in 90–120 μl of $ddH_2O$.
19. Perform PCR-based genotyping as described below.
20. Pellet the remaining one third of cells by centrifugation at $300 \times g$ for 5 min. Aspirate the supernatant and resuspend the cell pellet in 1 ml of RPMI medium supplemented with 10% FCS, penicillin/streptomycin and 800 μg/ml of G418.
21. Seed the cells in a 12-well format and select for 7 days. Changing the culture medium every second day and split the cells if they become too dense. Optimal cell density for K562 cells is $2–10 \times 10^5$ cells/ml.
22. After 7 days of selection, pellet the cells at $300 \times g$ for 5 min at RT. Aspirate the supernatant and resuspend the pellet in 1 ml of RPMI medium containing 10% FCS and penicillin/streptomycin. Let the cells recover for 2–5 days (see Note 13) before extraction of genomic DNA and flow cytometric analysis (Fig. 3a) (as described in steps 14–19). Save one third of cells for single clone expansion (see below).

### 3.3. Single Cell Expansion and Genotyping

1. Genotyping by PCR is quick and easy. The PCR setup described below can be used to verify ZFN-mediated gene targeting in the bulk culture before and after selection. Moreover, the protocol will be applied to characterize cell lines derived from a single clone and to determine the frequency of targeted integration (see Notes 14 and 15).
2. Set up the PCR for the detection of the donor DNA: 60 ng of genomic DNA, 10 mM dNTPs, 0.5 μl of 10 pmol/μl each of primers DD-F and DD-R, 5 μl of 5× Phire Buffer, 0.125 μl of Phire Hot Start II DNA Polymerase, and $ddH_2O$ to 25 μl.

Run PCR using the following cycling conditions: 1 cycle at 98°C for 3 min, 25 cycles of [30 s at 98°C; 30 s at 59°C; 50 s at 72°C], 1 cycle at 72°C for 7 min, and last cycle at 4°C until further use.

3. Set up the PCR for the detection of targeted integration: 100 ng of genomic DNA, 10 mM dNTPs, 0.5 μl of 10 pmol/μl each of primers TI-F and TI-R, 5 μl of 5× Phire Buffer, 0.125 μl of Phire Hot Start II DNA Polymerase, and $ddH_2O$ to 25 μl. Run PCR using the following cycling conditions: 1 cycle at 98°C for 3 min, 30 cycles of [30 s at 98°C; 30 s at 59°C; 70 s at 72°C], 1 cycle at 72°C for 7 min, and last cycle at 4°C until further use.
4. Add 4.2 μl of 6× loading dye to the PCR reaction and load entire sample on a 1% agarose gel. Expected product sizes for the detection of donor DNA or targeted integration event are 2.0 kb and 2.8 kb (Fig. 3b), respectively.
5. For clonal single-cell expansion by limiting dilution, use the sample with the most promising PCR results obtained for

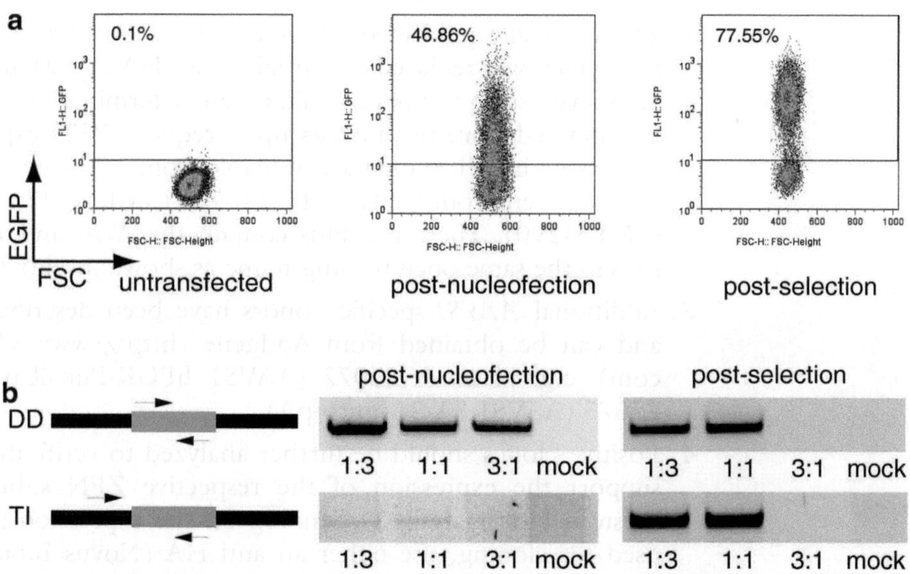


Fig. 3. Enrichment of targeted cells. (**a**) Flow cytometric analysis. K562 cells were nucleofected with 5 μg of plasmid DNA in a 1:3 ratio of ZFN expression vectors to donor. The *plots* indicate percentage of EGFP-positive cells. *Left panel*, untransfected cells; *middle panel*, cells 48 h post-nucleofection; *right panel*, cells after 7 days of selection with G418 and 4 days of recovery. (**b**) PCR-based genotyping. Detection of the donor DNA in the genomic DNA was performed with primer pair DD-F/DD-R (*black triangles*), both of them binding to a sequence within the homology arms (*light grey box*). The expected product of 2 kb was detected in cells transfected with ZFN:donor ratios of 1:3, 1:1 and 3:1 two days post-nucleofection (*middle panel*) and after selection (*right panel*). The lack of a signal in cells transfected with the 3:1 ratio is suggestive of ZFN-associated toxicity. Detection of the targeted integration event was performed with primer TI-F, which is binds to the genomic DNA outside the homology arms, and primer TI-R, which recognizes a sequence in the neomycin resistance gene. The expected PCR product of 2.8 kb was detected in cells transfected with ZFN:donor ratios of 1:3 and 1:1 after selection (*right panel*). The lack of a signal in cells transfected with the 3:1 ratio is suggestive of ZFN-associated toxicity.

targeted integration after selection. Count the cells and prepare a cell suspension in 50 ml RPMI medium with a density of 1 cell per 300 μl of culture medium.

6. Use a multi-dispenser pipette and add 100 μl of cell suspension per well of a 96-well plate. We recommend to set up a total of 480 wells, i.e. five 96-well plates (see Note 16).
7. Thirty minutes after seeding inspect wells meticulously to identify wells containing a single cell (see Note 17).
8. Check the wells regularly for growth of the cells and transfer them to a larger well size when necessary. Cultivate the cells until they grow dense in a 6-well format ($\sim 2 \times 10^6$ cells in 2 ml) and perform genotyping as described above in order to identify positive clones and to calculate the targeting frequency.

## 4. Notes

1. The *AAVS1*-specific zinc-finger arrays were synthesized by GENEART (Regensburg, Germany).
2. The originally published ZFNs contain a N-terminal FLAG tag, which was replaced in our lab by an HA tag. Mammalian ZFN expression vectors containing an N-terminal HA tag can be obtained from the authors upon request, ZFN expression vectors with a FLAG tag are available from Addgene (http://www.addgene.com): plasmids 21873 (pMLM292) 21872 (pMLM290). These plasmids contain the *Xba*I and *Bam*HI sites in the same open reading frame as shown in Fig. 2.
3. Additional *AAVS1*-specific donors have been described (13) and can be obtained from Addgene (http://www.addgene.com): e.g. plasmids 22072 (AAVS1 hPGK-PuroR-pA) and 22075 (AAVS1 SA-2A-puro-pA).
4. Positive clones should be further analyzed to verify that they support the expression of the respective ZFN subunit by Western blot analysis. Depending on the expression plasmid used for cloning, use either an anti-HA (Novus Biologicals, Littleton, CO, USA) or anti-FLAG (DYKDDDK tag antibody; Cell Signaling Technology, Danvers, MA, USA) antibody for detection.
5. The quality of the plasmid DNA is a major determinant of transfection efficiency by nucleofection. Run an undigested aliquot of your plasmid preparation on an agarose gel to verify that most of the plasmid is present in the supercoiled form. The $OD_{260/280}$ ratio should be at least 1.8.
6. The donor can also contain other selection markers, such as genes conferring resistance to puromycin (13) or hygromycin (17) for example.

7. Two *loxP* sites, which flank the region enclosing the poly(A) signal of the EGFP cassette as well as the entire HSVtk-Neo$^R$ cassette, are present in our donor DNA (Fig. 1c). This allows for an optional excision of all "intronic" poly(A) signals in the endogenous *PPP1R12C* locus by transient expression of *Cre* recombinase.
8. The cell density is crucial for high transfection efficiency. For instance, the transfection efficiency dropped twofold when employing $2 \times 10^6$ cells instead of the recommended $1 \times 10^6$ cells. The optimal cell number for nucleofection should be determined using the pmaxGFP plasmid (included in kit) that harbors an EGFP gene under control of a CMV promoter.
9. We have compared programs T-003 and T-016. Although transfection efficiency was similar for both programs, propidium iodide staining revealed that program T-016 was more cytotoxic, producing ~50% of dead K562 cells as compared to about 16% when using program T-003.
10. Sometimes white foam floating on top of the medium can be noticed when adding medium after the electric pulse. This does not necessarily mean that the nucleofection did not work or that your cells are dead.
11. Propidium iodide (PI) staining can be optionally used to determine the percentage of dead cells during the flow cytometric analysis. The percentage of EGFP and PI-positive cells, respectively, are indicators of transfection efficiency and nucleofection-associated toxicity.
12. Using the conditions given above, we usually obtain 20–40% EGFP-positive cells and 14–19% of PI-positive cells. For comparison, transfection of 2 μg of pmaxGFP using the same cell numbers and the same program results in more than 80% EGFP-positive cells one day post-nucleofection. The lower percentage of EGFP-positive cells after nucleofection of K562 cells with our donor plasmid likely reflects the weaker promoter activity of PGK versus CMV.
13. In case the cells have not recovered within 2 days such that flow cytometry and genomic DNA extraction can be performed, let the cells grow for some more days. Especially cells transfected with high amounts of ZFN expression plasmids grow slowly, probably reflecting ZFN-associated toxicity.
14. A Southern blot with an appropriately designed probe can be performed to confirm the targeted integration event.
15. In order to exclude clones that harbor randomly integrated donor plasmid or ZFN expression plasmids, a PCR with primers directed against the *amp*$^R$ gene should be performed (forward primer 5′-gtcacagaaaagcatcttacg, reverse primer: 5′-gttgccattgctacaggcatcg).

16. Round bottom 96-well plates are more suitable then flat bottom wells to identify single cells, as the cells sink to the middle of the well and are therefore easier to spot.
17. If you do not obtain enough wells with a single cell, try different dilutions ranging from 0.3 to 3 cells/100 μl.

## References


1. Capecchi, M. R. (2005) Gene targeting in mice: functional analysis of the mammalian genome for the twenty-first century, *Nat Rev Genet 6*, 507–512.
2. Vasquez, K. M., Marburger, K., Intody, Z., and Wilson, J. H. (2001) Manipulating the mammalian genome by homologous recombination, *Proc Natl Acad Sci USA 98*, 8403–8410.
3. Urnov, F. D., Rebar, E. J., Holmes, M. C., Zhang, H. S., and Gregory, P. D. (2010) Genome editing with engineered zinc finger nucleases, *Nat Rev Genet 11*, 636–646.
4. Jasin, M. (1996) Genetic manipulation of genomes with rare-cutting endonucleases, *Trends Genet 12*, 224–228.
5. Cathomen, T., and Joung, K. J. (2008) Zinc-finger nucleases: the next generation emerges, *Mol Ther 16*, 1200–1207.
6. Carroll, D. (2008) Progress and prospects: zinc-finger nucleases as gene therapy agents, *Gene Ther 15*, 1463–1468.
7. Kim, Y. G., Cha, J., and Chandrasegaran, S. (1996) Hybrid restriction enzymes: zinc finger fusions to Fok I cleavage domain, *Proc Natl Acad Sci U S A 93*, 1156–1160.
8. Smith, J., Bibikova, M., Whitby, F. G., Reddy, A. R., Chandrasegaran, S., and Carroll, D. (2000) Requirements for double-strand cleavage by chimeric restriction enzymes with zinc finger DNA-recognition domains, *Nucleic Acids Res 28*, 3361–3369.
9. Lombardo, A., Genovese, P., Beausejour, C. M., Colleoni, S., Lee, Y. L., Kim, K. A., Ando, D., Urnov, F. D., Galli, C., Gregory, P. D., Holmes, M. C., and Naldini, L. (2007) Gene editing in human stem cells using zinc finger nucleases and integrase-defective lentiviral vector delivery, *Nat Biotechnol 25*, 1298–1306.
10. Kotin, R. M., Linden, R. M., and Berns, K. I. (1992) Characterization of a preferred site on human chromosome 19q for integration of adeno-associated virus DNA by non-homologous recombination, *Embo J 11*, 5071–5078.
11. Ogata, T., Kozuka, T., and Kanda, T. (2003) Identification of an insulator in AAVS1, a preferred region for integration of adeno-associated virus DNA, *J Virol 77*, 9000–9007.
12. Smith, J. R., Maguire, S., Davis, L. A., Alexander, M., Yang, F., Chandran, S., ffrench-Constant, C., and Pedersen, R. A. (2008) Robust, persistent transgene expression in human embryonic stem cells is achieved with AAVS1-targeted integration, *Stem Cells 26*, 496–504.
13. Hockemeyer, D., Soldner, F., Beard, C., Gao, Q., Mitalipova, M., DeKelver, R. C., Katibah, G. E., Amora, R., Boydston, E. A., Zeitler, B., Meng, X., Miller, J. C., Zhang, L., Rebar, E. J., Gregory, P. D., Urnov, F. D., and Jaenisch, R. (2009) Efficient targeting of expressed and silent genes in human ESCs and iPSCs using zinc-finger nucleases, *Nat Biotechnol 27*, 851–857.
14. DeKelver, R. C., Choi, V. M., Moehle, E. A., Paschon, D. E., Hockemeyer, D., Meijsing, S. H., Sancak, Y., Cui, X., Steine, E. J., Miller, J. C., Tam, P., Bartsevich, V. V., Meng, X., Rupniewski, I., Gopalan, S. M., Sun, H. C., Pitz, K. J., Rock, J. M., Zhang, L., Davis, G. D., Rebar, E. J., Cheeseman, I. M., Yamamoto, K. R., Sabatini, D. M., Jaenisch, R., Gregory, P. D., and Urnov, F. D. (2010) Functional genomics, proteomics, and regulatory DNA analysis in isogenic settings using zinc finger nuclease-driven transgenesis into a safe harbor locus in the human genome, *Genome Res 20*, 1133–1142.
15. Szczepek, M., Brondani, V., Buchel, J., Serrano, L., Segal, D. J., and Cathomen, T. (2007) Structure-based redesign of the dimerization interface reduces the toxicity of zinc-finger nucleases, *Nat Biotechnol 25*, 786–793.
16. Söllü, C., Pars, K., Cornu, T. I., Thibodeau-Beganny, S., Maeder, M. L., Joung, K. J., Heilbronn, R., and Cathomen, T. (2010) Autonomous zinc-finger nuclease pairs for targeted chromosomal deletion, *Nucleic Acids Res 38*, 8269–8276.
17. Zou, J., Maeder, M. L., Mali, P., Pruett-Miller, S. M., Thibodeau-Beganny, S., Chou, B. K., Chen, G., Ye, Z., Park, I. H., Daley, G. Q., Porteus, M. H., Joung, J. K., and Cheng, L. (2009) Gene targeting of a disease-related gene in human induced pluripotent stem and embryonic stem cells, *Cell Stem Cell 5*, 97–110.

# Chapter 9

# RNA-Based Networks: Using RNA Aptamers and Ribozymes as Synthetic Genetic Devices

**Julia E. Weigand, Alexander Wittmann, and Beatrix Suess**

## Abstract

Within the last few years, a set of synthetic riboswitches has been engineered, which expands the toolbox of genetic regulatory devices. Small molecule binding aptamers have been used for the design of such riboswitches by insertion into untranslated regions of mRNAs, exploiting the fact that upon ligand binding the RNA structure interferes either with translation initiation or pre-mRNA splicing in yeast. In combination with self-cleaving ribozymes, aptamers have been used to modulate RNA stability. In this chapter, we discuss the applicability of different aptamers, ways to identify novel genetic devices, the pros and cons of various insertion sites and the application of allosteric ribozymes. Our expertise help to apply synthetic riboswitches to engineer complex genetic circuits.

**Key words:** Aptamer, Riboswitch, Ribozyme, Conditional gene expression, Synthetic biology

## 1. Introduction

A key objective in Synthetic Biology focuses on the engineering of complex biological systems and the reprogramming of cellular functions and networks. Within the last years, various tools, so called synthetic genetic devices, have been developed allowing the design of specific human-defined functions (1). Genetic switches are a class of these devices used to influence and/or modulate genetic circuits. Genetic devices can be composed of several components that perform basic biological functions. A sensor is in charge of signal detection, a regulator is responsible for the control of a specific process, and a transmitter sends/converts the signal from the sensor to the regulator.

An advantage of using RNA for engineering genetic devices is the possibility to integrate all the described functions within one molecule. RNA can act as a sensor – e.g. an aptamer domain can

Wilfried Weber and Martin Fussenegger (eds.), *Synthetic Gene Networks: Methods and Protocols*,
Methods in Molecular Biology, vol. 813, DOI 10.1007/978-1-61779-412-4_9, © Springer Science+Business Media, LLC 2012

directly monitor the concentration of small molecules with high affinity and specificity or can sense a physical signal like the temperature. RNA exhibits structural flexibility, which allows it to adopt different conformations in response to an external signal. Hence RNA-based genetic devices can be made of RNA only, without the need of any auxiliary factors, thereby integrating sensor and regulator in one molecule.

The easiest approach to create a genetic control device is the use of small molecule binding RNA aptamers. Aptamers are in vitro-selected RNA molecules that bind a ligand with high affinity and specificity (2, 3) and therefore are perfect sensors. They can be selected against a variety of targets, which allow the construction of sensor domains against nearly any target of choice.

RNA aptamers often adopt their final tertiary fold only in the presence of the ligand (4). Such ligand-induced conformational changes have been exploited to control gene expression in a variety of different approaches. Aptamers inserted into the 5′ untranslated region can control translation initiation in eukaryotes and bacteria. They can interfere with pre-mRNA splicing or control viral replication. In combination with self-cleaving ribozymes they can be exploited to modulate RNA stability. Within the last few years, a versatile set of RNA-based devices called engineered riboswitches, have been created in bacteria, eukaryotes, and even for RNA viruses and are reviewed elsewhere (5).

In this chapter we will focus on current aptamer-based conditional gene expression systems. Thereby we discuss the applicability of different aptamers, ways to identify novel aptamer devices and the pros and cons of various insertion sites and reporter constructs. Furthermore, we briefly discuss the application of allosteric ribozymes for conditional gene expression.

## 2. Materials

### *2.1. Small Molecule Binding Aptamers and Ribozymes*

1. Tetracycline aptamer

   Sequence: 5′-GGCCUAAAACAUACCAGAUCGCCACCCG CGCUUUAAUCUGGAGAGGUGAAGAAUACGACC ACCUAGGCC-3’ (see Note 1).

   Affinity: $K_D$(tetracycline) = 0.77 ± 0.11 nM (6).

   Specificity: $K_D$(doxycycline) = 118 ± 12 nM (see Note 2).

2. Theophylline aptamer

   Sequence: 5′-GGUGAUACCAGCAUCGUCUUGAUGCCC UUGGCAGCACC-3′.

   Affinity: $K_D$(theophylline) = 320 ± 130 nM (7).

   Specificity: $K_D$(caffeine) = 3,500 ± 1,500 nM (see Note 2).

3. Neomycin aptamer

   Sequence: 5′-GACUGCUUGUCCUUUAAUGGUCCAGU C-3′.

   Affinity: $K_D$(neomycin) = 9.2 ± 1.3 nM (8).

   Specificity: Neomycin but not paromomycin mediates regulation (see Note 2).

4. N79 hammerhead ribozyme

   Sequence: 5′-UGAGAUGCAGGUACAUCCAGCUGAUGA GUCCCAAAUAGGACGAAACGCGCUUCGGUGC GUCCUGGAUUCCACUGCUAUCCA-3’.

5. sTRSV hammerhead ribozyme

   Sequence: 5′-CUGUCACCGGAUGUGCUUUCCGGUCU GAUGAGUCCGUGAG GACGAAACAG-3′.

### 2.2. Suitable Vectors and Reporter Genes

1. High copy vectors:
   GFP expressed from a high copy vector (2μ origin of replication) is a suitable reporter gene in yeast. It allows sensitive and fast measurements without the need for cell lysis. Destabilized variants can be used to improve regulation (see Note 3) at the expense of sensitivity.
2. Single copy vector:
   Expression of reporter genes from single copy vectors minimizes cell-to-cell variability. In addition, lower mRNA amounts (and therefore aptamers) in relation to the ligand (see Note 4) can lead to higher regulation efficiencies. Here, GFP measurements are not sensitive enough. Luminescence measurements using luciferase as reporter gene have to be applied, which requires cell lysis and protein quantification as internal standard in yeast.
3. Chromosomal integration:
   Tetracycline aptamers can be exploited to regulate the expression of essential genes in yeast (see Subheading 3.5). Universal insertion cassettes for the tagging of any yeast gene (accession numbers P30598 and P30600) are deposited at the EUROSCARF yeast strain collection (http://web.unifrankfurt.de/fb15/mikro/euroscarf/index.html).

## 3. Methods

### 3.1. How to Choose the Right Aptamer?

Despite the fact that the majority of RNA aptamers are described to adopt their tertiary fold only in the presence of the ligand, only a few of them have the potential to control gene expression.

1. The malachite green aptamer was used for cell cycle control by inhibiting the translation of *CLB2* in yeast (9). No further studies made use of this aptamer, most probably due to the high toxicity of the ligand. Two antibiotic-binding aptamers recognizing tetracycline and neomycin efficiently control translation initiation and pre-mRNA splicing in yeast (see Subheadings 3.3 and 3.4). Biochemical and structural studies allowed a detailed look into the mechanism of regulation. The tetracycline aptamer (see Fig. 1a) adopts a highly complex structure, which is preformed in large parts (10–12). Ligand binding results in a stabilization of the binding pocket and the closing stem (13). The neomycin aptamer is the smallest regulating aptamer identified so far (27 nt, see Fig. 1c). It is largely unstructured in the absence of the ligand (14). Neomycin binding stabilizes the aptamer leading to dramatic conformational changes (8). The degree of conformational change differs between the tetracycline and the neomycin aptamer; however, both aptamers exhibit an extremely high binding affinity, which seems to be a basic necessity for regulating aptamers.
2. Rarely, small molecule binding aptamers have been described for conditional control of gene expression in higher eukaryotes. A pioneering study by Werstuck and Green uses two aptamers (H10 and H19) in tandem recognizing the Hoechst dye H33342 to control the translation of a luciferase reporter gene (15). Similar to the malachite green aptamer, the Hoechst dye-binding aptamers have not been used any more. Maybe the high concentration needed (10 mM) does not allow broad application. A biotin aptamer has been shown to be active in Xenopus oocytes (16).
3. In bacteria, the tetracycline and the neomycin aptamers cannot be used due to the antibiotic nature of their ligands. Instead, the theophylline aptamer (see Fig. 1b) has been used widely in bacteria, in the most cases only as sensor in combination with an additional regulator domain. Here, rational design combined with screening resulted in synthetic riboswitches which mask the ribosomal binding site (17–19).
4. Aptamers controling the activity of a ribozyme, so called aptazymes, have been used to control RNA integrity (see Subheading 3.6). Here, the theophylline aptamer and the aptamer domain of the thiamin riboswitch have been successfully used in *E. coli* (20–22). For application in eukaryotes the theophylline and tetracycline aptamers have been used (23–26).

In summary: The tetracycline aptamer is the most versatile aptamer for conditional gene expression in yeast. For other species the theophylline aptamer should be the first choice.

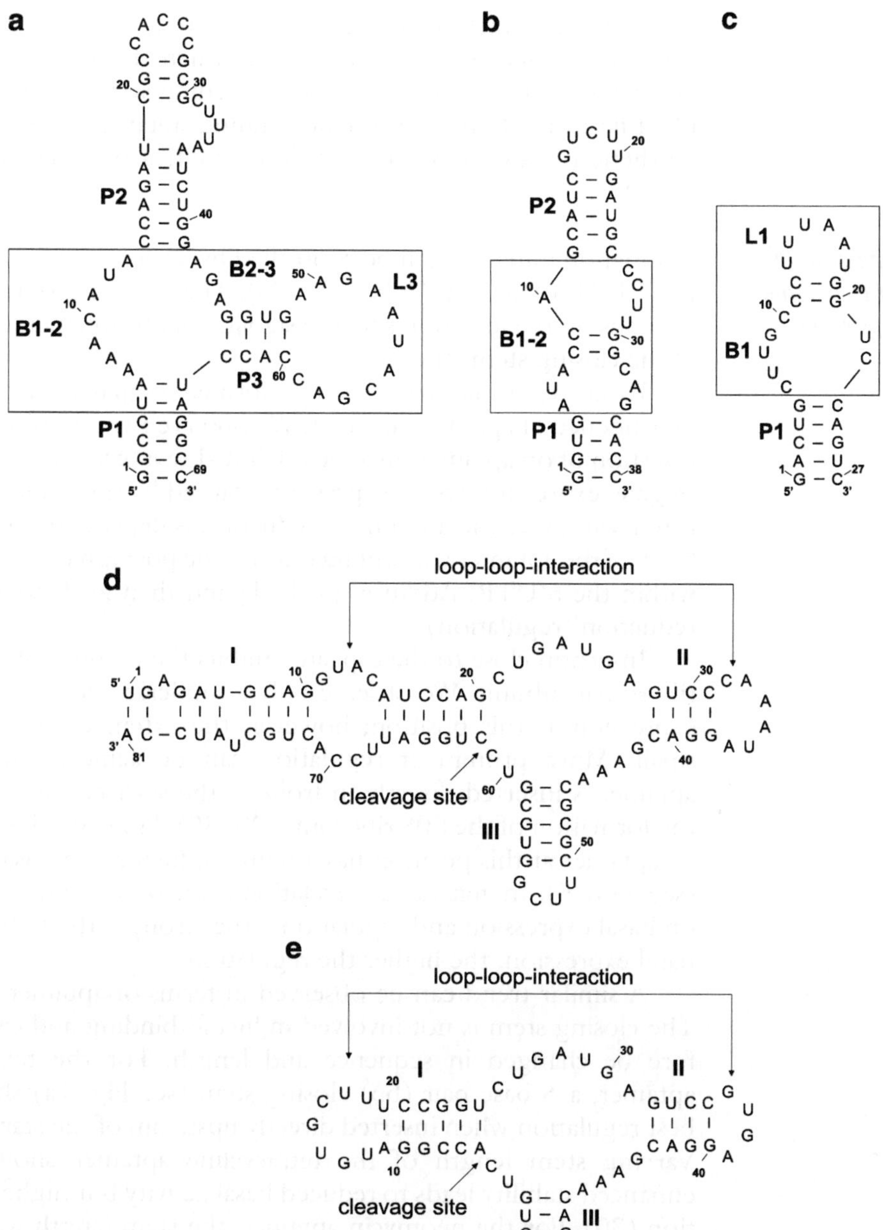


Fig. 1. Secondary structure of small molecule binding aptamers and ribozymes that have been used as engineered riboswitches. (**a**) Tetracycline aptamer. (**b**) Theophylline aptamer. (**c**) Neomycin aptamer. Invariable parts are *boxed*. (**d**) Sm1 hammerhead ribozyme N79 from *Schistosoma mansonii*. (**e**) sTRSV hammerhead ribozyme. P = stem, B = bulge, L = loop, hammerhead stems are *marked with* I, II, and III; cleavage site is indicated with an *arrow*.

### 3.2. How to Identify Novel Regulating Aptamers?

If there is a need to identify a regulating aptamer against a ligand of choice the first step is an in vitro selection procedure to isolate aptamers that bind the ligand with high affinity and specificity. However, it is necessary to add an additional screening step to isolate the subfraction of aptamers that are able to regulate gene expression. Detailed protocols exist, which describe simple and powerful yeast- and bacteria-based screening systems for the rapid identification of small molecule binding aptamers, which act as synthetic riboswitches for conditional control of gene expression (27, 28).

### *3.3. Aptamers for the Control of Translation Initiation in Yeast*

Translation initiation can be controlled by an aptamer inserted into the 5′UTR of an mRNA (see Fig. 2a). The efficiency of regulation depends on the insertion site, the stability of the aptamer, especially of the closing stem P1.

Extensive studies have been performed to analyze the position dependence of aptamers in yeast. We observe two opposite effects: insertion of on aptamer into the 5′UTR always leads to a reduction of gene expression (basal expression) caused by the aptamer structure itself. Here the amount of reduction is dependent on the stability of the aptamer without ligand and the position of the aptamer within the 5′UTR. Addition of the ligand then leads to a further reduction (regulation).

Insertion close to the cap site inhibits the binding of the small ribosomal subunit. We observe only a moderate decrease in basal expression at this position; however, the extent of regulation is small. More prominent regulation can be achieved when the aptamer is inserted directly in front of the start codon inhibiting the formation of the 80S ribosome (29, 30). However, insertion of an aptamer at this position has a higher influence on basal activity (see Note 5). In general, a correlation exists between the influence on basal expression and regulation – the stronger the reduction of basal expression, the higher the regulation.

A similar trend can be observed in terms of aptamer stability. The closing stem is not involved in ligand binding and can therefore be changed in sequence and length. For the tetracycline aptamer, a 6-base pair (bp) closing stem (see Fig. 1a) shows the best regulation when inserted directly upstream of the start codon. Varying stem length of the tetracycline aptamer showed that enhanced stability leads to reduced basal activity but higher regulation (30). For the neomycin aptamer, the stem length was varied between 5 and 10 bp. Here, an 8-bp long stem showed highest

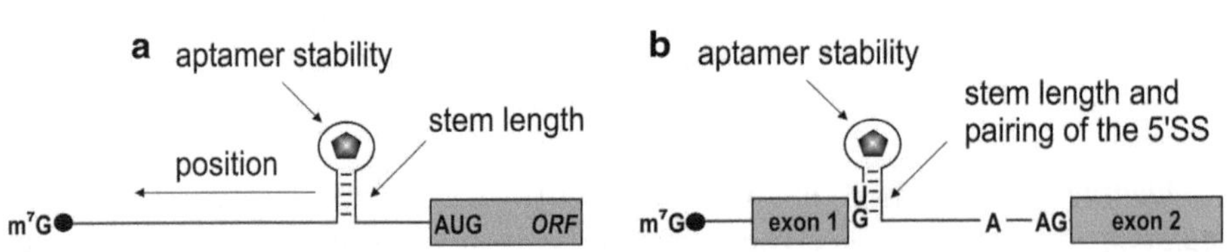


Fig. 2. Aptamer-mediated regulation of translation initiation (**a**) and pre-mRNA splicing (**b**) in eukaryotes. Indicated are possibilities for optimization of regulation and basal expression.

regulatory efficiency (31). It is recommended to analyze stem stabilities dependence when working with new aptamers.

In addition, loop L2 of the tetracycline aptamer can be changed by a tetraloop leading to a more stable aptamer variant (minimer). Insertion of the minimer into the 5′UTR led to a dramatic reduction of basal expression (to around 10%) and a decreased regulation (unpublished data), still variations of the overall aptamer stability can be favorable in other settings (see Subheading 3.4).

In general, it depends on the need of the experiment if you take a loss in basal activity for enhanced regulation.

### 3.4. Aptamers for the Control of Pre-mRNA Splicing

Both the tetracycline and the neomycin aptamer can be used to control pre-mRNA splicing in yeast ((13) and unpublished data). Here, the 5′ splice site (SS) is inserted into the closing stem of the aptamers (see Fig. 2b). This probably interferes with the binding of the U1 snRNP to the 5′SS. Breathing of the stem in the absence of the ligand seems to be sufficient for U1 recognition. The stabilization of the closing stem upon ligand binding prevents it.

A series of insertion sites have been tested with the tetracycline aptamer. Here, either stem P1 was shifted in relation to the 5′SS or P1 was stabilized by addition of further base pairs (see Note 6). The most active constructs harbor the complete 5′SS within an 11-bp long closing stem. Further stabilization of the aptamer using the minimer considerably enhanced regulation without affecting the basal expression.

The neomycin aptamer controls pre-mRNA splicing with the 5′SS completely inserted into an 8-bp long closing stem. Elongation of the closing stem leads to higher regulation efficiency, albeit with a slight drop in basal expression to around 75% (unpublished data).

In general the more stable the construct the higher the regulation efficiencies, without pronounced influence on basal expression.

### 3.5. Multiple Aptamer Insertions Result in Regulation Sufficient for the Control of Essential Genes

In general, insertion of more than one aptamer increases regulation. The combination of two tetracycline aptamer-containing introns or a combination of translation and splicing regulation leads to increased regulation efficiencies, but extremely reduced basal expression (13).

The insertion of three tetracycline aptamer copies into the 5′UTR allow a very tight regulation with up to 40-fold regulation efficiency and a basal expression of around 20%. Importantly, aptamers have to be separated by a linker sequence to allow for their proper folding (see Note 7). Recently, a tetracycline aptamer insertion cassette has been described, which allows a simple and powerful PCR-based strategy for easy tagging of any target gene in yeast (32). It harbors three tetracycline aptamers constitutively expressed by either an *ADH1* or a *TDH3* promoter (see Note 8). In addition it codes for a kanamycin-resistance gene flanked by *loxP* sites to select for chromosomal integration. After integration the aptamer construct replaces the 5′UTR and promoter of the endogenous gene. This system can even be used for the conditional control of essential genes in yeast and was shown to rapidly switch off protein synthesis.

### 3.6. Allosteric Ribozymes for the Regulation of RNA Stability

Aptamers have also been employed to regulate the degradation rate of RNA. To this end, aptamers were linked to ribozyme domains to modulate their catalytic activity in a ligand-dependent manner (named allosteric ribozyme or aptazyme). Early approaches created aptazymes functional in vitro utilizing the minimal motif of the hammerhead ribozyme as self-cleaving domain (33, 34). The minimal hammerhead motif consists of a three-helix junction and a conserved catalytic core of 11 nucleotides (35). However, the cleavage rate of the minimal hammerhead motif turned out to be too low to cleave efficiently under physiologic $Mg^{2+}$ concentration.

Extended hammerhead motifs with additional structural elements have been identified that show cleavage under physiologic conditions (36). Here, the loops of stem I and II interact with each other what stabilizes the tertiary structure of the hammerhead ribozyme and thereby increases the cleavage rate. Two extended hammerhead motives have been used for in vivo functional aptazymes. One is a derivative of the Sm1 hammerhead from *Schistosoma mansoni* (N79) and the other one originates from the satellite RNA of tobacco ringspot virus (sTRSV) (see Fig. 1d, e) (36, 37). Even if they adopt a similar hammerhead-shaped three dimensional structure, the tertiary contacts between stem I and II differs due to sequence variations in the loops (38). As a consequence, the sTRSV hammerhead exhibits a higher cleavage efficiency than N79 meaning that its rate of religation is lower. Both ribozymes have successfully been used in in vivo applications. It remains unclear to what extent the cleavage/ligation ratio in vitro correlates with functionality of aptazymes in vivo.

The aptamer can be attached to one of the three hammerhead stems in order to modulate either the formation of the catalytic core or the tertiary interactions between stem I and II. If the aptamer is grafted to stem I or II, rather the loop–loop interaction is affected. Linking the aptamer to stem III targets the catalytic core. The loops of stem I and II and the catalytic core tolerate only little sequence variation, whereas the sequence of the three stems is variable. Therefore, rational design and in vitro selection have been applied to find suitable linker sequences (21, 24). Up to now, all three stems have been used to engineer in vivo functional allosteric hammerhead ribozymes. Three aptamer domains have been employed for the development of in vivo functional aptazymes. The theophylline aptamer has been used for the development of aptazymes in both bacteria and eukaryotes. The tetracycline aptamer worked in yeast and mammalian cells and the aptamer domain of the thiamin riboswitch in *E. coli*.

In addition to the extended hammerhead motif, self-splicing group-I introns have been used for aptazyme based conditional gene expression (39). However, no further studies utilized group-I introns, most probably due to their size in comparison to the much smaller hammerhead motifs (group-I: 265 nt, N79: 81 nt, sTRSV: 50 nt).

Aptazymes have been integrated into bacterial RNA in a way that the ribosome binding site is released upon cleavage (21, 22). In eukarya, the RNA degradation rate was regulated by insertion of the aptazymes into the 5′ and 3′ UTR of a specific gene (see Notes 9 and 10, 23–26). Furthermore, aptazymes were applied to control the stability of rRNAs (40) or interfere with the processing of tRNAs and miRNAs (41, 42).

## 4. Notes

1. The given sequences were used to control translation initiation in yeast (tetracycline and neomycin aptamer) or *E. coli* (theophylline aptamer). The sequence of stem P1 is variable for all three aptamers. In addition, the sequence of stem P2 is variable for the tetracycline and the theophylline aptamer. Both can be inserted into an mRNA via their P1 or P2 stem. The terminal loop L1 in the neomycin aptamer is important for ligand binding and cannot be replaced. Invariable parts of the aptamers are indicated in Fig. 1 (boxed).
2. For all aptamers close derivatives of their cognate ligands are indicated, which are bound with lower affinity. They can be used as a negative control in biological assays when substituted for the cognate ligand. Doxycycline: similar to tetracycline with one OH-group shifted from the R6β to the R5α position. Caffeine: similar to theophylline with one additional methyl-group at position 7. Paromomycin: similar to neomycin with an OH-group instead of amino-group on position 6 in ring I.
3. Destabilized GFP variants in yeast harbor a so-called PEST sequence. Originates from the cyclin *CLN2,* which is a naturally short-lived protein (43). Reporter genes with shorter half-life reduce the background originated from residual expression of the inhibited reporter gene.
4. Ligand concentration is normally limited by the small molecule itself. At high concentrations (e.g. up to 10 mM for theophylline) every ligand shows side effects leading to reduced cell growth or even cell death. The limit is cell-type specific and should be tested in viability assays for every new aptamer/species. Suitable concentrations for tetracycline are up to 250 μM in yeast and 50 μM in HeLa cells.

5. Values for insertion of the tetracycline aptamer into a 5′UTR (30):
   (a) cap: 75% basal expression; 2.4-fold reduction upon ligand binding.
   (b) AUG: 20% basal expression; 5.8-fold reduction upon ligand binding.
6. Shift of P1 relative to the 5′SS showed that the more the 5′SS is included in P1 the higher the regulation. Further elongation of P1 also led to further inclusion of the 5′SS and higher regulation. Interestingly all constructs showed a similar basal expression of around 60% (13).
7. When inserting multiple aptamer copies into the 5′UTR it is necessary to separate them by unstructured spacers (e.g. 5–9 repeats of CAA) to allow correct folding of the aptamer domains.
8. The two promoters differ in strength and therefore allow adjustment in expression according to the target gene.
9. Unstructured spacer elements as mentioned before should be added upstream and downstream of an aptazyme to separate it from the surrounding RNA context.
10. When inserting aptazymes in the 5′UTR one has to be aware of the additional start codon within the catalytic core of the hammerhead ribozyme. On the one hand, insertion can be done in frame to the genuine start codon. However, the additional amino acids at the N-terminus may interfere with protein activity. On the other hand, there is one nucleotide within the start codon of the catalytic core that can be mutated without inhibiting hammerhead cleavage (N79: U25C, sTRSV: U29C). However, even if not completely inhibited, hammerhead cleavage is about twofold slower (37).

## Acknowledgments

This work was supported by the Aventis Foundation, the Deutsche Forschungsgemeinschaft (SU402/4-1 and the Excellence Cluster: Macromolecular Complexes) and the Else-Kröner-Fresenius Stiftung.

### References

1. Endy, D. (2005) Foundations for engineering biology. *Nature* **438**, 449–453
2. Ellington, A.D. and Szostak, J.W. (1990) In vitro selection of RNA molecules that bind specific ligands. *Nature* **346**, 818–822
3. Tuerk, C. and Gold, L. (1990) Systematic evolution of ligands by exponential enrichment: RNA ligands to bacteriophage T4 DNA polymerase. *Science* **249**, 505–510
4. Hermann, T. and Patel, D.J. (2000) Adaptive recognition by nucleic acid aptamers. *Science* **287**, 820–825
5. Weigand, J.E. and Suess, B. (2009) Aptamers and riboswitches: perspectives in biotechnology. *Appl. Microbiol. Biotechnol.* **85**, 229–236
6. Muller, M., Weigand, J.E., Weichenrieder, O. and Suess, B. (2006) Thermodynamic characterization of an engineered tetracycline-binding

riboswitch. *Nucleic Acids Res.* **34**, 2607–2617

7. Jenison, R.D., Gill, S.C., Pardi, A. and Polisky, B. (1994) High-resolution molecular discrimination by RNA. *Science* **263**, 1425–1429
8. Weigand, J.E., Schmidtke, S.R., Will, T.J., Duchardt-Ferner, E., Hammann, C., Wohnert, J. and Suess, B. (2010) Mechanistic insights into an engineered riboswitch: a switching element which confers riboswitch activity. *Nucleic Acids Res.* 39(8), 3363–3372
9. Grate, D. and Wilson, C. (2001) Inducible regulation of the S. cerevisiae cell cycle mediated by an RNA aptamer-ligand complex. *Bioorg. Med. Chem.* **9**, 2565–2570
10. Hanson, S., Bauer, G., Fink, B. and Suess, B. (2005) Molecular analysis of a synthetic tetracycline-binding riboswitch. *RNA* **11**, 503–511
11. Wunnicke, D., Strohbach, D., Weigand, J.E., Appel, B., Feresin, E., Suess, B., Muller, S. and Steinhoff, H.J. (2011) Ligand-induced conformational capture of a synthetic tetracycline riboswitch revealed by pulse EPR. *RNA* **17**, 182–188
12. Xiao, H., Edwards, T.E. and Ferre-D'Amare, A.R. (2008) Structural basis for specific, high-affinity tetracycline binding by an in vitro evolved aptamer and artificial riboswitch. *Chem. Biol.* **15**, 1125–1137
13. Weigand, J.E. and Suess, B. (2007) Tetracycline aptamer-controlled regulation of pre-mRNA splicing in yeast. *Nucleic Acids Res.* **35**, 4179–4185
14. Duchardt-Ferner, E., Weigand, J.E., Ohlenschlager, O., Schmidtke, S.R., Suess, B. and Wohnert, J. (2010) Highly modular structure and ligand binding by conformational capture in a minimalistic riboswitch. *Angew. Chem. Int. Ed. Engl.* **49**, 6216–6219
15. Werstuck, G. and Green, M.R. (1998) Controlling gene expression in living cells through small molecule-RNA interactions. *Science* **282**, 296–298
16. Harvey, I., Garneau, P. and Pelletier, J. (2002) Inhibition of translation by RNA-small molecule interactions. *RNA* **8**, 452–463
17. Desai, S.K. and Gallivan, J.P. (2004) Genetic screens and selections for small molecules based on a synthetic riboswitch that activates protein translation. *J. Am. Chem. Soc.* **126**, 13247–13254
18. Suess, B., Fink, B., Berens, C., Stentz, R. and Hillen, W. (2004) A theophylline responsive riboswitch based on helix slipping controls gene expression in vivo. *Nucleic Acids Res.* **32**, 1610–1614
19. Topp, S., Reynoso, C.M., Seeliger, J.C., Goldlust, I.S., Desai, S.K., Murat, D., Shen, A., Puri, A.W., Komeili, A., Bertozzi, C.R. *et al.* (2010) Synthetic riboswitches that induce gene expression in diverse bacterial species. *Appl. Environ. Microbiol.* **76**, 7881–7884
20. Wieland, M., Benz, A., Klauser, B. and Hartig, J.S. (2009) Artificial ribozyme switches containing natural riboswitch aptamer domains. *Angew. Chem. Int. Ed. Engl.* **48**, 2715–2718
21. Wieland, M. and Hartig, J.S. (2008) Improved aptazyme design and in vivo screening enable riboswitching in bacteria. *Angew. Chem. Int. Ed. Engl.* **47**, 2604–2607
22. Wieland, M., Gfell, M. and Hartig, J.S. (2009) Expanded hammerhead ribozymes containing addressable three-way junctions. *RNA* **15**, 968–976
23. Auslander, S., Ketzer, P. and Hartig, J.S. (2010) A ligand-dependent hammerhead ribozyme switch for controlling mammalian gene expression. *Mol. Biosyst.* **6**, 807–814
24. Win, M.N. and Smolke, C.D. (2007) A modular and extensible RNA-based gene-regulatory platform for engineering cellular function. *Proc. Natl. Acad. Sci. USA* **104**, 14283–14288
25. Win, M.N. and Smolke, C.D. (2008) Higher-order cellular information processing with synthetic RNA devices. *Science* **322**, 456–460
26. Chen, Y.Y., Jensen, M.C. and Smolke, C.D. (2010) Genetic control of mammalian T-cell proliferation with synthetic RNA regulatory systems. *Proc. Natl. Acad. Sci. USA* **107**, 8531–8536
27. Lynch, S.A., Topp, S. and Gallivan, J.P. (2009) High-throughput screens to discover synthetic riboswitches. *Methods Mol. Biol.* **540**, 321–333
28. Suess, B. and Weigand, J.E. (2009) Aptamers as artificial gene regulation elements. *Methods Mol. Biol.* **535**, 201–208
29. Hanson, S., Berthelot, K., Fink, B., McCarthy, J.E. and Suess, B. (2003) Tetracycline-aptamer-mediated translational regulation in yeast. *Mol. Microbiol.* **49**, 1627–1637
30. Suess, B., Hanson, S., Berens, C., Fink, B., Schroeder, R. and Hillen, W. (2003) Conditional gene expression by controlling translation with tetracycline-binding aptamers. *Nucleic Acids Res.* **31**, 1853–1858
31. Weigand, J.E., Sanchez, M., Gunnesch, E.B., Zeiher, S., Schroeder, R. and Suess, B. (2008) Screening for engineered neomycin riboswitches that control translation initiation. *RNA* **14**, 89–97
32. Kotter, P., Weigand, J.E., Meyer, B., Entian, K.D. and Suess, B. (2009) A fast and efficient translational control system for conditional

expression of yeast genes. *Nucleic Acids Res.* **37**, e120

33. Soukup, G.A. and Breaker, R.R. (1999) Engineering precision RNA molecular switches. *Proc. Natl. Acad. Sci. USA* **96**, 3584–3589
34. Tang, J. and Breaker, R.R. (1997) Rational design of allosteric ribozymes. *Chem. Biol.* **4**, 453–459
35. Hammann, C. and Lilley, D.M. (2002) Folding and activity of the hammerhead ribozyme. *Chembiochem* **3**, 690–700
36. Khvorova, A., Lescoute, A., Westhof, E. and Jayasena, S.D. (2003) Sequence elements outside the hammerhead ribozyme catalytic core enable intracellular activity. *Nat. Struct. Biol.* **10**, 708–712
37. Yen, L., Svendsen, J., Lee, J.S., Gray, J.T., Magnier, M., Baba, T., D'Amato, R.J. and Mulligan, R.C. (2004) Exogenous control of mammalian gene expression through modulation of RNA self-cleavage. *Nature* **431**, 471–476
38. Scott, W.G., Martick, M. and Chi, Y.I. (2009) Structure and function of regulatory RNA elements: ribozymes that regulate gene expression. *Biochim. Biophys. Acta* **1789**, 634–641
39. Thompson, K.M., Syrett, H.A., Knudsen, S.M. and Ellington, A.D. (2002) Group I aptazymes as genetic regulatory switches. *BMC Biotechnol.* **2**, 21
40. Wieland, M., Berschneider, B., Erlacher, M.D. and Hartig, J.S. (2010) Aptazyme-mediated regulation of 16S ribosomal RNA. *Chem. Biol.* **17**, 236–242
41. Berschneider, B., Wieland, M., Rubini, M. and Hartig, J.S. (2009) Small-molecule-dependent regulation of transfer RNA in bacteria. *Angew. Chem. Int. Ed. Engl.* **48**, 7564–7567
42. Kumar, D., An, C.I. and Yokobayashi, Y. (2009) Conditional RNA interference mediated by allosteric ribozyme. *J. Am. Chem. Soc.* **131**, 13906–13907
43. Mateus, C. and Avery, S.V. (2000) Destabilized green fluorescent protein for monitoring dynamic changes in yeast gene expression with flow cytometry. *Yeast* **16**, 1313–1323

# Chapter 10

# MicroRNA Circuits for Transcriptional Logic

**Madeleine Leisner, Leonidas Bleris, Jason Lohmueller, Zhen Xie, and Yaakov Benenson**

## Abstract

One of the longstanding challenges in synthetic biology is rational design of complex regulatory circuitry with multiple biological inputs, complex internal processing, and physiologically active outputs. We have previously proposed how to address this challenge in the case of transcription factor inputs. Here we describe the methods used to construct these synthetic circuits, capable of performing logic integration of transcription factor inputs using microRNA expression vectors and RNA interference (RNAi). The circuits operate in mammalian cells and they can serve as starting point for more complex synthetic information processing networks in these cells.

**Key words:** Transcriptional regulation, RNA interference, Synthetic biology, RNAi computing, Mammalian cells, Molecular logic

## 1. Introduction

Research in the fields of in vivo molecular computing (1) and specifically, RNA-based molecular computing (2) has grown significantly in the past decade. Progress in synthetic biology (3–7) has contributed to this growth, to a point when the distinction no longer exists in most cases. A popular benchmark challenge in the community is the rational design of complex regulatory networks controlled by transcriptional signals, namely transcription factors (8). A large number of approaches have been tested (3, 9–12), leading to increasingly sophisticated regulatory networks. We have recently proposed and tested a novel approach that relies on RNAi logic (13, 14) to integrate transcriptional signals in a rational fashion (15). Our approach (Fig. 1) is in theory indefinitely scalable and highly flexible with respect to the input integration program.

Wilfried Weber and Martin Fussenegger (eds.), *Synthetic Gene Networks: Methods and Protocols*,
Methods in Molecular Biology, vol. 813, DOI 10.1007/978-1-61779-412-4_10, © Springer Science+Business Media, LLC 2012

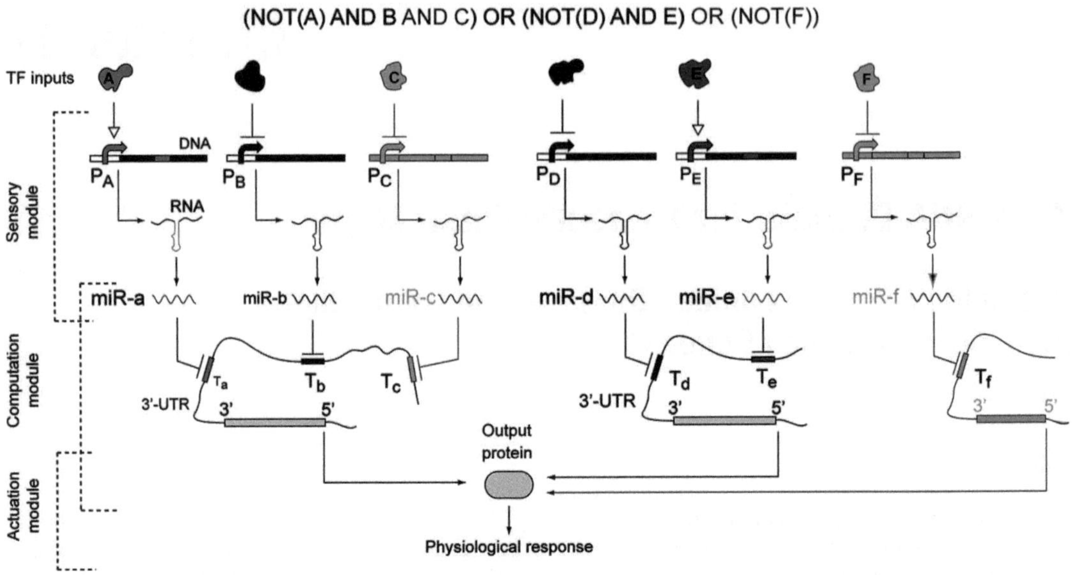


Fig. 1. Transcription regulatory programs with RNAi logic circuits. (**a**) The schematics of a logic circuit with multiple transcription factor (TF) inputs and a fluorescent ZsYellow protein output. Different organizational modules of the circuit are indicated. In the sensory module, each TF input controls one synthetic "sensor" microRNA gene via simple responsive promoter. Processed mature microRNA converge on their targets embedded in the 3′-UTR of the output transcript/s encoding the same protein. TF inputs A–F, promoters $P_A$–$P_F$, miRs miR-a to miR-f, and output-encoding mRNA transcripts containing miR targets $T_a$–$T_f$ are indicated. *Arrows* with *arrowheads* denote activation, and *blunt arrows* represent repression. Elements in *gray* denote potential opportunities for circuit scale-up. (Reproduced from Leisner et al., Nature Nanotechnology **5**, 666–670 (2010).).

The methods described here apply to a few example circuits we implemented but they can be easily adapted to novel designs. Currently the circuits serve as model systems, but they will be used in the future as components of rationally designed cell-controlling machinery for biotechnology and medicine (16, 17).

## 2. Materials

### *2.1. Cloning*

1. microRNA backbone plasmid: Plasmid 11665, pPRIME-CMV-Neo-FF3 (Addgene).
2. Vectors pTRE-Tight-BI, pZsYellow-C1, pAmCyan-C1, pDsRed-Expess-C1 (Clontech), pNEBR- R1 and pNEBR-X1-Hygro (New England Biolabs).
3. Inoculating loop.
4. 15 mL culture tubes.
5. Plasmid miniprep kit (Qiagen).
6. PCR purification kit (Qiagen).

7. Gel purification kit (Qiagen).
8. Agarose.
9. DNA oligomers.
10. XhoI, EcoRI, SapI, NotI, SpeI, SmaI, BamHI restriction enzymes (New England Biolabs).
11. Alkaline Phosphatase, Calf Intestinal (CIP), T4 polynucleotide kinase (PNK) (New England Biolabs), T4 DNA ligase.
12. Pfu Ultra II Fusion HS DNA polymerase (Agilent Technologies).
13. dNTP mixture (Invitrogen).
14. NaCl solution in water, 0.5 M.
15. RT-PCR grade water (Ambion).
16. Max Efficiency DH5 *E. coli* cells (Invitrogen) or equivalent chemically-competent *E. coli* bacteria.
17. Buffers NEB1, NEB2, NEB3, NEB4 (New England Biolabs).
18. Selective LB broth.
19. Agar plates.
20. DNA size ladders (New England Biolabs).
21. Camera for taking gel pictures.

### 2.2. Cell Culture

1. HEK293-H Cells (Invitrogen).
2. Cell growth medium (all reagents from Invitrogen): Dubecco Modified Eagle Medium (DMEM), high glucose with 10% Fetal Bovine Serum (FBS), 0.045 units/mL of penicillin and 0.045 μg/mL streptomycin.
3. PBS (Invitrogen).
4. 0.25% Trypsin-EDTA (Invitrogen).
5. T75 Flasks (Corning).
6. 50 mL test tubes.
7. Lipofectamine and PLUS transfection reagents (Invitrogen).
8. 37°C/5% $CO_2$ incubator.

### 2.3. Microscopy

1. Inverted microscope equipped with appropriate filters for measuring Cyan, Yellow, and Red fluorescent proteins simultaneously.
2. Incubation chamber.
3. 12-well plates.

### 2.4. Flow Cytometry

Flow cytometer with appropriate lasers and filters for simultaneous measurement of Cyan, Yellow, and Red fluorescent proteins.

## 3. Methods

### *3.1. Cloning of microRNA Expression Vectors*

1. Acquire the pPRIME-CMV-Neo-FF3. The plasmid is provided in the form of bacterial stab. The stabs are stable at 4°C for 2 weeks. To recover the bacteria, use a sterile inoculating loop to gather bacteria from the bacterial stab. Spread the bacteria onto an agar plate with chloramphenicol (Cam) and streak for single colonies. Expand a number of single colonies in 2 mL of LB-Cam medium and purify plasmid DNA using a MiniPrep kit.
2. Design a set of synthetic non-endogenous microRNA sequences and their targets. For detailed instructions, see ref. 18. Design microRNA-coding insert to express a mature microRNA of interest. Use the rules described in the pPRIME papers (19, 20) to generate sense and antisense strands of about 100 nt each such that the anticipated annealing product has an XhoI-compatible sticky end on the 5′-end and EcoRI-compatible sticky end on the 3′-end. Order without purification using 50 nmol scale. Resuspend in TE buffer to a final concentration of 1 nmol/μL and store as stock solutions. Dilute 1:10 with TE to 100 pmol/μL. Mix 2.5 μL of diluted sense and antisense strands with 2 μL of 0.5 M NaCl and 13 μL TE buffer to a total volume of 20 μL. Anneal in a PCR machine block first heating up to 95°C and cooling down to 10°C at a rate of about −1.5°C/min. The resulting solution will contain 250 pmol of the annealed duplex (because the oligomers are not purified, the actual concentration of fully formed duplex is substantially lower). Examples of oligomer pairs that have been used include:

   miR-FF3

   ```
   TCGAGTGCTGTTGACAGTGAGCGCACGATATGGGC
     TGAATACAAATAGTGAAGCCACAGATGTATTTGT
     ATTCAGCCCATATCGTTTGCCTACTGCCTC
     GGAG (sense)
   ```

   ```
   AATTCTCCGAGGCAGTAGGCAAACGATATGGGCTG
     AATACAAATACATCTGTGGCTTCACTATTTGT
     ATTCAGCCCATATCGTGCGCTCACTGTCA
     ACAGCAC (antisense)
   ```

   miR-FF4

   ```
   TCGAGTGCTGTTGACAGTGAGCGCCGCTTGA
     AGTCTTTAATTAAATAGTGAAGCCACAGAT
     GTATTTAATTAAAGACTTCAAGCGGTGCC
     TACTGCCTCGGAG (sense)
   ```

   ```
   AATTCTCCGAGGCAGTAGGCACCGCTTGAA
     GTCTTTAATTAAATACATCTGTGGCTTCACTATTT
     AATTAAAGACTTCAAGCGGCGCTCACTGTCA
     ACAGCAC (antisense)
   ```

miR-FF5

TCGAGTGCTGTTGACAGTGAGCGAAGCACTCTGATTTGACAATTATAGTGAAGCCACAGATGTATAATTGTCAAATCAGAGTGCTTTGCCTACTGCCTCGGAG (sense)

AATTCTCCGAGGCAGTAGGCAAAGCACTCTGATTTGACAATTATACATCTGTGGCTTCACTATAATTGTCAAATCAGAGTGCTTCGCTCACTGTCAACAGCAC (antisense)

miR-FF6

TCGAGTGCTGTTGACAGTGAGCGAACCAAAGAGATTCCTCATAAATAGTGAAGCCACAGATGTATTTATGAGGAATCTCTTTGGTTTGCCTACTGCCTCGGAG (sense)

AATTCTCCGAGGCAGTAGGCAAACCAAAGAGATTCCTCATAAATACATCTGTGGCTTCACTATTTATGAGGAATCTCTTTGGTTCGCTCACTGTCAACAGCAC (antisense)

3. Digest pPRIME-CMV-Neo-FF3 with XhoI and EcoRI as follows: Digest 6 μg of plasmid DNA with 200 units (10 μL) of XhoI in 100 μL of NEB buffer 2 and 1× BSA for 1 h at 37°C. Elute the product using PCR purification kit into 40 μL EB buffer. Redigest with 200 units (10 μL) of EcoRI in 100 μL of NEBU buffer. Gel-purify the resulting product from 1% agarose gel using Gel Purification kit. Expected yield is between 1 and 2 μg double-digested DNA.
4. Dephosphorylate the digested vector: add 25 μL of the gel-purified DNA to 1 μL of CIP enzyme, 4 μL of 10× NEB3 buffer and 10 μl water. Incubate 1 h at 37°C. Purify with PCR purification kit. Expected concentration is about 10 ng/μL.
5. Phosphorylate the annealed insert: combine 12 μL of the annealed duplex solution with 5 μL of 10 mM ATP, 5 μL of 10× PNK buffer, 1.5 μL of PNK enzyme, and 26.5 μL of water to a total volume of 50 μL. Incubate for 30 min at 37°C.
6. Subclone the insert into the vector: use about 30 ng of the dephosphorylated vector and increasing amounts of the phosphorylated insert from zero (negative control) up to 2 μL (nominal amount: 6 pmol) in 10 μL of 1× T4 DNA ligase buffer with 400 units (1 μL) of T4 DNA ligase. Ligate at 16°C overnight; however, much shorter times often work just as well.
7. Transform 50 μL of competent *E. coli* cells with the entire ligation mixture using heat shock. Plate on selective agar plates.
8. Examine the plates next morning. Make sure the control plate has no or very few colonies while the ligation reactions have colonies. Verify insertion by colony PCR or proceed to colony

expansion in 2 mL selective medium, followed by Miniprep isolation and sequencing. Sequence at least five colonies, as due to using non-purified oligomers the chance of mutations in the insert is substantial.

9. MicroRNA genes can be further subcloned under various tunable promoters using common cloning strategies. However, we found that while pPRIME constructs elicit strong knock-down, transplanting the Neo-miR fragments into other host vectors results in significant reduction of the miR activity, suggesting that other elements in the pPRIME vector contribute to efficient miR processing. This can be overcome by subcloning miR-30-miR-XX fusions as introns (see below).

### 3.2. Cloning of microRNA Target Vectors

1. Obtain pZsYellow-C1 plasmid. The plasmid is delivered in DNA form, 500 ng/mL. Transform competent *E. coli* cells with a small amount of plasmid DNA, ~1 ng. Plate the cells on agar plates with Kanamycine selection marker. Pick colonies and transfer to 2 mL of LB-Kan medium, shake overnight and isolate DNA using a MiniPrep Kit.
2. Design the sense and the antisense DNA oligomers for the target insert. The insert should include BamHI and XhoI-compatible sticky ends on 5′- and 3′-ends, respectively, and the 5′-end of the sense strand should start with [TCGA→BamHI sticky end←]GC**TTA**ACAAGCTTCGAXXXX, where TTA is a stop codon and XXXX represents the sequence of small RNA target. We recommend using triple tandem repeats of the same target to increase sensitivity to microRNA. Order and process the oligomers according to steps described in Subheading 3.1, all the way through phosphorylation.
3. Perform a double digest of pZsYellow-C1 plasmid with BamHI and XhoI as follows. Combine about 5 μg of purified plasmid with 10 μL of 10× NEB2 buffer, 10 μL of 10× BSA solution, 1 μL (20 units) of BamHI and 5 μL (100 units) of XhoI and complete to 100 μL with water. Incubate for 1 h at 37°C, then gel-purify from 1% agarose gel using Gel Purification kit. Expected concentration of the purified product is about 30 ng/μL, 1.5 μg total yield. The isolate can be optionally dephosphorylated as in step 4 in Subheading 3.1.
4. Calculate molar concentration of the gel-purified vector and the annealed insert. Perform series of ligations with ~30 ng of the vector and varying amount of the insert, from no insert (negative control) to about fourfold molar excess. Large dilution of the annealed and phosphorylated insert will be required to get to this ratio. Proceed according to steps 7 and 8 in Subheading 3.1 to obtain sequenced constructs with RNAi targets in the 3′-UTR of ZsYellow protein.

5. (optional) Subclone the reporter under a constitutive promoter of interest. pZsYellow is driven by a CMV promoter. However, other promoters such as EF1a can perform better as our experience has shown.
6. (optional) Replace CMV promoter as follows: amplify the EF1a promoter using PCR from pEF4/V5-His A plasmid (Invitrogen) using the primers GGAGTGGGAATTGGCTCCG and TCACGACACCTGAAATGGAAGA. The amplicon was then TOPO-cloned into pGLOW-TOPO (Invitrogen) to create pEF1a-GLOW. PCR-amplify the EF1a promoter again from pEF1a-GLOW PCR using primers CCAAGGGCGGCCGCCAGTGTGGTGGAATTGCCCT and GGTTCCACTAGTATTCTAGATCTGCAGAATTGCCC containing NotI and SpeI sites, respectively. PCR-amplify pZsYellow furnished with microRNA targets in order to eliminate the CMV promoter and to add SpeI and NotI sites using the primers CCAAGGACTAGTGTCAGATCCGCTAGCGCTAC and GGTTCCGCGGCCGCGCATGGCGGTAATACGGTTATC. Digest the promoter-less vector and the promoter insert with SpeI and NotI and ligate with T4 DNA ligase to render the required construct.

### 3.3. Preparing Cells for Transfection

1. A day prior to transfection, examine your culture flask to verify that the cells are healthy at no more than 80% confluence. Remove growth medium from the cells, wash once with 2 mL PBS and add 2 mL of 0.25% Trypsin-EDTA solution. Incubate the flask at 37°C for 5–10 min and confirm by microscopic examination that the cells have detached.
2. Add 8 mL of fresh growth medium to the cells and pipet up and down a few times to break the clumps. Confirm by visual examination that the cells are not clumping. Count the cells using hemocytometer and determine cell concentration in the flask. Calculate the volume required to obtain 100,000 cells. Multiply this volume by the number of experiments you are planning to arrive at the volume of resuspended cells used for plating. Normally 1 mL of medium is used for one well in a 12-well plate. The number of experiments equals the number of wells.
3. In a 50 mL test tube combine the required volume of cell suspension and complete with fresh growth medium to the total volume whose size, in milliliter, equals the number of wells – for example, for 36 experiments you will use 36 wells, and the total volume of cell suspension with fresh medium should be 36 mL. Mix the tube well by inverting up and down a few times. Split the solution into the wells of 12-well plates, 1 mL per well, using a single sterile pipet. Incubate the plates in the cell culture incubator for ~24 h.

### 3.4. Confirming the Activity of microRNA Expression Vectors

1. Prepare cells for transfection (see Subheading 3.3) a day before the experiments.
2. A typical test contains a mixture of 100 ng transfection marker pAmCyan-C1, 100 ng of ZsYellow reporter fused with the appropriate target, and 200 ng of the pPRIME-based microRNA expression cassette. Two control experiments are recommended: a mixture of pAmCyan-C1 and pZsYellow fused with a scrambled target, 100 ng each, and 200 ng of DNA "filler", such as pUBI-linker-NOS (21); and another where the filler is replaced with the pPRIME-microRNA construct. The first control provides the baseline expression level of ZsYellow protein and the second provides the "off-target" activity of a given microRNA. In addition, the microRNA construct can be compared to an siRNA with the same active sequence. In this case, 100 ng of transfection marker and 100 ng of ZsYellow with the appropriate target are co-transfected with 2.5 pmol of siRNA duplex. Lastly, individual plasmids expressing fluorescent marker should be transfected to provide controls for image collection and flow cytometry.
3. Transfect the cells using Lipofectamine and PLUS reagents. Shortly before transfection, replace the growth medium in plated cells with 1 mL DMEM without supplements with a single medium wash step.
4. Prepare transfection mixtures by mixing all nucleic acids, including the plasmids and the siRNAs into 40 μl of DMEM. Add 2.4 μl of the PLUS reagent to the final mix and incubate for 20 min at room temperature. In parallel, mix 1.6 μl Lipofectamine (Invitrogen) with 40 μl DMEM. Next, mix PLUS- and lipofectamine-containing solutions and incubate for 20 more min at room temperature before application to the cells. Add the transfection mixture (~90 μl) to the wells and mix with the medium by gentle shaking. Three hours after transfection, add 120 μl FBS to the wells and incubate the cells for 48 h at 37°C/5% $CO_2$ before the analysis.
5. Take microscopy images of the live cells. Any standard equipment will be adequate. As an illustration, one can use Zeiss Axiovert 200 microscope equipped with Sutter filter wheels, Prior mechanized stage and an environmental chamber (Solent) held at 37°C during measurements. Collect the images using Orca ERII camera cooled to −60°C, in the high precision (14 bit) mode using a 20× PlanApochromat NA 0.8, PH2 objective. Use the following settings: S500/20× (excitation) and S535/30 m (emission) filters for ZsYellow; and S430/25× (excitation) and S470/30 m (emission) for AmCyan. A dichroic mirror 86004v2bs (Chroma) can be used for both fluorophores. Verify the settings by using "single-color" controls and make sure the spill-over of the signal into the unintended channels is negligible while the intended channel shows bright fluorescence

(for example, AmCyan-only sample should exhibit strong signal in AmCyan channel but not in ZsYellow channel, and vice versa).

6. Images can be collected using standard image processing software such as Metamorph 7.0 (Molecular Devices). Care should be taken not to saturate the camera. For this purpose, exposure times have to be set using the brightest sample in the experiment (for example, a control mixture where the transfection marker is cotransfected with a ZsYellow plasmid lacking a microRNA target sequence).
7. The purpose of the images is to provide qualitative impression of the experimental data. To obtain statistically significant set of images, collect at least four snapshots from randomly chosen locations. Alternatively, stage scan can be performed and 3 × 3 or 4 × 4 image sets collected. Occasionally the images can be used for quantitative measurement of the knock-down efficiency by integrating the total ZsYellow signal from multiple frames (after background subtraction) and normalizing it by the total AmCyan signal from the same frames. The ratio should be normalized to similar ratio from the control sample where ZsYellow is not repressed.
8. Following microscopy measurements, the cells are prepared for flow cytometry analysis (FACS). Cells in each well are trypsinized with 0.5 mL 0.25% trypsin-EDTA, and the cell suspension is centrifuged at 90 g for 5 min. Carefully remove supernatant and resuspend cell pellet by short vortexing in 0.5 mL PBS buffer.
9. Any standard flow cytometry instrument with appropriate lasers can be used for measurements. As an example, one could use a BD LSRII machine with 405 nm and 488 nm lasers. The settings are 488 nm laser and a 530/30 emission filter for ZsYellow, and a 405 nm laser and 450/50 emission filter for AmCyan. The sensitivity of the detectors (PMT values) should be set based on the intensity of the unrepressed samples. "Single color" samples, as well as negative control cells should be measured to determine autofluorescence and the degree of detector cross-talk. In general, cross-talk can be expected in the system, in particular from AmCyan to ZsYellow channel. In this case appropriate compensation should be performed before quantitative analysis.
10. Knock-down efficiency of a microRNA or siRNA is calculated as the ratio of total ZsYellow signal normalized by the total AmCyan signal, subsequently normalized by the same ratio in the control sample with non-repressed ZsYellow:

$$\frac{\text{mean(ZsYellow)*\%ZsYellow}^{+}}{\text{mean(AmCyan)*\%AmCyan}^{+}} \Big/ \frac{\text{mean(ZsYellow,negctrl)*\%ZsYellow}^{+}\text{(negctrl)}}{\text{mean(AmCyan,negctrl)*\%AmCyan}^{+}\text{(negctrl)}}$$

where mean(label) is the mean intensity of a label in those cells that express this label above autofluorescence, and %Label$^+$ is the frequency of label-expressing cells among all cells.

11. It is important that the measured efficiency be high if the microRNA is considered as a candidate for large circuits. Another consideration is the correlation between microRNA and the equivalent siRNA activity. If siRNA activity is high but that of the pPRIME microRNA is not, some improvement might be achieved when the microRNA is transferred to an intron (see below). However if both the siRNA and the microRNA are inefficient, this specific sequence should probably be discarded or modified to achieve at least 95% knock-down with siRNA and at least 85% knock-down with pPRIME-based microRNA.

### 3.5. Transplanting microRNA into Intron-Based Vectors

1. The procedures described below are specific protocols used in actual vector construction. However, they are easily adaptable to any comparable cloning project.

   An intron-encoded microRNA can be regulated by an inducible or repressible promoter and in addition it can be co-expressed with a fluorescent marker that serves as an indicator of microRNA level in cells. When the intron is embedded in the protein-coding frame, splicing and microRNA processing must take place to allow fluorescent signal (unlike when the intron is embedded in 3′-UTR or 5′-UTR).

2. miR-30-miR-X pPRIME-based fusion needs to be converted to an intron-like structure with the help of known splicing signals (22). The fluorescent protein of interest has to contain a suitable insertion site, characterized by the pair of nucleotides AG at the 3′-end of the first exon and G at the 5′-end of the second exon. In our case we inserted the intron between the amino acids Glu-Gly (GAG-GGC).

3. We use "seamless" cloning approach to implement one-pot ligation of multiple PCR products, digested with an SII-type restriction enzyme such as SapI. These enzymes create arbitrarily prescribed, non-palindromic sticky ends that can be exploited for properly oriented, ordered ligation of multiple-digested products in one step. To plan these cloning experiments, the desired final sequence is first constructed. It is examined for the presence of existing DNA sequences that can be PCR-amplified from other sources, and novel DNA sequence that has to be introduced via PCR primers. Junction sites of length 3 (in case of SapI-like enzymes) are chosen in such a way that each segment contains an existing DNA flanked by novel DNA. In addition all sticky ends must be distinct and ideally have high CG content. Consequently, PCR primers will include SapI, spacer of one nucleotide, the sticky end precursor

that matches the desired junction site, and any novel DNA that has to be inserted.

4. Below we describe the construction of two intron-embedded regulated microRNA genes: pTRE-DsRed-miR-FF3 and Rheo-AmCyan-miR-FF4.
5. pTRE-DsRed-miR-FF3: obtain the following commercial template plasmids: pTRE-Tight-BI, pDsRed-Express-C1; using protocol 3.1, create a plasmid pPRIME-miR-30-miR-FF3.
6. PCR-amplify pTRE-Tight (unidirectional) promoter using primers CCGGGCTCTTCCCGTACACGCCACCTCGACA and GGCCGCTCTTCGGCGGCGATCTGACGGTTCACT. Use the following protocol: 10× Pfu reaction buffer (5 μL), dNTP mix 10 mM each (1.2 μL), template plasmid at 5 ng/μL (1 mL), primers at 10 mM (1 μL each), Pfu Ultra fusion HS II enzyme (1 μL), and water (39.8 μL) to a total volume of 50 μL. Follow manufacturer protocol for PCR program with elongation time of 15 s at 72°C and annealing temperature of 58°C for 30 cycles.
7. PCR-amplify DsRed exon-I using pDsRed-Express-C1 as a template and primers CCGGGCTCTTCCCGCTAGCGCTACCGGTCG and GGCCGCTCTTCGCACCTCGCCCTCGCCCTCGA. Follow the same protocol as above.
8. PCR-amplify miR-30-miR-FF3 fusion using pPRIME-miR-30-miR-FF3 as a template and primers CCGGGCTCTTCCGTGAGTATGTGCTCGCTTCGGCAGC and GGCCGCTCTTCGGCCCTGAGGAAAAAAAAGGAAACAATTGAAAAAAGTGATTTAATTTATACCA. Follow the above protocol.
9. PCR-amplify the second exon of DsRed using pDsRed-Express-C1 as a template and primers CCGGGCTCTTCCGGCCGCCCCTACGAGG and GGCCGCTCTTCGTGAGACGTTGGAGTCCACGTTCTTTAATAG. Follow the above protocol.
10. Verify that the PCR reactions yielded single band products of expected length and purify them using PCR purification kit (Qiagen). Measure DNA concentration of the eluates.
11. Digest 136 ng of pTRE promoter, 56 ng of DsRed Exon I, 147 ng of miR30-miR-FF3 fusion, and 407 ng of DsRed Exon II separately with SapI (NEB) at a final concentration of 2 U/μL (use 10 U/μL stock) in 10 μL of NEB4 buffer at 37°C for 1 h.
12. Following SapI digest, purify the reactions using PCR purification kit and elute them into 50 μL EB buffer. Mix 4 μL of each digested fragment in 1× T4 DNA ligase buffer (NEB) and ligate in the presence of 40 units of T4 DNA ligase (NEB) for 1 h at 16°C in total volume of 20 μL. Resolve the resulting

mixture on 1% agarose gel and identify the correct four-way ligation product based on its size. Gel-purify the band using Gel purification kit (Qiagen) and elute into 50 μL EB buffer.

13. Subject 10 μL of that mixture to a second round of PCR using primers CCGGGCTCTTCCCGTACACGCCACCTCGACA and GGCCGCTCTTCGTGAGACGTTGGAGTCCACGTTCTTTAATAG (first primer of the promoter amplicon and the last primer of the DsREd-Exon II amplicon) using annealing temperature of 56°C and elongation time of 30 s. The PCR reaction should render a strong band of the correct side alongside shorter by-products.
14. Gel-purify the correct product again (expected concentration of the eluate about 10 ng/μL) and phosphorylate with PNK (NEB) as follows: use 25 μL of purified fragment solution, 5 μL of PNK×10 buffer, 5 μL of 10 mM ATP and 14 μL water. Add PNK after incubating the mixture at 70°C for 10 min. Incubate 30 min at 37°C.
15. Obtain a suitable host vector (*see next paragraph for clarification regarding the choice of the backbone!*). For example, pBluescript II KS (−) can be digested with SmaI to create blunt end ligation sites. Digest about 1 mg of the vector with SmaI in 10 μL of 1× NEB4 buffer for 1.5 h at 25°C. Add 0.5 μL CIP (NEB) to the reaction and incubate for another hour at 37°C to dephosphorylate the blunt ends.
16. After transformation, analyze colonies by colony PCR and restriction analysis. Sequence promising clones. The construction is complete. Figures 2 and 3 show a detailed description of the amplicons and the relation between the primers and the templates for different fragments, as well as the annotation of the final construct. Note that pBluescript contains LacO site. If LacI is planned for use in the circuit, use a different backbone plasmid. Alternatively the DsRed-microRNA fusion construct can be subcloned into another pTRE-regulated vector, for example pTRE-Tight-BI, using primers AGTGAACCGTCAGATCGCCG and CACTCAACGATATCTCGGTCTATTCTT and restriction enzymes NheI and EcoRV.
17. The construction of Rheo-AmCyan-FF4 fusion follows essentially the same protocols and in steps 7–17 with a different set of templates and primers. The templates are pNEBR-X1-Hygro for the Rheo promoter, pAmCyan-C1 for AmCyan exons and pPRIME-miR-30-miR-FF4 that needs to be constructed according to 3.1 prior to this cloning project. The primers for the Rheo promoter are CCAAGGGCTCTTCCAATGTGAGTTAGCTCACTCATTAGGCACCC and CCAAGGGCTCTTCGGCGATGAGCTCTGGTAGTAGCTGACGGCCCCCATTATAT. The primers for the first AmCyan exon are CCAAGGGCTCTTCCCGC

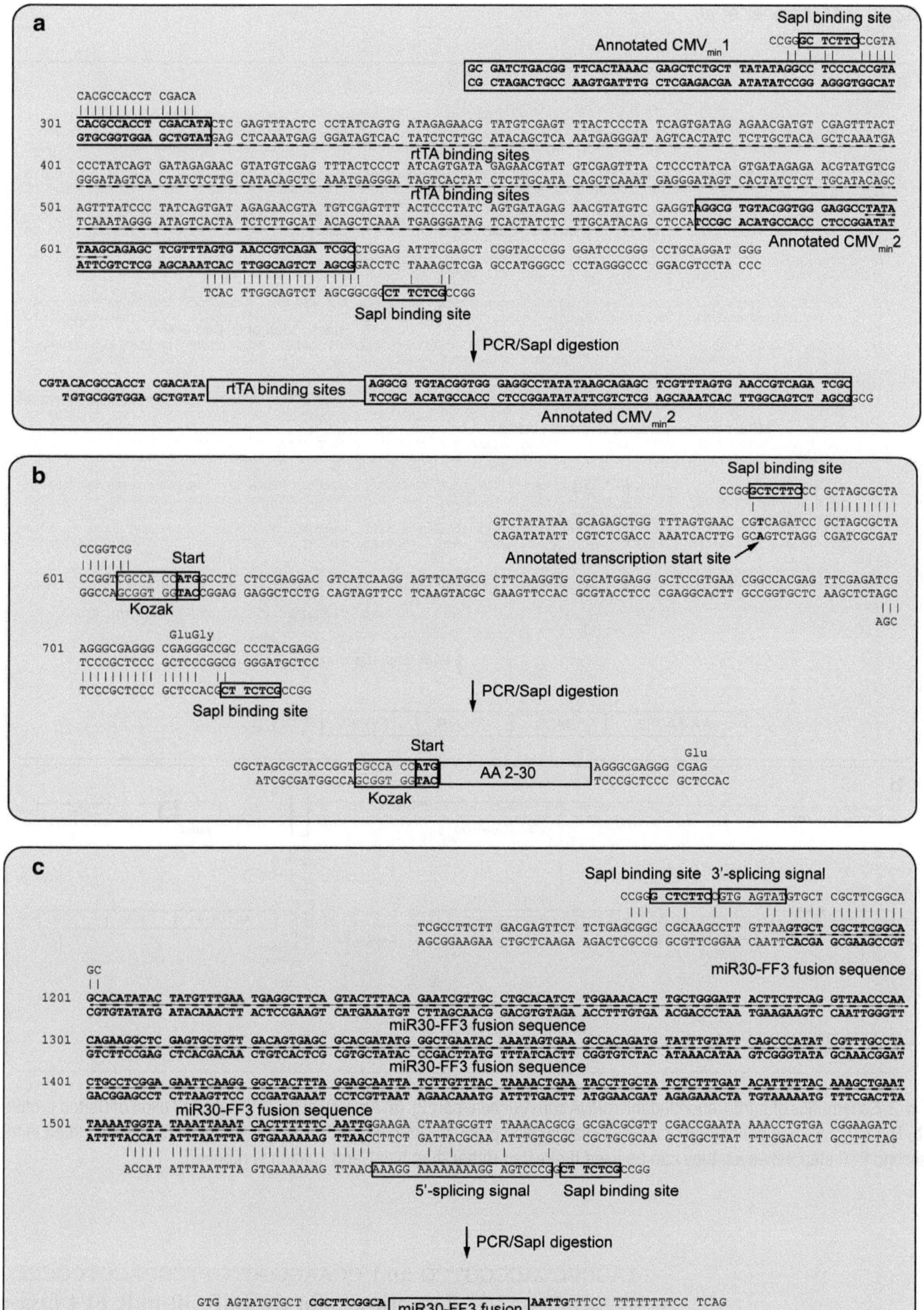


Fig. 2. Schematics of intron-encoded microRNA construction (part 1). (**a**) Amplification of the pTRE-Tight promoter and the resulting post-digestion fragment. PCR primers are shown at their hybridization locations. Different functional elements are indicated. (**b**) Amplification of the first exon of DsRed protein. (**c**) Amplification of the microRNA sequence and introduction of splicing signals.

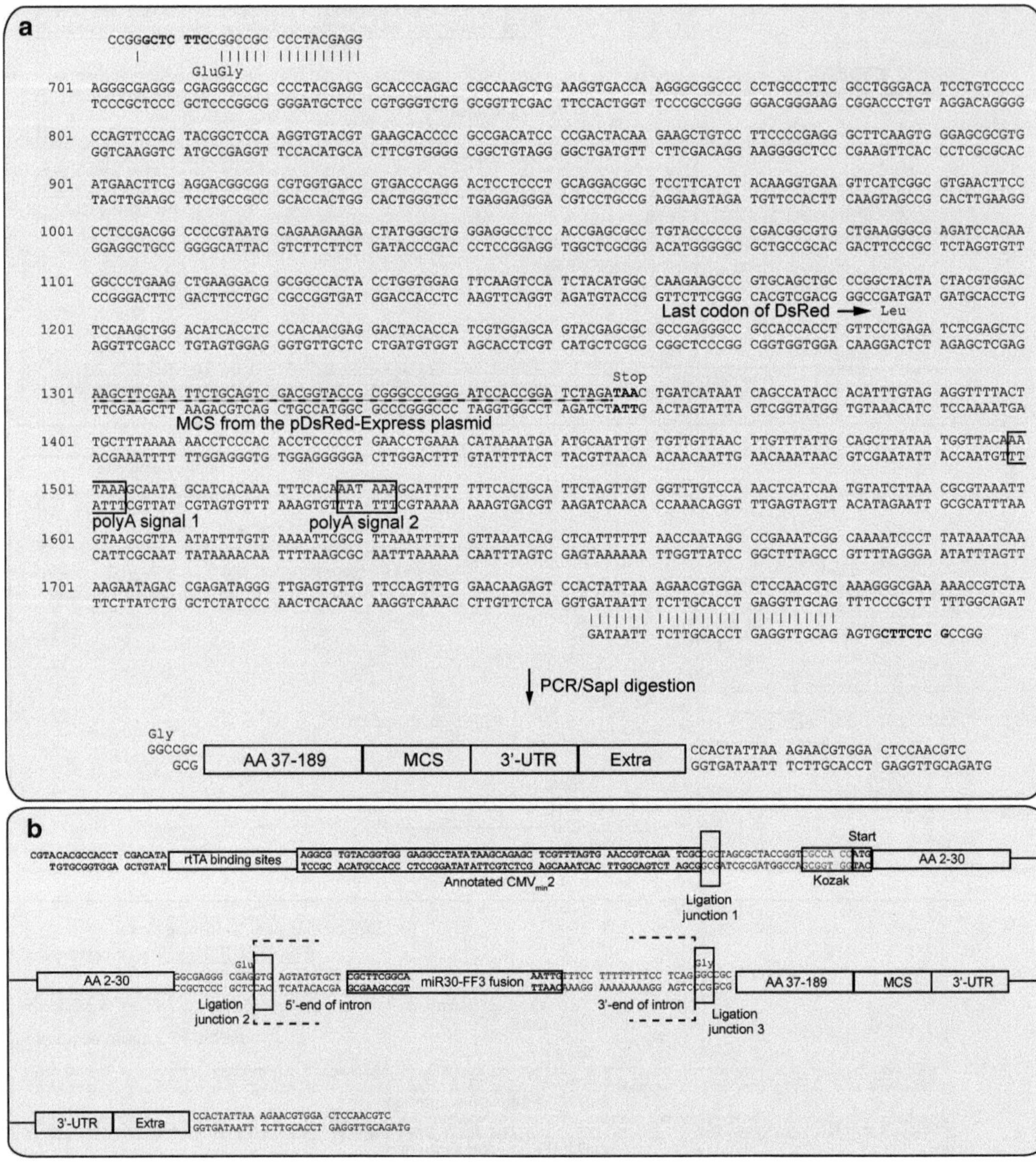


Fig. 3. Schematics of intron-encoded microRNA construction (part 2). (**a**) Amplification of the second exon of DsRed protein. (**b**) The structure and annotation of the four-way ligation product. Note that the 5′- and 3′-sticky ends are eliminated in the second PCR step. However, they can be used if directed rather than blunt-end ligation is required.

TAGCGCTACCGGTCG and CCAAGGGCTCTTCGCACCTCGCCCT TCACGGTGAAG. The primers for the miR-30-miR-FF4 fusion are the same as for the miR-30-miR-FF3 fusion above, and the primers for the second AmCyan exon are CCAAGGGCTCTTC CGGCAGCGGCAAGCCC and CCAAGGGCTCTTCGTACACCTCG ACCCCAAAAAACTTGA.

18. Digest PCR-purified products with SapI as described above and elute using PCR purification kit into 50 μL EB buffer each. Ligate 10 μL of each digested fragment simultaneously in 50 μL of 1× T4 DNA ligase buffer in the presence of 200 units of T4 DNA ligase for 1 h at 16°C.

19. Resolve the mixture by 1% agarose gel, and isolate a band corresponding to the full-length ligated product. Amplify the entire product and subclone into a suitable acceptor as described above. Note that pBluescript contains a LacO site; if LacI regulator is planned for use in the circuit together with the Rheo-controlled construct, avoid using pBluescript. Alternatively, the LacO site can be removed by amplifying two separate regions of the pBluescript-based plasmid using the primers TTTACATGTGATTACGCCAAGCTACGGG and CAATTTACGCGTTAAGATACATTGATGAG; and TTTACGCGTTGAGCAAAAGGCCAGCAAAAG and TTTACATGTCCGTCGTTTTACAACGTCGTG. Digest both PCR products with PciI and MluI and religate using T4 Ligase, removing the LacO site.

### 3.6. Calibration of Intron-Encoded Microrna Knock-Down

1. In general, the approach is similar to protocol 3.4. Two calibrations are performed: first, one needs to find the dosage of miR-encoding plasmid that elicits strong knockdown under fully induced conditions. Second, one needs to confirm that the construct is sensitive to the regulator and find the optimal regulator dosage relative to the microRNA plasmid.

2. Prepare transfection mixtures with the excess of a regulator plasmid. For example, in the case of two miR constructs described below, one could use as much as 500 ng of pTET-On-Advanced (rtTA-expressing plasmid) or pNEBR-R1 (Rheo activator – expressing plasmid) with their cognate cofactors, Dox and RSL1 respectively. Add fixed amount of a fluorescent reporter (e.g., ZsYellow) furnished with the appropriate microRNA target (triple tandem repeat of the same target is strongly recommended) and transfection markers. Finally, add increasing amount of the microRNA gene. *Do not balance with filler DNA.* One could measure in the range between zero and 400 ng of a plasmid. The choice of the transfection marker is dictated by the marker co-expressed with the microRNA; the combination of AmCyan, ZsYellow and DsRed can generally be resolved with microscopy and flow cytometry. Perform a series of corresponding control experiments where the cognate microRNA target is replaced by a scrambled or unrelated target.

3. Transfect cells and analyze data as described in the protocols 3.3 and 3.4. In addition, monitor expression of the fluorescent marker co-expressed with the microRNA. In general one expects that increased dosage of microRNA gene will result in higher marker expression and lower levels of the reporter.

In the control experiments, the levels of the reporter should remain constant. This is not always the case and some changes in reporter levels can be observed even when the microRNA target is scrambled. Specific activity of a microRNA can be measured by normalizing reporter levels with cognate targets to reporter levels with scrambled targets. Non-specific changes in the reporter level of more than twofold over the entire range of microRNA concentrations should be a cause for concern.

4. Based on previous analysis, find minimal microRNA plasmid dosage that elicits efficient reporter knock-down (experience shows that the knock-down will become constant above certain microRNA levels). Perform additional series of measurements with fixed reporter and microRNA levels but with changing amount of the regulator plasmid in the range between zero and 500 ng. Set up two series of control experiments: in the first series use a reporter with scrambled target. In the second series transfect either filler DNA instead of the regulator plasmid or the regulator plasmid without adding small molecule cofactor (if required). These controls will help quantify the specific versus non-specific effects of the regulators on reporter knock-down. Measure knock-down and find the minimal regulator dosage that elicits knock-down that is close to saturation.

### *3.7. Transcriptional Logic Computations with Tunable Intron-Encoded microRNAs*

1. Assemble the following plasmids: constitutive regulator-expressing vectors; tunable microRNA-expressing vectors; and reporter (output) plasmid/s with properly arranged targets as required by a specific logic computation to be performed. In a specific example with the vectors whose construction has been described above, the plasmids are pTET-On-Advanced, pNEBR-R1, pTRE-DsRed-miR-FF3, pRheo-AmCyan-FF4, and pZsYellow-tFF3-tFF4. As a rule, use triple repeats of the same target. For control experiments use ZsYellow with scrambled target. Cloning strategy for multi-target reporters has been described elsewhere (13, 18).
2. Based on calibration experiments in the previous paragraph, determine the amounts of microRNA and the regulator plasmids. The computation circuit contains the output and the microRNA genes. The inputs are the regulators (transcription factors, TFs). In different computation experiments we test for different combinations of inputs while the circuit is always the same. In a logic function, the input is either on or off. For $N$ inputs, there are total of $2N$ input combinations where each input is either on or off. Circuit characterization requires that all these combinations be tested. With two inputs there are four combinations, namely (1) rtTA:off, Rheo:off; (2) rtTA:on, Rheo:off; (3) rtTA:off, Rheo:on; (4) rtTA:on; Rheo:on. The off state is achieved by not adding a TF plasmid to the

transfection mixture. The on state is achieved by adding the plasmid at a concentration found in above calibrations. For each measurement, perform a control where the output is replaced by a reporter with scrambled targets. In all measurements add small molecule cofactors at saturating concentrations because they are not the inputs to the circuit.

3. Transfect cells and measure fluorescent reporters as described above. Two reporters indicate microRNA levels and the third one is the output. Be careful to properly compensate the channels and measure single-reporter controls in all experiments. Calculate specific circuit response by normalizing output level in each computation to its level in the corresponding control with the same input combination.

## 4. Notes

1. The primers with SapI site should ideally include overhangs of about 10 nt long to the left of the SapI binding site. With 4 or 6 nts as used in this report, large concentrations of SapI are required.
2. SapI can be replaced with BspQI that has the same specificity but higher activity.
3. If one-pot ligations do not seem to work, examine your substrates by ligating pairs of adjacent fragments in control reactions. At times the digestion is not full or the sticky ends are not compatible.
4. pPRIME vectors elicit efficient RNAi, but the Neo-MicroRNA fragments somewhat lose their activity when subcloned to other backbones.
5. In case of circuits with more than two inputs, one could use Infra-red fluorescent protein (23) as a fourth color. In principle there is no requirement to co-express fluorescent protein with the microRNA, and the exons can be replaced with non-coding sequence.
6. Promoter that drives the output has crucial effect on circuit performance. We found that EF1a is significantly better than CMV, and additional effects are likely to be found as more promoters are tested.
7. There are no transfection markers in the large-scale computation experiments, because all the labels are used as reporters. The sample-to-sample variability can be minimized by careful handling and by multiple replicas of the same measurement, three at least for both the experimental and the control samples.

## References


1. Benenson, Y. (2009) Biocomputers: from test tubes to live cells, *Molecular Biosystems 5*, 675–685.
2. Benenson, Y. (2009) RNA-based computation in live cells, *Current Opinion in Biotechnology 20*, 471–478.
3. Weiss, R., Homsy, G. E., and Knight, T. F. (1999) Toward in vivo digital circuits, In *Evolution as Computation: DIMACS Workshop* (Landweber, L. F., and Winfree, E., Eds.), pp 275–295, Springer.
4. Gardner, T. S., Cantor, C. R., and Collins, J. J. (2000) Construction of a genetic toggle switch in Escherichia coli, *Nature 403*, 339–342.
5. Elowitz, M. B., and Leibler, S. (2000) A synthetic oscillatory network of transcriptional regulators, *Nature 403*, 335–338.
6. Isaacs, F. J., Dwyer, D. J., and Collins, J. J. (2006) RNA synthetic biology, *Nature Biotechnology 24*, 545–554.
7. Canton, B., Labno, A., and Endy, D. (2008) Refinement and standardization of synthetic biological parts and devices, *Nature Biotechnology 26*, 787–793.
8. Weber, W., and Fussenegger, M. (2002) Artificial mammalian gene regulation networks – novel approaches for gene therapy and bioengineering, *Journal of Biotechnology 98*, 161–187.
9. Guet, C. C., Elowitz, M. B., Hsing, W. H., and Leibler, S. (2002) Combinatorial synthesis of genetic networks, *Science 296*, 1466–1470.
10. Cox, R. S., Surette, M. G., and Elowitz, M. B. (2007) Programming gene expression with combinatorial promoters, *Molecular Systems Biology 3*, 11.
11. Ellis, T., Wang, X., and Collins, J. J. (2009) Diversity-based, model-guided construction of synthetic gene networks with predicted functions, *Nature Biotechnology 27*, 465–471.
12. Kramer, B. P., Fischer, C., and Fussenegger, M. (2004) BioLogic gates enable logical transcription control in mammalian cells, *Biotechnology and Bioengineering 87*, 478–484.
13. Rinaudo, K., Bleris, L., Maddamsetti, R., Subramanian, S., Weiss, R., and Benenson, Y. (2007) A universal RNAi-based logic evaluator that operates in mammalian cells, *Nature Biotechnology 25*, 795–801.
14. Xie, Z., Liu, S. J., Bleris, L., and Benenson, Y. (2010) Logic integration of mRNA signals by an RNAi-based molecular computer, *Nucleic Acids Research 38*, 2692–2701.
15. Leisner, M., Bleris, L., Lohmueller, J., Xie, Z., and Benenson, Y. (2010) Rationally designed logic integration of regulatory signals in mammalian cells, *Nature Nanotechnology 5*, 666–670.
16. Shapiro, E., and Benenson, Y. (2006) Bringing DNA computers to life, *Scientific American 294*, 44–51.
17. Baker, D., Group, B. F., Church, G., Collins, J., Endy, D., Jacobson, J., Keasling, J., Modrich, P., Smolke, C., and Weiss, R. (2006) Engineering life: Building a fab for biology, *Scientific American 294*, 44–51.
18. Benenson, Y. (2011) Engineering RNAi circuits, *Methods in Enzymology 497*, 187-205.
19. Chang, K., Elledge, S. J., and Hannon, G. J. (2006) Lessons from Nature: microRNA-based shRNA libraries, *Nature Methods 3*, 707–714.
20. Stegmeier, F., Hu, G., Rickles, R. J., Hannon, G. J., and Elledge, S. J. (2005) A lentiviral microRNA-based system for single-copy polymerase II-regulated RNA interference in mammalian cells, *Proceedings of the National Academy of Sciences of the United States of America 102*, 13212–13217.
21. Zhang, Z. L., Xie, Z., Zou, X. L., Casaretto, J., Ho, T. H. D., and Shen, Q. X. J. (2004) A rice WRKY gene encodes a transcriptional repressor of the gibberellin signaling pathway in aleurone cells, *Plant Physiology 134*, 1500–1513.
22. Wagner, E. J., Baines, A., Albrecht, T., Brazas, R. M., and Garcia-Blanco, M. A. (2004) Imaging alternative splicing in living cells, *Methods in Molecular Biology*, 29–46.
23. Shu, X. K., Royant, A., Lin, M. Z., Aguilera, T. A., Lev-Ram, V., Steinbach, P. A., and Tsien, R. Y. (2009) Mammalian Expression of Infrared Fluorescent Proteins Engineered from a Bacterial Phytochrome, *Science 324*, 804–807.

# Chapter 11

# Light-Regulated Gene Expression in Yeast


## Laszlo Kozma-Bognar, Anita Hajdu, and Ferenc Nagy



## Abstract

An important basic requirement of synthetic genetic networks is the option of external control of gene expression. Although several chemically inducible systems are available, all of these suffer from the common problem: the chemical inducers are difficult to remove so that to terminate the response. We have described a regulatory expression system for yeast, which employs light as inducer. This light switch translates light-controlled protein–protein interactions into the transcription of selected genes in a dose-dependent and reversible manner.



**Key words:** Yeast, Light, Photoreceptors, Control of gene expression, Two-hybrid assay


## 1. Introduction

Design of regulatable expression systems with controlled induction or repression of target gene function was of intense scientific interest for many years. A variety of approaches has been reported, generally representing the external stimuli working in complex with responsive promoter to provide the rapid, reversible, and specific response to dosed perturbation (1).

Phytochromes (Phys) are plant photoreceptors that exist in two inter-convertible conformational states. Upon red light ($\lambda_{max}$ = 660 nm) illumination the biologically inactive form (Phy Pr) is converted to the active, far-red light ($\lambda_{max}$ = 730 nm) absorbing conformer (Phy Pfr), which initiates signalling cascades via physical interaction of specific regulatory proteins. The Pfr → Pr conversion is triggered by far-red light irradiation that results in the dissociation of the Phy interacting proteins (IPs) and termination of signalling (2). This unique light-dependent protein–protein interaction/dissociation mechanism combined with the basic principles of yeast


Wilfried Weber and Martin Fussenegger (eds.), *Synthetic Gene Networks: Methods and Protocols*,
Methods in Molecular Biology, vol. 813, DOI 10.1007/978-1-61779-412-4_11, © Springer Science+Business Media, LLC 2012

two-hybrid assays has led to the construction of artificial light switches controlling gene expression in yeast (3, 4). In the model plant *Arabidopsis thaliana*, phytochromes are represented by a small protein family with five closely related receptors designated PhyA, B, C, D, and E. Based on the same molecular principles, PhyA or PhyB and their corresponding IPs were employed as artificial light switches in yeast so far (3, 4). In these typical yeast two-hybrid systems, Phys and IPs were expressed as chimera proteins, fused to DNA-binding (GBD) or the transcriptional activator (GAD) domain of the GAL4 transcription factor, respectively. The *cis* component of the switch was a reporter gene controlled by a GAL4-responsive promoter. In darkness, Phy-GBD binds the promoter, but does not induce transcription. Red light illumination converts Phy into the Pfr form, therefore facilitates Phy–IP interaction, which recruits IP-GAD to the GAL4-dependent promoter resulting in activation of transcription. Subsequent far-red light illumination coverts Phy Pfr to Pr, which results in the dissociation of the Phy-GBD–IP-GAD complex and abrogation of transcription.

Functional phytochrome receptors consist of the apoprotein and the covalently linked chromophore called phytochromobilin (5). Since the chromophore is not synthesized in yeast, an analogous compound, phycocyanobilin (PCB) purified from cyanobacteria is added to the media. PCB is readily uptaken by yeast cells and is autoligated by phytochrome apoproteins resulting in photochemically functional phytochrome photoreceptors (6).

By comparing different combinations of components, we showed that optimal induction of gene expression (low background levels and high fold-changes) can be achieved by the light switch composed of the PhyA-GBD and the FAR-RED ELONGATED HYPOCOTYL 1 (FHY1)-GAD fusion proteins and the GAL4-responsive *GAL1* promoter in yeast cells grown on solid media (4). Moreover, we demonstrated that an unidentified compound normally present in yeast cells can act as chromophore for phytochromes (4).

## 2. Materials

### 2.1. Plasmids and Gene Constructs

1. The GAL4-based yeast two-hybrid plasmids expressing the PhyA-GBD and FHY1-GAD fusion proteins are available upon request (4). In case of use of different protein interaction detection system (e.g. LexA-based) requiring re-cloning of the components, these plasmids can serve as reliable sources of the corresponding cDNAs.
2. The light-dependent expression of the target gene is driven by a GAL4-responsible promoter. We recommend using the authentic *GAL1* promoter, which can be amplified using

the following primers: pG1 Fwd: 5′-ATTACCACCATATAC ATATCC-3′ and pG1 Rev: 5′-TCTCCTTGACGTTAAA GTAT-3′ (see Note 1). In case of non-GAL4-based systems, the promoter must be selected and tested accordingly.

### 2.2. Yeast Culturing and Transformation

1. Host yeast strain with appropriate auxotrophic markers to enable transformation of the plasmids expressing the components of the light switch.
2. Synthetic drop-out medium (SD) without specific amino acids corresponding to the selectable markers of the plasmids; 3.35 g Difco™ yeast nitrogen base (Becton, Dickinson and Company, Sparks, MD), 1 g complete supplement mixture without the appropriate amino acids (MP Biomedicals United States, Solon, OH), 10 g glucose in 500 mL water. Add 10 g agar if solid medium is required. Sterilize by autoclaving.
3. YPD medium; 20 g peptone, 20 g glucose, 10 g yeast extract in 1,000 mL water. Add 20 g agar if solid medium is required. Sterilize by autoclaving.
4. 10× Tris–EDTA (TE) buffer pH 7.5: 100 mM Tris–HCl pH 8.0, 10 mM EDTA. Set pH to 7.5 with acetic acid. Sterilize by autoclaving.
5. 10× lithium-acetate (LiAc) solution pH 7.5; 10.2 g LiAc in 100 mL water, set pH to 7.5 with acetic acid. Filter sterilize.
6. 1× TE/LiAc buffer pH 7.5: prepare freshly by mixing equal amounts of 10× TE buffer and 10× LiAc solution. Add sterile water to obtain 1× working concentration for each component.
7. 50% (w/v) PEG 3350 solution; 50 g PEG 3350 in 100 mL water. Sterilize by autoclaving.
8. Carrier DNA: 10 mg/mL in water, salmon sperm DNA (Sigma, St. Louis, MO).

### 2.3. Light Treatments and Measurement of Output Processes

1. Phycocyanobilin (PCB) stock solution, optional (see Note 2).
2. 2.5 mM luciferin solution in water (D-Luciferin Firefly, potassium salt from Biosynth AG, Staad, Switzerland).
3. Green safety light can be produced by wrapping green fluorescent tubes with plastic filters (7) or by using light emitting diodes (LEDs) with emission maxima at 525 nm.
4. Red and far-red light conditions are best produced by LED arrays (Snap-Lite LED modules from Quantum Devices, Barneveld, WI).
5. We have used a liquid nitrogen-cooled CCD camera (Visitron Systems GmbH, Munich, Germany) to detect, and the Metamorph software (Molecular Devices, Downingtown, PA) to quantify luminescent signals.

## 3. Methods

### 3.1. Construction of Chimeric Genes

1. Standard molecular cloning methods are used to create the components of the light switch (8).

### 3.2. Yeast Transformation

#### 3.2.1. Preparation of Competent Yeast Cells

1. Inoculate 5 mL liquid YPD medium with yeast cells and incubate for 16 h at 30°C with agitation.
2. Transfer the 5 mL culture to a dish with 45 mL liquid YPD medium and incubate for 5–6 h at 30°C with agitation.
3. Transfer the culture to a sterile centrifuge tube and sediment yeast cells by centrifugation at 3,000 × *g* for 5 min at room temperature.
4. Discard the supernatant and wash the pellet with 20 mL sterile distilled water.
5. Spin the tube at 3,000 × *g* for 5 min at room temperature.
6. Discard the supernatant and wash the pellet with 20 mL 1× TE/LiAc buffer.
7. Spin the tube at 3,000 × *g* for 5 min at room temperature.
8. Discard the supernatant and suspend yeast cells in 1–1.5 mL TE/LiAc. Cells can be stored at 4°C for 24 h without significant loss in transformation competence.

#### 3.2.2. Yeast Transformation

Volumes and amounts are given for a single transformation.

1. Incubate 50 μL of competent yeast cells at 30°C for 15 min.
2. Heat-denature 15 μL carrier DNA by incubation at 100°C for 10 min. Chill on ice.
3. Prepare transformation master mix by mixing 30 μL 10× TE pH 7.5 buffer, 30 μL 10× LiAc pH 7.5 solution, and 240 μL 50% (w/v) PEG 3350 solution.
4. Add the denatured carrier DNA to the transformation master mix and vortex immediately.
5. Add 50 μL competent yeast cells (from step 1) and vortex immediately.
6. Add 5 μg plasmid DNA (see Note 3) to be transformed and vortex immediately.
7. Incubate cells with plasmid DNA at 30°C for 30 min.
8. Incubate sample at 42°C for 20 min (heat shock).
9. Spin for 1 min in a microcentrifuge at full speed (about 16,000 × *g*).
10. Discard supernatant and re-suspend the pellet in 100–200 μL sterile water.
11. Spread on solid selective medium and incubate plates at 30°C for 3 days until transformed colonies appear (see Note 4).

### 3.3. Switching Transcription On and Off: Light Treatments and Measurement of Output Processes

In this section we provide a protocol for measuring light-regulated reporter gene expression (luciferase activity) in vivo in yeast cells grown on solid medium. In this particular configuration the cDNA of the firefly luciferase gene was cloned in the effector construct controlled by the *GAL1* promoter and, therefore, by the PhyA-FHY1 light switch (4). Depending on the particular aim of the application, the effector construct is modified and so is the method for detecting the light-induced changes of target gene expression.

1. Inoculate 2 mL of selective SD media yeast cells and incubate for 16 h at 30°C with agitation (see Note 4).
2. Transfer 20 μL droplets of the cultures to SD agar plates and incubate for 48 h at 30°C in darkness. Prepare identical plates according to the number of different light treatments planned. All further manipulations are conducted under green safety light.
3. During the incubation, yeast cells form merged colonies (or patches) with 5–8 mm diameter. Pipette 20 μL of 2.5 mM L-luciferin solution at the centre of each patch and transfer the plates to the imaging chamber (see Note 5).
4. Start taking images using a CCD camera designed for detecting bioluminescence.
5. 16–18 h after the application of luciferin (see Note 6), pause imaging, remove separate plates for different light treatments, then return plates in the camera and resume imaging. The routine test to verify proper phytochrome response includes red light, far-red light, red followed by far-red light treatments, and the dark control. The response of the switch is dose dependent (3). We have used 10 min irradiation at 70 $\mu mol/m^2/s^1$ fluence rate, independent of the wavelength. These are considered as saturating conditions.
6. Quantify luminescence signals using appropriate software and analyse results.

## 4. Notes

1. In contrast to our results, light-dependent interaction of phytochromes with other proteins were not demonstrated in yeast without addition of exogenous chromophore (PCB) (3, 9). It should be noted, however, that those experiments used the activation of different reporter genes as a measure of protein–protein interactions. Reporter genes in yeast strains routinely used for GAL4-based yeast two-hybrid assays are usually composed of an artificial/engineered promoter with binding sites for the GAL4 transcription factor that controls the expression

of the reporter proteins, which confer prototrophy (HIS3, ADE2) or measurable enzyme activity (beta-galactosidase). In our test system a different promoter (unmodified *GAL1* promoter) and a different reporter protein (luciferase) were used. We propose that phytochromes can employ an unidentified compound from yeast as chromophore, but light absorbing efficiency of the constituted receptor is less than that of the receptor binding PCB. As a result, red light induces less amount of phytochrome Pfr, which results in a lower level of the phytochrome/other protein complex and less-efficient induction of transcription. This weak activating effect could be under the threshold of detection in case of the "factory-made" reporter genes, but is clearly seen by our *GAL1* promoter-driven reporter.

2. We showed that activity of the *GAL1* promoter displayed clear, high level of induction in response to red light without the application of PCB. We obtained only threefold higher induction if PCB was added to the medium. Therefore, we suggest that PCB is not required if the *GAL1* promoter is used for output regulation. However, if other promoters and/or other protein interaction detection systems are used, addition of PCB to the medium is advised at 10 μM final concentration. PCB is not available commercially, but has to be purified from cyanobacteria (7).

3. Plasmid DNA should be free of any traces of organic solvents and optimally has a concentration of 1–3 μg/μL. Plasmids expressing components of the light switch are co-transformed; add 5 μg of each.

4. Apply conditions (i.e. particular drop-out medium) to select for the presence of the plasmids transformed, but not for the interaction of the proteins expressed.

5. Imaging chambers (or dark boxes) of ultra-sensitive CCD cameras are usually cannot be set to keep 30°C. We found that yeast cells can tolerate room temperature over several days. Although growth is restricted, the function of the light switch is retained.

6. Even in the absence of any positive effect (i.e. in darkness), luciferase protein is produced from the readout construct and accumulates to a certain level. When luciferin is first applied to these yeast cells, the accumulated enzyme molecules react with the substrate and become inactivated promptly. It follows that the detected luciferase activity reports the rate of transcription of the readout construct only, if the accumulated luciferase proteins are eliminated. According to our findings, this state of the system is reached 16–18 h after the application of luciferin.

## References


1. Banaszynski LA, Wandless TJ (2006) Conditional control of protein function. Chem Biol. 13:11–21
2. Bae G, Choi G (2008) Decoding of light signals by plant phytochromes and their interacting proteins. Annu Rev Plant Biol. 59:281–311
3. Shimizu-Sato S, Huq E, Tepperman JM, Quail PH (2002) A light-switchable gene promoter system. Nat Biotechnol 20:1041–1044
4. Sorokina O, Kapus A, Terecskei K, Dixon LE, Kozma-Bognar L, Nagy F, Millar AJ (2009) A switchable light-input, light-output system modeled and constructed in yeast. J Biol Eng 3:15 doi:10.1186/1754-1611-3-15
5. Emborg TJ, Walker JM, Noh B, Vierstra RD (2006) Multiple heme oxygenase family members contribute to the biosynthesis of the phytochrome chromophore in Arabidopsis. Plant Physiol. 40:856–868
6. Li L, Lagarias JC (1994) Phytochrome assembly in living cells of the yeast Saccharomyces cerevisiae. Proc Natl Acad Sci USA. 91:12535–12539
7. Tepperman, J. http://openwetware.org/images/2/27/Y2H_documentation.pdf
8. Sambrook J, Russel DW (2001) Molecular Cloning: A Laboratory Manual. Cold Spring Harbor, New York
9. Hiltbrunner A, Viczián A, Bury E, Tscheuschler A, Kircher S, Tóth R, Honsberger A, Nagy F, Fankhauser C, Schäfer E (2005) Nuclear accumulation of the phytochrome A photoreceptor requires FHY1. Curr Biol. 15:2125–2130

# Chapter 12

# Light-Controlled Gene Switches in Mammalian Cells

**Fuzhong Zhang, Kristian M. Müller, G. Andrew Woolley, and Katja M. Arndt**

## Abstract

Remote control of cells is a desirable feature in synthetic biology. We established a light-switchable interfering peptide (iPEP) which controls gene expression by modulating the activity of a transcription factor. For photo-switching, the iPEP is cross-linked with a *cis-trans* isomerizable cross-linker in such a way that the light-activated *cis* form enables inhibitor folding rendering it active, whereas the dark-adapted *trans* form forces the inhibitor into an inactive form. Switching can be repeated in both directions. The iPEP acts as dominant-negative inhibitor targeting c-Jun and c-Fos of the transcription factor activator protein-1 (AP-1). Light-activated peptides exhibited much stronger inhibition of AP-1:DNA complexes and interference with gene transcription than their nonactivated counter parts. In this chapter, we provide protocols for cross-linking, peptide purification, observation of structural changes upon photo-switching, DNA binding analyses as well as gene expression studies in mammalian cells.

**Key words:** *cis-trans* Isomerization, Electrophoretic mobility shift assay, Interfering peptide, Light-switchable, Inducible gene expression

## Abbreviations

| | |
|---|---|
| AFosW | Dominant-negative peptide with acidic extension targeting cJun |
| AP-1 | Activator protein-1 |
| BSBCA | 3,3′-bis(sulfonato)-4,4′-bis(chloroacetamide)-azobenzene |
| bZIP | Basic domain Leucine Zipper |
| CD | Circular dichroism |
| DMEM | Dulbecco's modified Eagle's medium |
| EMSA | Electrophoretic mobility shift assay |
| FBS | Fetal bovine serum |
| FosW | Dominant-negative peptide targeting cJun |
| iPEP | Interfering peptide |
| ONPG | O-Nitrophenyl-β-galactoside |
| PMA | Phorbol-12-myristate 13-acetate |
| TCEP | Tris(2-carboxyethyl)phosphine |
| X-iPEP | Cross-linked version of iPEP |

Wilfried Weber and Martin Fussenegger (eds.), *Synthetic Gene Networks: Methods and Protocols*,
Methods in Molecular Biology, vol. 813, DOI 10.1007/978-1-61779-412-4_12, © Springer Science+Business Media, LLC 2012

## 1. Introduction

Remote control of electronic devices is a standard feature nowadays. A similar remote control system would be advantageous for programming or controlling cells in a contact-free manner. We describe in this chapter a general method for establishing a light-switchable system by photo-control of proteins. Protocols are given for the design, characterization, and application of a light-switchable coiled-coil forming peptide targeting the transcription factor activator protein-1 (AP-1) (1).

A dominant-negative interfering peptide (iPEP) targeting AP-1 was developed using a combination of semi-rational design and in vivo selection by randomizing the coiled-coil interface residues (2, 3) of the natural selection partner and selecting for peptides with high affinity using a protein-fragment complementation assay (4–6). Resulting iPEPs were shown to bind to their target c-Jun and c-Fos with considerably higher stability than the wild-type interaction partner (7, 8). Importantly, protocols can easily be transferred and such peptides can be generated for different targets (9–14). While these peptides target solely the coiled-coil dimerization domain, they are already efficient inhibitors of DNA binding (1, 12). For further increase of binding, these peptides can be combined with an acidic extension capturing the basic DNA binding domain of AP-1 (15).

For light-switching, iPEP were coupled with a small photo-switchable cross-linker, such as 3,3′-bis(sulfonato)-4,4′-bis(chloro acetamide)-azobenzene (BSBCA) (16). These cross-linkers react site-specifically with proteins via Cys residues and undergo *trans*-to-*cis* isomerization when exposed to a light pulse. These pulses are of a wavelength and energy that does not cause significant direct effects on cells and thus can be used to switch the conformation of modified proteins in the context of living cells. Such compounds have been shown to be able to control dimerization of the coiled-coil domain of transcription factors and can be used for many different peptides or protein sequences (17–21). Additionally, the cross-linkers can be adopted for specific uses (22–25).

Light-induced switching of the linker causes folding or unfolding of the peptides (Fig. 1), and switching can be repeated in both directions. Importantly, light-activated peptides exhibited much stronger inhibition of AP-1:DNA complexes and interference with gene transcription than their nonactivated counter parts.

In this chapter, we give detailed protocols for the cross-linking and purification of a photo-switchable dominant-negative peptide inhibitor (Subheading 3.1). Furthermore, detailed protocols are given for the analysis of structural changes upon photo-isomerization using UV/VIS and circular dichroism spectroscopy (Subheading 3.2). We demonstrate that the photo-switchable dominant-negative

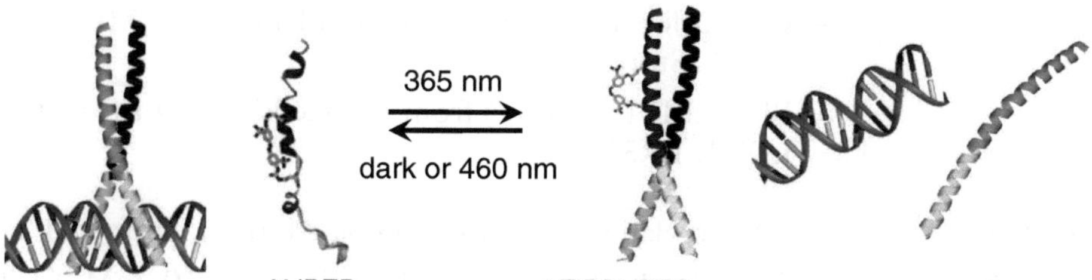


Fig. 1. Photo-control of bZIP protein DNA binding activity by a BSBCA cross-linked dominant negative interfering peptide (X-iPEP). When X-iPEP is irradiated at 365 nm, the BSBCA cross-linker isomerizes to the *cis* form and enables folding of X-iPEP. The X-iPEP$_{cis}$ binds to the target bZIP protein forming a bZIP/X-iPEP$_{cis}$ complex that cannot bind DNA. When X-iPEP$_{cis}$ is exposed to a 460 nm light or relaxed to *trans* thermally, the BSBCA cross-linker unfold its helical structure, causing the target bZIP proteins recovering their DNA binding activity. The bZIP structure is derived from the crystal structure (PDB ID: 1FOS) of an AP-1 heterodimer-DNA complex – cFos/cJun/DNA.

inhibitor is able to interfere with DNA binding of AP-1 in a light-controlled, reversible manner as seen in electrophoretic mobility shift assays (Subheading 3.3). The inhibitors are able to block AP-1-mediated gene expression in mammalian cells, and we provide protocols for peptide transfection (Subheading 3.4) and reporter gene assays (Subheading 3.5).

## 2. Materials

### 2.1. Design of Dominant Negative Mutation Sites and Preparation of Cross-linked Peptide

1. Sodium phosphate buffer: 1 M sodium phosphate at pH 8.5. Store at room temperature.
2. BSBCA solution: Dissolve BSBCA powder in water at 10 mM. Prepare aliquots and store at −20°C (see Note 1).
3. TCEP solution: 100 mM tris(2-carboxyethyl)phosphine water solution. Prepare aliquots and store at −20°C.
4. Zorbax SB-C18 column.
5. HPLC buffer A: Water containing 0.1% TFA. Store at room temperature.
6. HPLC buffer B: Acetonitrile containing 0.1% TFA. Store at room temperature.

### 2.2. Structural Changes Measured by UV-VIS and Circular Dichroism Spectrometry

1. UV-VIS buffer: 10 mM sodium phosphate at pH 7.0. Store at room temperature.
2. High intensity 365 nm UV LED (200 mW) from Opto Technology, Inc. Wheeling, IL, USA.
3. Metal halide Tri-Lite lamp (70 W) from World Presion Instruments, coupled to a 370 ± 10 nm band-pass filter from Harvard Apparatus, Canada.

4. CD buffer: 10 mM sodium phosphate at pH 7.0 with 0.5 mM TCEP. Store at 4°C.
5. Blue LED (~34 mW) from LumiledsTM, San Jose, CA, USA, LXHL-LR3C.

#### 2.3. DNA Binding Properties Measured by EMSA

1. 40% acrylamide solution (Sigma-Aldrich). Store at 4°C.
2. 5× TBE buffer: 450 mM boric acid, 10 mM EDTA and 500 mM Tris-HCl at pH 8.3. Store at room temperature.
3. Ammonium persulfate: Prepare 10% (w/v) ammonium persulfate in water and use the same day (see Note 2).
4. *N,N,N,N*-Tetramethyl-ethylenediamine (TEMED; Sigma-Aldrich). Store at room temperature.
5. 10× EMSA loading buffer: 900 mM boric acid, 20 mM EDTA, 1 mg/mL BSA, 1 mg/mL nonspecific salmon sperm DNA, and 1 M Tris-HCl at pH 8.3. Prepare aliquots and store at −20°C.
6. 6× EMSA loading dye: 60% glycerol, 0.006% bromophenol blue, 10 mM Tris-HCl at pH 8.3.
7. Fluorescent photo imager, such as Molecular Dynamics Storm 860 phosphorimager (Molecular Dynamics, Amersham Biosciences).

#### 2.4. Transfection of Mammalian Cells with Cross-linked Peptides

1. Dulbecco's modified Eagle's medium (DMEM) containing 10% fetal bovine serum (FBS), 0.1 mg/mL streptomycin, and 200 U/mL penicillin. Store at 4°C, equilibrate to room temperature before use.
2. Solution of trypsin/EDTA: 0.25% trypsin and 1 mM EDTA. Store at 4°C, equilibrate to room temperature before use.
3. Poly-L-lysine solution (15 ng/mL): Mix 7 mL of 100 ng/mL stock poly-L-lysine solution with 39.7 mL of distilled water. Store at 4°C.
4. PBS buffer: 137 mM NaCl, 2.7 mM KCl, and 10 mM sodium phosphate at pH 7.4. Store at room temperature.
5. OptiM media.
6. Lipofectamine 2000 (Invitrogen).
7. 1,000× PMA solution: Dissolve phorbol-12-myristate 13-acetate (PMA; Sigma-Aldrich) at 10 μg/mL (1,000×) in DMSO and store in single use aliquots at −80°C.

#### 2.5. Reporter Gene Assays

1. 1× lyse buffer: 50 mM NaCl, 2 mM EDTA, 1 mM $MgSO_4$, 5 mM DTT, 1% Triton-x-100, and 40 mM Tricine at pH 7.8. Store at −20°C.
2. Luciferase Assay Buffer A (Promega, E152A).
3. Luciferase Assay Buffer B (Promega, E152B).
4. Luciferase Assay Substrate (Promega).

5. ONPG phosphate solution: Dissolve *o*-nitrophenyl-β-galactoside (ONPG, Sigma-Aldrich) at 4 mg/mL in 100 mM sodium phosphate buffer at pH 7.0. Store at −20°C.
6. 100× magnesium chloride solution: 100 mM $MgCl_2$, 4.5 M β-mercaptoethanol. Store at −20°C freezer.
7. ELISA plate reader, such as TECAN Plate Reader with XRead Plus software (TECAN Group Ltd.).

## 3. Methods

### 3.1. Design of Mutation Sites and Preparation of Cross-linked Light-Switchable Peptides

Analogs of the target bZIP protein containing the leucine zipper domain but lacking the basic DNA binding domain can be used as dominant negative peptides, albeit often with relatively low inhibition potential (1). Alternatively, dominant negative peptides targeting coiled-coil proteins, such as bZIP-type transcription factors, have been designed theoretically (26) and evolved through semi-rational selection approaches (8, 13). This approach, yielding the so-called interfering peptides can be combined with the addition of negatively charged heptad repeats to the N-terminus capturing the basic region of the target protein to yield inhibitors with even higher efficacy (15).

The photo-switchable small molecule, BSBCA, is introduced into a dominant-negative peptide by covalently cross-linking at two engineered cysteine residues. BSBCA is an azobenzene derivative and adopts the *trans* conformation in the dark. Photo-irradiation at 365 nm isomerizes BSBCA to its *cis* confirmation. In order to activate a helical dominant-negative peptide, such as iPEPs targeting bZIP-type transcription factors by photo-irradiation, the light-activated (*cis*-form) of the cross-linker should be compatible with helix formation, whereas the *trans* form should promote unfolding of the inhibitor to keep it in an inactive state. To achieve this, the BSBCA cross-linker should be introduced at two residues spacing seven amino acids apart ($i$ and $i+7$ positions) (21). To do so, two consecutive $f$ position residues in the middle of a dominant-negative peptide are mutated to cysteines. To promote peptide stability and folding into α-helical structure without any charge effects at the termini, the N-terminus of the peptide is acetylated and the C-terminus is capped with amide. The designed peptide can be synthesized or purchased commercially.

#### 3.1.1. Preparation of Cross-linked Peptide

1. Dissolve the di-cysteine mutant iPEP in water to a concentration of 1 mM (see Note 3).
2. Mix 500 μL of peptide solution, 50 μL of sodium phosphate buffer, 200 μL of BSBCA solution and 100 μL TCEP solution in a 5 mL round bottom flask. Bring the volume to 1 mL by adding 150 μL water.

3. Add a magnetic stir bar to the flask and cap the flask with a rubber septum.
4. Cut a 3 mL disposable plastic syringe (with the plunger removed) at the 1 mL mark. Attach an empty balloon to the cut end using Parafilm.
5. Create a nitrogen inlet and vent system on the 5-mL round bottom flask by inserting two disposable needles into the septum. Blow nitrogen gas for 2 min to purge oxygen from the flask.
6. Attach the syringe-balloon on to the vent needle. Fill the balloon with nitrogen gas to maintain positive nitrogen pressure in the flask.
7. Incubate the flask at 37°C with stirring at 200 RPM for 10 h (see Note 4).
8. Remove the septum. Quench the reaction by adding 100 μL 10% TFA (see Note 5).

#### *3.1.2. Purification of Cross-linked Peptides*

The cross-linked peptides are purified by reversed-phase HPLC with a Zorbax SB-C18 column and a UV-VIS detector monitoring at 370 nm.

1. Pre-equilibrate the column with 95% HPLC buffer A and 5% HPLC buffer B.
2. Inject 250 μL of the reaction mixture each time into HPLC.
3. Elute using a linear gradient changing from 95% HPLC buffer A + 5% HPLC buffer B to 40% HPLC buffer A + 60% HPLC buffer B over 15 min.
4. Collect fractions of the cross-linked peptide.
5. Transfer the collected peptide solution into a clean 5-mL round bottom flask. Evaporate the acetonitrile from HPLC buffer B using a rotary evaporator.
6. Transfer the peptide solution into a 15-mL disposable plastic tube, flash freeze with liquid nitrogen. Freeze dry the sample. Store the cross-linked peptide in dark.
7. Redissolve a small fraction of the dried peptide sample in distilled water. Measure molecular weight of the cross-linked peptide using MALDI mass spectrometry.

### ***3.2. Structural Changes Measured by UV-VIS and Circular Dichroism Spectrometry***

#### *3.2.1. UV-VIS Spectroscopy*

1. Dissolve peptides in UV-VIS buffer at a peptide concentration of 10 μM.
2. Transfer the sample into a UV cuvette with 1 cm pathlength.
3. Set up a UV-VIS spectrophotometer.
4. Preheat the cuvette holder to 25°C.
5. Preset the scan range to 600–220 nm.

6. Place the sample cuvette into the spectrophotometer and let the sample equilibrate for 5 min to reach the temperature.
7. Start the UV-VIS scan.
8. To obtain a cross-linked peptide UV-VIS spectrum in the irradiated state, expose the sample cuvette to a 365 nm UV light for 2 min. A variety of light sources can be employed, such as a high intensity 365 nm UV LED or a metal halide Tri-Lite lamp coupled to a 370 ± 10 nm band-pass filter. To obtain even irradiation on the sample, expose the cuvette to the light source for 1 min. Rotate the cuvette by 180° to irradiate from the other side (see Note 6).
9. After irradiating the sample, immediately place the cuvette back to the UV-VIS spectrometer and start the scan.

#### *3.2.2. CD Spectroscopy*

1. Dissolve peptides in CD buffer at a peptide concentration of 50 μM (see Note 7).
2. Transfer the sample into a CD cuvette of 1 mm pathlength.
3. Set up a CD spectrometer. Set the scanning range from 260 nm to 190 nm with 1 nm step and an integration time of 5 s at each wavelength.
4. Place the cuvette inside the CD spectrometer. Start the scan.
5. Irradiate the sample by exposing the CD cuvette to a high intensity 365 nm UV LED for 2 min. Rotate the cuvette by 180° and irradiate from the other side for another 2 min. Place the cuvette inside the CD spectrometer and start a CD scan immediately.
6. To relax the 365 nm light-irradiated peptide, expose the sample in the CD cuvette to a 460 nm blue light using a blue LED for 1 min from each side (see Note 8). Place the cuvette inside the CD spectrometer and start the CD scan immediately. An example of the results is shown in Fig. 2.

### 3.3. DNA Binding Properties Measured by EMSA

For EMSA, a 1.5 mm thick, 8% polyacrylamide gels are used.

#### *3.3.1. Preparation of Polyacrylamide Gels for EMSA*

1. Mix 2 mL of 40% acrylamide solution, 0.5 mL of 5× TBE buffer, 0.42 mL of 60% glycerol, 7.03 mL distilled water.
2. Add 40 μL of 10% ammonium persulfate and 12 μL of TEMED.
3. After mixing, pour the gel and insert the comb immediately. The gel should polymerize in about 30 min.
4. Prepare the running buffer by diluting 100 mL of the 5× TBE buffer with 900 mL of water in a measuring cylinder. Cover with Para-Film and invert to mix. Pour this solution into an electrophoretic tank.
5. Once the polyacrylamide gel is polymerized, carefully remove the comb and assemble the gel in the electrophoretic tank. Make sure the entire gel has been submerged with the running buffer.

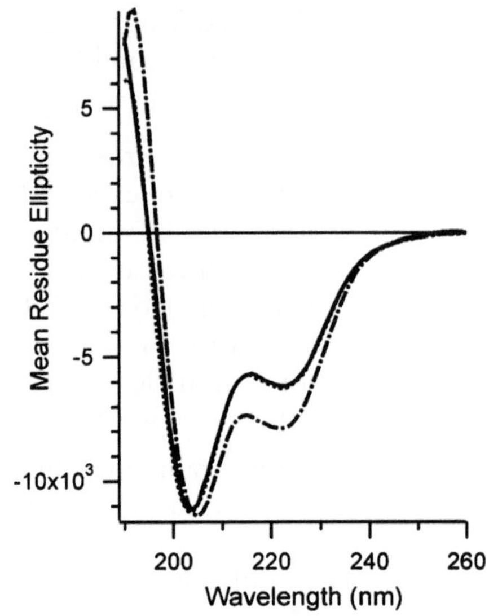


Fig. 2. CD spectra of BSBCA cross-linked AFosW (a dominant negative iPEP of cFos) in CD buffer. Dark-adapted (—), irradiated at 365 nm (· — · — ·–), then irradiated at 460 nm (· · · · · ·). Irradiation at 365 nm increases helical content; irradiation at 460 nm, restores helical content to dark-adapted levels. (Figure adapted from the supplementary data of ref. 1).

6. Wash the wells with running buffer using a 100 μL pipette or a Hamilton syringe.

*3.3.2. Measurement of Binding Affinities Between Transcription Factor and DNA*

1. Mix different concentrations of bZIP proteins (in the form of either homodimer or equimolar heterodimer) with 0.48 pmol of fluorescein-labeled DNA probe and 3 μL of 10× EMSA loading buffer in a total volume of 30 μL in an 0.6 mL Eppendorf tube (see Note 9).
2. Incubate each sample at 25°C for 5 min (see Note 10).
3. Add 6 μL EMSA loading dye and load samples on the gel.
4. Connect the electrophoretic tank to a power supply and run the gel at 270 V for 15 min.
5. Carefully transfer the gel to a fluorescent photo imager.
6. Scan the gel with a 450 nm blue LED.
7. DNA bands can be quantified in the scanned image with Image Quant software.

*3.3.3. Evaluation of the Inhibitory Effect of iPEP on DNA Binding*

1. Mix a fixed concentration of target bZIP proteins with DNA probe and EMSA loading buffer (see Note 11).
2. Add different amounts of dominant-negative iPEP to each sample with a final volume of 30 μL.
3. Incubate samples at 25°C for 15 min (see Note 12).

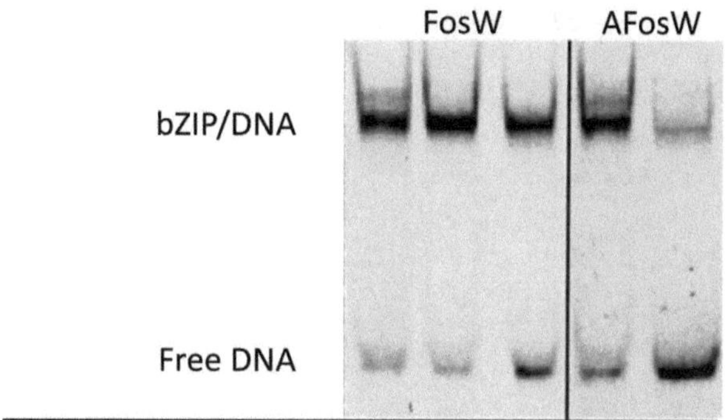


| | | | | | |
|---|---|---|---|---|---|
| cFos/cJun conc. [μM] | 0.3 | 0.3 | 0.3 | 0.3 | 0.3 |
| X-iPEP conc. [μM] | - | 3 | 15 | 0.3 | 1 |

Fig. 3. Electrophoretic mobility shift assay (EMSA) of two iPEP competing with the cFos/cJun heterodimer binding to DNA. Each lane contains cFos/cJun with or without iPEP at the indicated concentration. FosW is a dominant negative of cFos selected from a semi-rational library for binding to cJun (8). AFosW is an analog of FosW created by the addition of two acidic heptad repeat motifs to the N-terminus of FosW (1).

4. Load on an 8% polyacrylamide gel for EMSA.
5. Scan the gel as described above.
6. An example of the results is shown in Fig. 3.

#### *3.3.4. Evaluation of the Effect of Photo-Isomerization on DNA Binding*

1. Add different amounts of cross-linked dominant-negative interfering peptides (X-iPEP) to mixtures of target bZIP proteins, DNA probe, and EMSA loading buffer.
2. Transfer samples onto ice and chill for 5 min before photo-irradiation. We recommend performing this experiment at 4°C in order to maximize the amount of *cis* isomer (see Note 13).
3. Irradiate sample tubes with 365 nm UV light pulses (5 min on and 5 min off) for 1 h (see Note 14).
4. To isomerize the *cis* isomer back to the *trans* form, expose some samples to a 460 nm blue LED for another 2 min.
5. Load all samples (light- and dark-adapted form) onto a 4°C prechilled 8% polyacrylamide gel.
6. Attach a power supply and run the gel at 4°C, 270 V for 15 min.
7. An example of the results is shown in Fig. 4.

### ***3.4. Transfection of Mammalian Cells with Cross-linked Peptides***

1. Passage human embryonic kidney 293 T (HEK293T) cells when approaching confluence with trypsin/EDTA on 100-mm culture dishes with DMEM (containing 10% FBS, 0.1 mg/mL streptomycin, and 200 U/mL penicillin) at 37°C to provide new maintenance cultures.

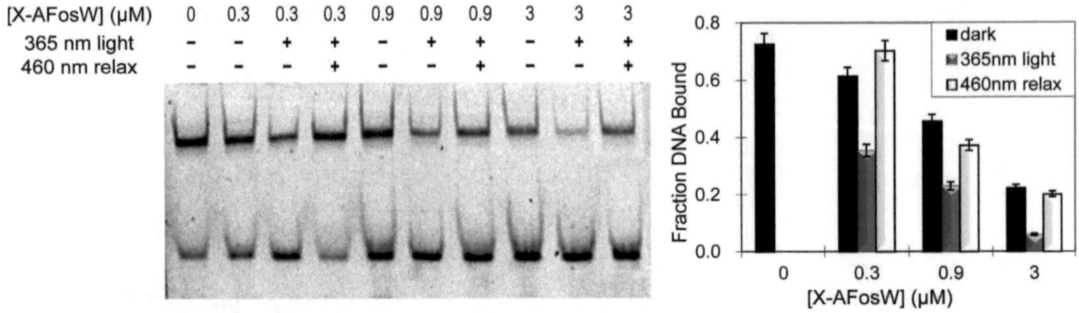


Fig. 4. EMSA of a mixture containing 0.1 μM cFos/cJun and 16 nM DNA in the absence (*first lane*) or presence of different amounts of XAFosW (the cross-linked version of AFosW). Samples were dark-adapted or irradiated with 365 nm light with or without subsequent irradiation at 460 nm as indicated. Irradiation was performed at 4°C. Band intensities were integrated and fraction DNA bound was calculated for each lane. Dark-adapted samples are shown in *black*, 365 nm light irradiated samples are shown in *gray* and 460 nm light relaxed samples are shown in *white*. (Figure adapted from the supplementary data of ref. 1).

2. Add 300 μL of 15 ng/mL poly-L-lysine into each well of a 96-well plate.
3. Incubate at room temperature for 1 h.
4. Remove the solution with vacuum. Wash each well with 300 μL sterile distilled water twice, remove the water, and dry the plate under sterile bench (see Note 15).
5. When HEK293T cells reach confluency, remove the media from 100-mm culture dishes.
6. Wash cells twice with 10 mL PBS buffer.
7. Add 2 mL trypsin, incubate the dish at 37°C for 5 min and then add 10 mL of DMEM to dilute the trypsin.
8. Transfer cells to a sterile 50-mL tube and spin at 1,500 × *g* for 5 min.
9. Remove the supernatant and resuspend the cells in 5 mL DMEM.
10. Use 10 μL of cell culture to measure cell density.
11. Dilute the resuspended cell culture with DMEM to $2.5 \times 10^5$ cells/mL and add 100 μL to each well of a poly-lysine coated 96-well plate.
12. Incubate the 96-well plate overnight at 37°C. At the next morning, cells reach approximately 70% confluency.
13. Replace DMEM media on the 96-well plate with 50 μL of OptiM media.
14. Incubate at 37°C for 1 h.
15. Mix 1 μg of reporter plasmid DNA, different amounts of dominant negative peptides and 1 μL of lipofectamine 2000 with OptiM media in a total volume of 50 μL in sterile 0.6-mL Eppendorf tubes.

16. Incubate at room temperature for 20 min.
17. Add the 50 μL solution to each well of cells in the 96-well plate.
18. Incubate at 37°C for 6 h (see Note 16).
19. Replace the transfection solution with 100 μL of DMEM.
20. External stimuli can be applied to induce expression of native bZIPs if they are not expressed under normal condition. For example, DMEM containing 10 ng/mL of PMA can be used to induce native AP-1 protein expression (see Note 17).

In order to obtain maximal photo-isomerization efficiency, we recommend photo-irradiating the 96-well plate in a confined space lined with aluminum foil.

1. Sterilize a paper box used for cryopreservation as well as several pieces of aluminum foil.
2. Line the paper box with aluminum foil.
3. Place a piece of sterile mirror at the bottom of the paper box.
4. Carefully place the 96-well plate on the top of the mirror.
5. Tape a 365 nm UV LED to the inside of the paper box cover. Close the cover.
6. Once the LED is turned on, the 365 nm light will shine right onto the top of the 96-well plate. Penetrating light will be reflected back to the plate by the mirror at the bottom.
7. Place the whole set-up into a 37°C incubator.
8. Turn on the LED for 3 min followed by 27 min incubation in the dark. This 30 min irradiation cycle is repeated 20 times (see Note 18). Such an irradiation pattern can be controlled by controlling the power supply using a clock timer.
9. Parallel with this experiment, incubate another plate of cells treated the same way at 37°C but kept in the dark as control.

### 3.5. Reporter Gene Assays

The reporter gene luciferase is controlled by the transcription factor under investigation, and dominant-negative peptides directed against the transcription factor will decrease luciferase expression. Consequently, photo-switching of BSBCA cross-linked peptides can be monitored by measuring luciferase activity. As a control for equal transfection and/or expression, a plasmid with β-galactosidase under a constitutive promotor is used, and obtained luciferase values can be normalized taking the corresponding β-galactosidase values into account (Fig. 5).

1. Cells can be analyzed 45 h after transfection.
2. Remove the media using vacuum.
3. Wash each well twice with 100 μL of PBS buffer.
4. Remove the PBS buffer.

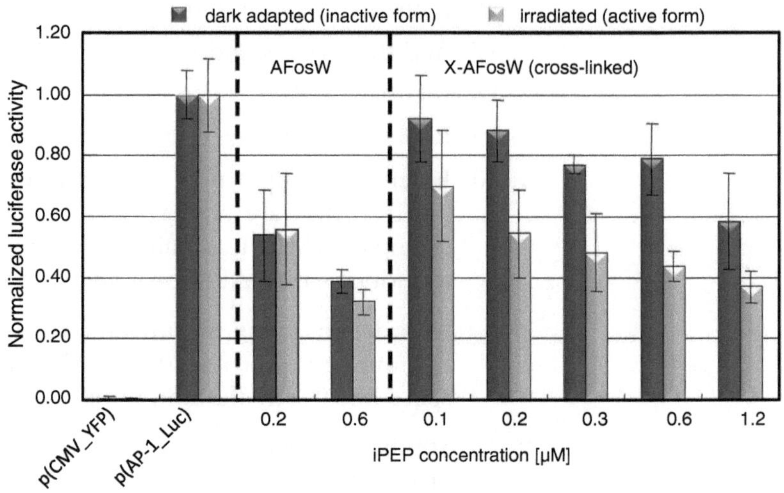


Fig. 5. Effect of photo-irradiation on cellular AP-1 activity in X-AFosW transfected HEK293T cells. Cells were co-transfected with 1 μg of pAP1-Luc (a plasmid containing a firefly luciferase gene under the control of an AP-1 promoter), 0.5 μg pRSV-Gal (a plasmid containing a β-galactosidase gene under the control of a constitutional CMV promoter) and dominant negative peptides (uncross-linked: AFosW or BSBCA cross-linked: X-AFosW) at indicated concentrations. pRSV-Gal serves as an internal control. The first two columns represent cells transfected with 1 μg of pCMV-YFP (a plasmid containing the YFP gene under the control of the CMV promoter) as a negative control for luciferase activity and as an indicator to monitor transfection efficiency. Specific inhibition on AP-1 activity is evaluated by the ratio of luciferase activity over ß-galactosidase activity.

5. Add 50 μL of 1× lyse buffer, incubate at 37°C for 5 min.
6. Use a 100 μL pipetman to pipette the cell lysate up and down for at least six times.
7. Transfer lysate from each well to a 1.5 mL Eppendorf tube.
8. Centrifuge the cell lysate in Eppendorf tubes at 1,500 × *g* for 2 min.
9. Transfer the supernatant to a new 1.5 mL Eppendorf tube for analysis. This cell lysate can be stored at −80°C for 5 days without significant changes in reporter signal.

*3.5.1. Luciferase Assay*

1. Prepare luciferase substrate solution by adding 10 mL of luciferase assay buffer A and 100 mL of luciferase assay buffer B to a vial of lyophilized Luciferase Assay Substrate (Sigma-Aldrich). Dispense into working aliquots. Unused luciferase substrate solution can be stored at −80°C. Equilibrate the solution to room temperature each time before use.
2. Luciferase assay can be analyzed on a Micro Lumat EG&G luminescence photometer (Berthold Technologies Ltd.).
3. Transfer 20 μL of cell lysate to each well of a white 96-well plate.

4. Wash the luminescence photometer with PBS buffer followed by rinse with luciferase substrate solution.
5. Set the instrument to add 100 μL of luciferase substrate solution to 20 μL of cell lysate in each well of the 96-well plate.
6. Set the delay time (time after adding luciferase substrate solution and before recording luminescence) to 2 s and the luminescence integration time 10 s.
7. Start the instrument (see Note 19).

*3.5.2. β-Galactosidase Assay*

1. Prepare β-galactosidase substrate solution by mixing 2.2 mL of 4 mg/mL ONPG phosphate solution with 0.1 mL of 100× magnesium chloride solution and 7.7 mL of 100 mM phosphate buffer at pH 7.0. Dispense into working aliquots. Unused galactosidase substrate solution can be stored at −80°C. Equilibrate the solution to room temperature each time before use.
2. β-galactosidase assay can be analyzed on an ELISA plate reader at 405 nm.
3. Preset a kinetic run with intervals of 30 s with 5 s shaking between cycles.
4. Set the number of total kinetic cycles to 10 and the measuring temperature to 25°C.
5. Transfer 20 μL of the cell lysate to each well of a clear bottom transparent 96-well plate.
6. Add 130 μL of β-galactosidase substrate solution to the cell lysate. Immediately place the plate in the plate reader and start the kinetic measurement (see Note 20).

## 4. Notes

1. BSBCA is most stable in the solid form. Aqueous aliquots should be only prepared for short-term usage. Always store BSBCA in the dark to prevent photo side reactions.
2. In order to produce high-quality polyacrylamide gels, we suggest to prepare ammonium persulfate solution within the same day of use.
3. A high peptide concentration is key for obtaining high cross-linking efficiency. Peptide concentration lower than 0.1 mM requires a long reaction time to reach completion, during which unfavorable side reactions may happen.
4. BSBCA cross-linking of short helical peptide usually reacts smoothly with >90% yield. Common byproducts include peptides with only one cysteine reacted to one chloroacetyl group of the BSBCA molecule leaving the other cysteine unreacted, as

well as peptides with two cysteine side-chains linked to two BSBCA molecules. In both cases, the other chloroacetyl group is unreacted indicating hydrolysis of the chloroacetyl group is slow under experimental condition. One BSBCA molecule linking two peptides together was never observed, probably because peptide concentrations were never too high. If the cross-linked peptide is not observed after 10 h or if the reaction has a low yield, the following steps may increase the yield: (a) increase the reaction temperature to 45°C, (b) irradiate the reaction flask with a 365 nm UV LED, (c) add guanidinium chloride or urea to increase peptide backbone flexibility.

5. If the cross-linked peptide precipitates from aqueous solution, add ethanol (up to 50%) to dissolve it.
6. Under this condition, ~87% of *cis* isomer can be obtained.
7. When the concentration of the cross-linked peptide is too high, it is difficult to irradiate the whole sample cuvette evenly. If a high concentration is necessary under special situation, an extended irradiation time and mixing during irradiation is recommended.
8. BSBCA *cis*-to-*trans* isomerization is much more efficient than *trans*-to-*cis* isomerization. Irradiation with a 460 nm blue light for 2 min is enough to produce >98% trans form.
9. Alternatively, radio-labeled DNA probe can be used followed by scintillation counting.
10. The 5 min incubation time allows proper protein/DNA complexes to form.
11. In order to test the inhibition effect in the most sensitive manner, we recommend to use a target bZIP protein concentration at which 80–90% of DNA is in the bound form.
12. An extended incubation time may be needed if the competition of a dominant negative peptide with the target bZIP is slow.
13. Thermal relaxation of BSBCA *cis* isomer to trans is much slower at 4°C (half life of 7 h) than 37°C (half life of 17 min). We recommend 4°C in order to maintain the *cis* isomer.
14. The 1 h irradiation time is to compromise the potentially slow competition of dominant negative peptides with the target bZIP protein at 4°C.
15. Coating 96-well plates with poly-L-lysine significantly increases HEK293T cell attachment thus improves data reproducibility.
16. The 6-h incubation time allows efficient transfection to occur. Alternatively, more Lipofectamine 2000 can be added to enhance transfection efficiency. However, doing so often leads to decreased cell viability due to the toxicity of Lipofectamine 2000.
17. High concentration of PMA is toxic to HEK293T cells.

18. Due to the fast thermal relaxation rate of BSBCA at 37°C (half life 17 min), it is very difficult to produce a high percentage of the *cis* isomer and retain it for a longer period of time while not introducing phototoxicity or other side-effects from the irradiation light. The irradiation pattern we described in this protocol produces ~50% *cis* isomer during irradiation. Under this condition, there is no detectable phototoxicity nor UV exposure induced endogenous bZIP response.
19. If the luminescence is too strong, decrease the integration time or dilute the cell lysate, make sure the luciferase substrate is not all consumed during data recording time. If the luminescence is too weak, increase integration time and the volume of cell lysate. Otherwise, modify the reporter DNA construct.
20. If the slope of absorption-over-time is not linear, the activity of β-galactosidase is probably too strong. Dilute the cell lysate or decrease the kinetic interval to have more frequent time points. If there is little change in absorbance at 405 nm, the β-galactosidase activity is too weak. Add more cell lysate or increase the kinetic interval time.

## Acknowledgments

The authors wish to thank Katharina Timm and Barbara Sauer for technical assistance. This work has been supported by the Canadian Institutes for Health Research (F.Z.), the Natural Sciences and Engineering Research Council of Canada (NSERC, Post-Doctoral Fellowship, F.Z.), by the Emmy-Noether Program of the German Research Foundation (DFG) (Ar 373/1-3; K.M.A.), by the German Academic Exchange Service (DAAD; K.M.A.) and by the Excellence Initiative of the German Federal and State Governments (EXC 294; F.Z., and K.M.A.).

### References

1. Zhang, F., Timm, K. A., Arndt, K. M. & Woolley, G. A. (2010). Photocontrol of coiled-coil proteins in living cells. *Angew Chem Int Ed Engl* **49**, 3943–3946.
2. Mason, J. M., Müller, K. M. & Arndt, K. M. (2007). Considerations in the design and optimization of coiled coil structures. *Methods Mol Biol* **352**, 35–70.
3. Mason, J. M. & Arndt, K. M. (2004). Coiled coil domains: stability, specificity, and biological implications. *Chembiochem* **5**, 170–176.
4. Arndt, K. M., Pelletier, J. N., Müller, K. M., Plückthun, A. & Alber, T. (2002). Comparison of in vivo selection and rational design of heterodimeric coiled coils. *Structure* **10**, 1235–1248.
5. Arndt, K. M., Pelletier, J. N., Müller, K. M., Alber, T., Michnick, S. W. & Plückthun, A. (2000). A heterodimeric coiled-coil peptide pair selected in vivo from a designed library-versus-library ensemble. *J Mol Biol* **295**, 627–639.

6. Pelletier, J. N., Arndt, K. M., Plückthun, A. & Michnick, S. W. (1999). An in vivo library-versus-library selection of optimized protein-protein interactions. *Nat Biotechnol* **17**, 683–690.
7. Mason, J. M., Müller, K. M. & Arndt, K. M. (2008). iPEP: peptides designed and selected for interfering with protein interaction and function. *Biochem Soc Trans* **36**, 1442–1447.
8. Mason, J. M., Schmitz, M. A., Müller, K. M. & Arndt, K. M. (2006). Semirational design of Jun-Fos coiled coils with increased affinity: Universal implications for leucine zipper prediction and design. *Proc Natl Acad Sci USA* **103**, 8989–8994.
9. Jouaux, E. M., Timm, B. B., Arndt, K. M. & Exner, T. E. (2009). Improving the interaction of Myc-interfering peptides with Myc using molecular dynamics simulations. *J Pept Sci* **15**, 5–15.
10. Gaida, A., Hagemann, U. B., Mattay, D., Räuber, C., Müller, K. M. & Arndt, K. M. (2009). Selection of peptides interfering with protein-protein interaction. *Methods Mol Biol* **535**, 263–291.
11. Jouaux, E. M., Schmidtkunz, K., Müller, K. M. & Arndt, K. M. (2008). Targeting the c-Myc coiled coil with interfering peptides. *J Pept Sci* **14**, 1022–1031.
12. Hagemann, U. B., Mason, J. M., Müller, K. M. & Arndt, K. M. (2008). Selectional and mutational scope of peptides sequestering the Jun-Fos coiled-coil domain. *J Mol Biol* **381**, 73–88.
13. Mason, J. M., Müller, K. M. & Arndt, K. M. (2007). Positive aspects of negative design: simultaneous selection of specificity and interaction stability. *Biochemistry* **46**, 4804–4814.
14. Mason, J. M., Hagemann, U. B. & Arndt, K. M. (2007). Improved stability of the Jun-Fos Activator Protein-1 coiled coil motif: A stopped-flow circular dichroism kinetic analysis. *J Biol Chem* **282**, 23015–23024.
15. Olive, M., Krylov, D., Echlin, D. R., Gardner, K., Taparowsky, E. & Vinson, C. (1997). A dominant negative to activation protein-1 (AP1) that abolishes DNA binding and inhibits oncogenesis. *J Biol Chem* **272**, 18586–18594.
16. Burns, D. C., Zhang, F. & Woolley, G. A. (2007). Synthesis of 3,3'-bis(sulfonato)-4,4'-bis(chloroacetamido)azobenzene and cysteine cross-linking for photo-control of protein conformation and activity. *Nat Protoc* **2**, 251–258.
17. Morgan, S. A., Al-Abdul-Wahid, S. & Woolley, G. A. (2010). Structure-based design of a photocontrolled DNA binding protein. *J Mol Biol* **399**, 94–112.
18. Beharry, A. A., Sadovski, O. & Woolley, G. A. (2008). Photo-control of peptide conformation on a timescale of seconds with a conformationally constrained, blue-absorbing, photo-switchable linker. *Org Biomol Chem* **6**, 4323–4332.
19. Woolley, G. A. (2007). Channel-forming activity of alamethicin: effects of covalent tethering. *Chem Biodivers* **4**, 1323–1337.
20. Woolley, G. A., Jaikaran, A. S., Berezovski, M., Calarco, J. P., Krylov, S. N., Smart, O. S. & Kumita, J. R. (2006). Reversible photocontrol of DNA binding by a designed GCN4-bZIP protein. *Biochemistry* **45**, 6075–6084.
21. Woolley, G. A. (2005). Photocontrolling peptide alpha helices. *Acc Chem Res* **38**, 486-493.
22. Sadovski, O., Beharry, A. A., Zhang, F. & Woolley, G. A. (2009). Spectral tuning of azobenzene photoswitches for biological applications. *Angew Chem Int Ed Engl* **48**, 1484–1486.
23. Zhang, F., Sadovski, O. & Woolley, G. A. (2008). Synthesis and characterization of a long, rigid photoswitchable cross-linker for promoting peptide and protein conformational change. *Chembiochem* **9**, 2147–2154.
24. Chi, L., Sadovski, O. & Woolley, G. A. (2006). A blue-green absorbing cross-linker for rapid photoswitching of peptide helix content. *Bioconjug Chem* **17**, 670–676.
25. Zhang, Z., Burns, D. C., Kumita, J. R., Smart, O. S. & Woolley, G. A. (2003). A water-soluble azobenzene cross-linker for photocontrol of peptide conformation. *Bioconjug Chem* **14**, 824–829.
26. Grigoryan, G., Reinke, A. W. & Keating, A. E. (2009). Design of protein-interaction specificity gives selective bZIP-binding peptides. *Nature* **458**, 859–864.

# Chapter 13

# Expressed Protein Modifications: Making Synthetic Proteins


## Birgit Wiltschi



## Abstract

Techniques to manipulate cellular gene expression such that amino acid analogs not encoded by the genetic code are incorporated into a polypeptide chain have recently gained increasing interest. The so-called noncanonical amino acids often have unusual properties that can be translated into target proteins by reprogrammed ribosomal protein synthesis. Residue-specific substitution of a specific canonical amino acid by its analogs provokes global effects in the resulting protein congeners that include improved stability or catalytic activity, reduced redox sensitivity, as well as altered spectral properties. Thus, the approach holds great promise for the engineering of synthetic proteins.

This contribution describes a protocol for the incorporation of a noncanonical amino acid into a target protein expressed in an appropriate amino acid auxotrophic *E. coli* strain.



**Key words:** Aminoacyl-tRNA synthetase, Auxotrophy, Expressed protein modification, New minimal medium, Noncanonical amino acid, Protein engineering, Residue-specific incorporation, Substrate tolerance, Unnatural amino acid


## 1. Introduction

Gene expression in living cells can be manipulated at several steps. The time when a certain mRNA is transcribed from a coding DNA sequence as well as its abundance can be regulated by the use of inducible promoters of varying strength (1–3). Regulatory systems that rely on alternative mRNA structures, the so-called RNA switches (4–6), or internal ribosome entry sites (IRES) (7, 8) enable to manipulate the translation of the genetic information into proteins. Finally, the introduction of protein building blocks beyond the 20 canonical amino acids that are prescribed by the standard genetic code permits control at the topmost level of gene expression, directly at the ribosome.


Wilfried Weber and Martin Fussenegger (eds.), *Synthetic Gene Networks: Methods and Protocols*,
Methods in Molecular Biology, vol. 813, DOI 10.1007/978-1-61779-412-4_13, © Springer Science+Business Media, LLC 2012

Noncanonical amino acids (NCAAs) are not encoded by the standard genetic code, yet they are used as protein building blocks by the cellular translation machinery under strictly controlled conditions (9). The site-specific introduction of NCAAs in response to an amber stop codon provokes mostly local ramifications (10–13) while the residue-specific substitution of one or more (14, 15) canonical amino acids by their noncanonical analogs usually has global effects (16–21). For the first method, the insertion of an in-frame amber stop codon into the coding sequence of the target protein is vital. For the latter no DNA mutagenesis is necessary. The coding DNA sequence is transiently interpreted by an alternative genetic code and protein congeners are formed (22). In this way, modifications can be introduced into a target protein directly at the amino acid sequence level (expressed protein modifications).

For approval by the cellular translation machinery, NCAAs must be structurally similar to their canonical counterparts although their chemical properties may be quite different (20, 23). In fact, a whole plethora of NCAAs with interesting and useful chemical side chains occur naturally (24, 25). The substances usually fit well into the three-dimensional structure of a protein (26–29) yet expressed protein modifications elicit remarkable functional consequences. These encompass the manipulation of the fluorescence properties of proteins (30–38), of their redox properties (39), or of the pH sensitivity (40), as well as the improvement of protein stability (41–43), or the amelioration of the catalytic activity of enzymes (22, 44–49). Expressed protein modifications can be an extension to classical protein engineering that relies on mutagenesis within the canonical pool of amino acids. Due to the limited number and diversity of available protein congeners, however, the predictability of the effect(s) of certain NCAAs upon incorporation into a target protein is currently low.

For the residue-specific incorporation of NCAAs (see Fig. 1 for details), an inducible expression construct of the target protein is required. It may be fused to a tag for facilitated purification. The expression strain must be auxotrophic for the specific canonical amino acid whose analog will be incorporated into the target protein. This strain will not grow unless it is supplemented with the specific canonical amino acid, which permits to control its free intracellular pool available for protein translation. Cells do not grow on NCAAs; in fact, many NCAAs are cytotoxic (50). Thus, a two step procedure must be followed. In the first step, cell mass accumulates in the presence of the specific canonical amino acid. During the following second step, the target protein is expressed in the presence of the NCAA while the specific canonical amino acid is absent. During both steps, a strict control of the availability of the canonical amino acid is necessary, which requires the use of defined synthetic medium.

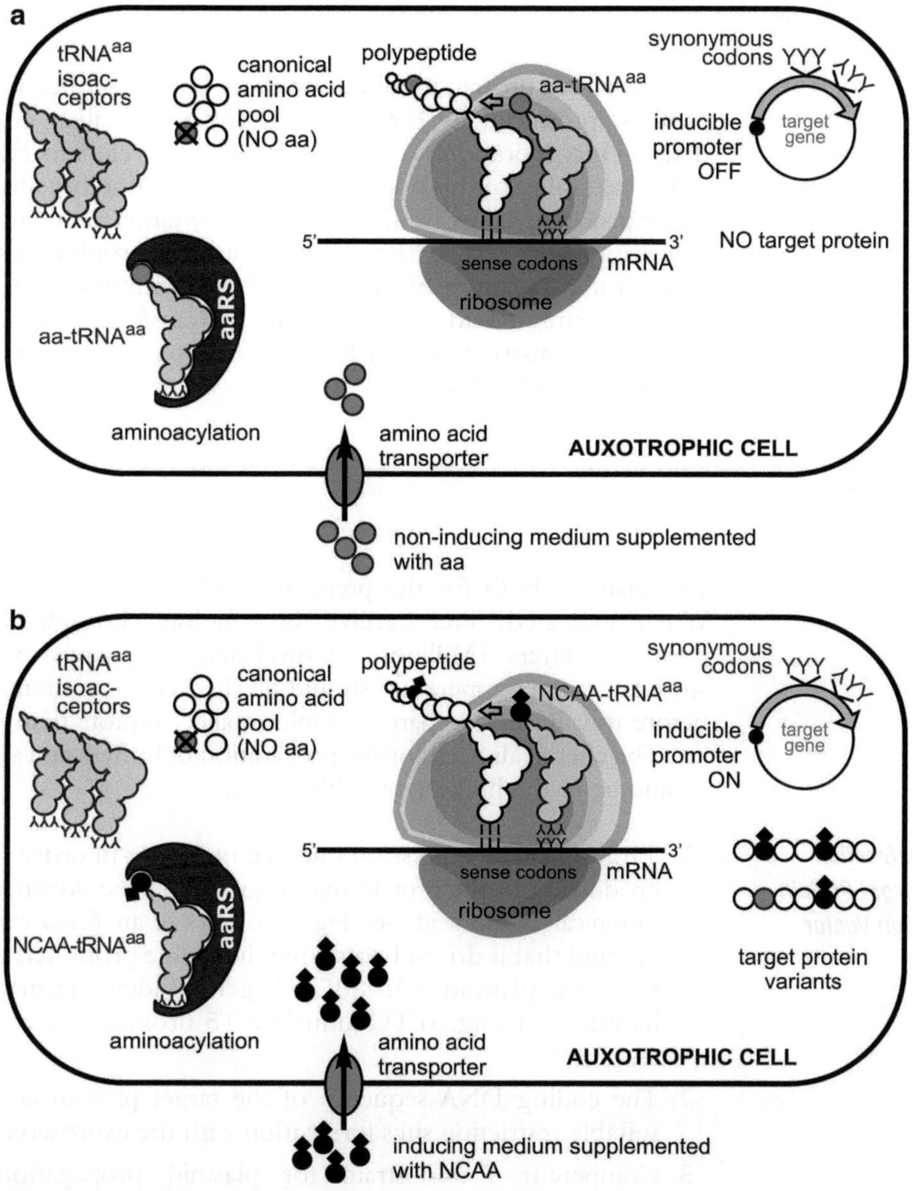


Fig. 1. Residue-specific incorporation of noncanonical amino acids. (**a**) Step one, accumulation of cell mass. The auxotrophic host cell cannot grow unless the required amino acid *aa* is supplemented in the medium since it is absent from the intracellular canonical amino acid pool (*NO aa*). Amino acid *aa* is taken up into the cell by an amino acid transporter and is charged onto its cognate *tRNA*$^{aa}$ by a specific aminoacyl-tRNA synthetase (*aaRS*). Whether only one or several isoacceptor tRNAs are available depends on the number of *synonymous codons* specifying *aa*. The aminoacylated *aa-tRNA*$^{aa}$ molecules participate in translation at the ribosome and *aa* is incorporated into the nascent polypeptide. The amount of *aa* in the medium is limited such that it gets depleted by the cells in mid-log. As soon as the canonical amino acid is depleted from the medium, translation stalls and the cells stop growing, i.e., the culture becomes stationary. (**b**) Step two, protein production. The medium is supplemented with an *NCAA* instead of *aa* and the expression of the target protein is turned on. The analog can be taken up into the cells by the relatively unspecific amino acid transporters (9, 60). *aaRS* will tolerate *NCAA* as its substrate if the analog is chemically and sterically not too dissimilar to *aa*. However, *aa* must accumulate intracellularly to a level that allows its aminoacylation to the *tRNA(s)*$^{aa}$ by *aaRS*. The mis-aminoacylated *NCAA-tRNA*$^{aa}$ molecules participate in ribosomal translation and *NCAA* is incorporated into the nascent polypeptide. The stringency of the culture conditions during analog incorporation impinges on the labeling efficiency of the target protein variants with *NCAA*. For instance, the turnover of cellular proteins synthesized during step one liberates *aa*. Since *aa* is a much better substrate for *aaRS* than *NCAA*, it will successfully compete with the analog for aminoacylation. This undesirable effect becomes more pronounced with prolonged induction periods.

Expressed protein modifications have been described for target proteins produced in *E. coli*, yeasts, and mammalian cells. The detailed incorporation protocol described here refers to *E. coli* due to the versatility of this popular protein production organism. The major procedures involve the cloning of the target gene into a suitable expression vector, the choice of an auxotrophic expression strain, the assessment of the minimal concentration of a specific canonical amino acid for this strain, its transformation with the expression construct and finally the two step procedure for incorporation of an NCAA.

## 2. Materials

Use distilled $H_2O$ for the preparation of all buffers and media. Where indicated, filter sterilize the solutions through 0.22 μM pore size filters (Millipore, Schwalbach, Germany). Solutions stored at room temperature should be checked for contaminations before use. Rather prepare multiple smaller aliquots than one big lot. The chemicals used for the preparation of buffers and solutions should be of the highest available purity.

### 2.1. Construction of the Target Protein Expression Vector

1. Target protein expression must be inducible in order to avoid production of the protein during growth in the presence of the canonical amino acid (see Fig. 1a). Choose an *E. coli* expression plasmid that is driven by a strong, inducible promoter. We routinely use plasmid pQE-80L (Qiagen, Hilden, Germany) that harbors a strong, IPTG-inducible T5 promoter (see Notes 1 and 2).
2. The coding DNA sequence of the target protein containing suitable restriction sites for ligation with the expression vector.
3. Competent *E. coli* strain for plasmid propagation, e.g., DH5alpha (Takara Bio Inc., Otsu, Japan), TOP10 (Invitrogen GmbH, Karlsruhe, Germany), or XLIBlue (Agilent Technologies, Palo Alto, CA).

### 2.2. Assessing the Limiting Concentration of a Specific Canonical Amino Acid for the Expression Strain of Choice

1. Choose an appropriate amino acid auxotrophic expression strain, e.g., if a methionine analog is to be incorporated, the strain must be auxotrophic for methionine (see Note 3).
2. LB medium (5 g/L yeast extract (BD Difco, Franklin Lakes, NJ), 10 g/L tryptone (BD Difco)); LB is a complex medium, it contains all canonical amino acids. Autoclave and store at room temperature.
3. NMM medium is synthetic and contains defined components; individual amino acids can be omitted. Always prepare fresh from sterile stock solutions (see Table 1), do not store.

**Table 1**
**Composition of the synthetic new minimal medium (NMM, (61)). Dissolve the components of the stock solutions in $H_2O$. To prepare the medium, calculate the required volume of $H_2O$ and autoclave. After cooling, add the stock solutions to yield the final concentration of each component and mix. All stock solutions are stored at room temperature in the dark. Check for contaminations by visual inspection each time before use**

| Component | Stock concentration | Stock ×-fold | Preparation of the stock solution | Final concentration |
|---|---|---|---|---|
| $(NH_4)_2SO_4$ | 1 M | 133× | 1 L; autoclave, store at room temperature | 7.5 mM |
| NaCl | 5 M | 588× | 1 L; autoclave, store at room temperature | 8.5 mM |
| $KH_2PO_4$ | 1 M | 45.5× | 1 L; autoclave, store at room temperature | 22 mM |
| $K_2HPO_4$ | 1 M | 20× | 1 L; autoclave, store at room temperature | 50 mM |
| $MgSO_4$ | 1 M | 1,000× | 1 L; autoclave, store at room temperature | 1 mM |
| D-Glucose monohydrate | 1 M | 50× | 1 L; autoclave, store at room temperature | 20 mM |
| Canonical L-amino acids (see Note 4) | 0.5 g/L each | 10× | 1 L; filter sterilize, store in 50 mL aliquots at room temperature | 50 mg/L |
| $CaCl_2$ | 1 mg/mL | 1,000× | 50 mL; filter sterilize, store at room temperature | 1 mg/L |
| $FeCl_2$ | 1 mg/mL | 1,000× | 50 mL; filter sterilize, store at room temperature | 1 mg/L |
| Trace elements ($CuSO_4$, $ZnCl_2$, $MnCl_2$, $(NH_4)_2MoO_4$) | 10 μg/mL each | 1,000× | 1 L; autoclave, store at room temperature | 10 μg/L |
| Thiamine | 10 mg/mL | 1,000× | 50 mL; filter sterilize, store at room temperature | 10 mg/L |
| Biotin (see Note 5) | 10 mg/mL | 1,000× | 50 mL; filter sterilize, store at room temperature | 10 mg/L |

4. Sterile baffled 250 mL and 500 mL flasks.
5. Sterile 50 mL tubes; sterile disposable 10 mL culture tubes (Greiner Bio-One GmbH, Frickenhausen, Germany); disposable plastic cuvettes; spectrophotometer.
6. 50, 5, and 0.5 mM solutions of the canonical amino acid to be tested in $H_2O$; filter sterilize; store at room temperature in the dark.

### 2.3. Transformation of the Expression Strain with the Target Protein Expression Vector

1. LB medium (see Subheading 2.2, item 2 for ingredients) and LB plates. To prepare the plates, add 20 g/L agar (BD Difco) to the LB medium and add the appropriate antibiotic (see Note 6) for the selection of transformants.
2. Sterile 100 mL and 1 L flasks.
3. Autoclaved centrifuge buckets.
4. 10% (v/v) glycerol; autoclave and store at 4°C. Prechill on ice before use.
5. Target protein expression plasmid in salt-free solution (see Note 7).
6. Sterile electroporation cuvettes, 0.1 cm gap; electroporator suitable for electroporation of bacteria.

### 2.4. Incorporation of an NCAA into the Target Protein

1. NMM medium containing a specific canonical amino acid at the limiting concentration; do not store, prepare fresh from stock solutions (see Table 1 for details).
2. 50 mM solution of the NCAA in $H_2O$, filter sterilized; store at room temperature in the dark.
3. 1M isopropyl β-D-1-thiogalactopyranoside (IPTG, Sigma, Steinheim, Germany) in $H_2O$, filter sterilize and store 1 mL aliquots at −20°C.

### 2.5. Analysis of the Analog Incorporation Efficiency

Mass analysis (51) is a convenient way to analyze the extent of the substitution of a specific canonical amino acid by the noncanonical analog. It takes advantage of the mass shift introduced by the NCAA. Instrumentation for mass analysis depends on the preferred method.

## 3. Methods

### 3.1. Construction of the Target Protein Expression Vector

1. Cut your expression vector of choice (see Notes 1, 2, and 8) with the appropriate restriction enzymes.
2. Cleave the coding DNA sequence of your target protein with the same restriction enzymes.
3. Gel purify the restricted DNA fragments using a commercially available DNA purification kit.
4. Ligate the purified vector backbone and insert. Typically, ligations a performed in a total volume of 10 μL containing 1 μL T4 DNA ligase and 10–100 ng DNA (vector plus insert). Use insert:vector molar ratios between 2:1 and 6:1.
5. Transform commercially available competent *E. coli* cells with the ligation sample (see Note 9).
6. Isolate the expression construct and verify by sequencing.

### *3.2. Assessing the Limiting Concentration of a Specific Canonical Amino Acid for the Expression Strain of Choice*

1. Grow an overnight culture of your expression strain of choice in LB medium (see Note 10).
2. Inoculate 100 mL NMM containing all canonical amino acids in a 250 mL baffled flask with 100 μL of the overnight culture (see Note 11).
3. Incubate at 37°C with vigorous shaking and record a growth curve by monitoring the optical density at 600 nm ($OD_{600}$) over time, until the culture enters the stationary growth phase (see Note 12).
4. From the growth curve, deduce the mid-log $OD_{600}$ and the time that the cells need to reach it (see Fig. 2a).
5. Inoculate 20 mL NMM lacking the desired canonical amino acid in a sterile 50 mL tube with 20 μL of the overnight culture from step 1 (see Note 13).
6. Dispense 2 mL of this cell suspension each into eight sterile 10 mL culture tubes.
7. Add the canonical amino acid to a final concentration of 0, 10, 20, 40, 80, 160, 300, and 500 μM (see Note 14).
8. Incubate overnight at 37°C with shaking.
9. Determine the final $OD_{600}$ of each culture the next day and plot the canonical amino acid concentration versus cell density ($OD_{600}$). A typical amino acid concentration versus $OD_{600}$ curve is shown in Fig. 2b.
10. Deduce the specific amino acid concentration that is necessary for your strain to reach mid-log (see Fig. 2b).

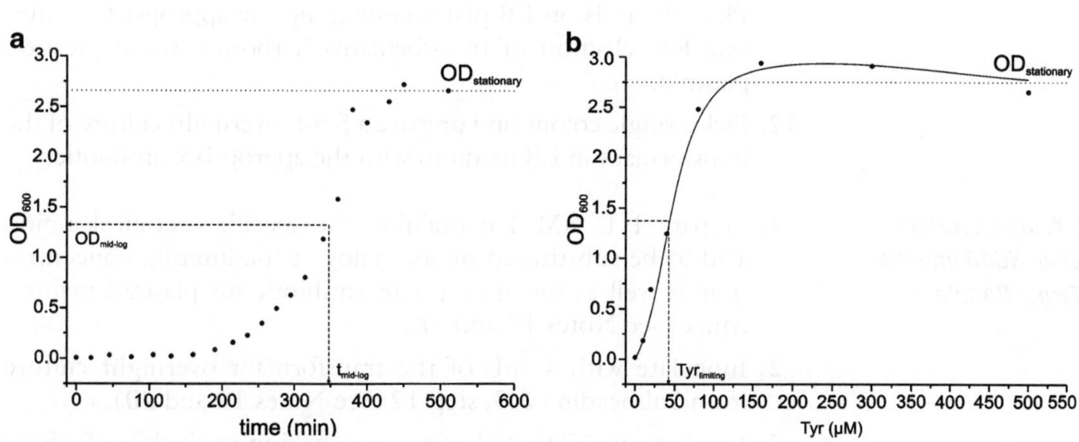


Fig. 2. Determination of the limiting concentration of a specific canonical amino acid. (**a**) Generic growth curve of a tyrosine auxotrophic *E. coli* strain in NMM containing all canonical amino acids. The optical densities at mid-log and in stationary phase ($OD_{mid\text{-}log}$ and $OD_{stationary}$), as well as the time that the auxotroph requires to reach mid-log ($t_{mid\text{-}log}$) are indicated. (**b**) A typical concentration *vs.* $OD_{600}$ curve for the assessment of the limiting concentration of a specific canonical amino acid, here tyrosine (Tyr), is shown. For facilitated determination of the limiting Tyr concentration, the data points were approximated by a curve. The limiting Tyr concentration ($Tyr_{limiting}$) is indicated.

11. Inoculate 20 mL NMM containing the specific canonical amino acid at the limiting concentration with 20 μl overnight culture from step 1 and grow at 37°C. The cells should stop growing at mid-log, after the time deduced from the growth curve.

### 3.3. Transformation of the Expression Strain with the Target Protein Expression Vector

1. In a 100 mL flask, grow a 20 mL overnight culture of your desired strain in LB medium at 37°C.
2. Inoculate 500 mL LB medium in a 1 L flask with 15 mL of the overnight culture.
3. Grow at 37°C with vigorous shaking to an $OD_{600}$ 0.7–0.8.
4. Harvest the cells by low speed centrifugation (5,000 × *g*) for 15 min at 4°C.
5. Discard the supernatant and gently wash the cells with 500 mL sterile, chilled 10% (v/v) glycerol (see Note 15).
6. Repeat the centrifugation and washing steps as indicated above.
7. After the second wash, resuspend the cells in the remaining 10% (v/v) glycerol and prepare 40 μL aliquots on ice.
8. Flash freeze the aliquots in liquid nitrogen and store at -80°C.
9. For electroporation, thaw an aliquot on ice and mix with 10–100 ng target protein expression construct (see Note 7). Transfer the mix to a prechilled electroporation cuvette.
10. Electroporate by following the instructions of the electroporator manual and recover cells by incubating in 1 mL LB medium at 37°C for 1 h with gentle shaking (see Note 16).
11. Plate the cells on LB plates containing the appropriate antibiotic for selection of transformants harboring the expression plasmid.
12. Pick a single colony and prepare a 5 mL overnight culture of the transformant in LB medium with the appropriate antibiotic.

### 3.4. Incorporation of an NCAA into the Target Protein

1. Prepare 1 L NMM containing the specific canonical amino acid to be substituted by its analog at the limiting concentration as well as the appropriate antibiotic for plasmid maintenance (see Notes 17 and 18).
2. Inoculate with 1 mL of the transformant overnight culture from Subheading 3.3, step 12 (see Notes 19 and 20).
3. Incubate at 37°C with vigorous shaking until the cells have grown to mid-log.
4. Determine $OD_{600}$ and repeat after 20 min; the canonical amino acid has been used up by the cells if the cell density remains constant (see Note 21).

5. Add the NCAA to a final concentration of 0.5 mM and induce the target protein expression by the addition of 1 mM IPTG.
6. Perform the target gene expression in the presence of the non-canonical analog at 30°C for 4 h with vigorous shaking (see Note 22).
7. Harvest the culture and freeze the cell pellet at −80°C until purification of the target protein (see Note 21).

### 3.5. Analysis of the Analog Incorporation Efficiency

We routinely perform liquid chromatography-coupled electrospray ionization mass spectrometry (LC-ESI-MS, (51)) to determine the extent of analog incorporation. This mass analysis technique is fast and conveniently permits the analysis of intact proteins.

Under tightly controlled incorporation conditions, the amino acid substitution by the analog is usually quantitative and the variant protein yield does not differ grossly from that of the unlabeled parent protein. However, the incorporation efficiency as well as the yield greatly depends on the target protein and on the NCAA. A comprehensive analysis in this regard can be found in ref. 22.

## 4. Notes

1. The functional *E. coli* RNA polymerase that is produced during the growth phase in the presence of the canonical amino acid is stable enough to drive the expression of large amounts (22) of target protein from the T5 promoter on pQE-80L during the NCAA incorporation phase. pQE-80L harbors the *lac* repressor *lacI*. If your expression plasmid of choice does not contain a *lacI* gene, co-transformation with a plasmid for constitutive *lacI* expression, such as pREP4 (Qiagen) is necessary to ensure a high level of *lac* repressor for tight repression of protein expression during the growth phase in the presence of the canonical amino acid.
2. Expression plasmids relying on the T7 system, such as pET vectors (Novagen Merck Chemicals Ltd., Nottingham, UK), are less suitable. pET vectors are employed with DE3 expression strains. Addition of lactose analogs, such as IPTG to cultures of these strains induces the expression of T7 RNA polymerase (52) under the *lacUV5* promoter on the DE3 lysogen in the chromosome of the host (53). The T7 RNA polymerase subsequently induces transcription of the target gene under control of the strong bacteriophage T7 promoter/*lac* operator in the pET plasmid (54). Induction of T7 RNA polymerase occurs simultaneously to that of the target gene, and thus, to the addition of the NCAA to the culture. The NCAA can be incorporated into all newly synthesized proteins in

addition to the target protein. Consequently, it is incorporated also into the induced T7 RNA polymerase, which may have deleterious effects on enzyme activity (55). This undesirable effect is not restricted to the T7 expression system but applies to any expression system, where an essential protein component is co-induced with the target protein in the presence of an NCAA.

3. The commercially available *E. coli* strain B834(DE3) (Novagen Merck Chemicals Ltd., Nottingham, UK) is a methionine auxotroph. Other amino acid auxotrophic strains are available from public strain collections, e.g., *E. coli* Strain National BioResource Project (http://www.shigen.nig.ac.jp/ecoli/strain/top/top.jsp), and the *E. coli* Genetic Resources at Yale (CGSC, The Coli Genetic Stock Center http://cgsc.biology.yale.edu/).

4. All listed amino acids are from Roth (Carl Roth GmbH, Karlsruhe, Germany).

   L-Alanine (≥99%), L-arginine (≥99%), L-asparagine monohydrate (≥99%), L-aspartic acid (≥98,5%), L-cysteine (≥98%), L-glutamic acid (≥99%), L-glutamine (≥99%), glycine (≥99%), L-histidine (≥98,5%), L-isoleucine (≥98,5%), L-leucine (≥98,5%), L-lysine hydrochloride (≥98,5%), L-methionine (≥99%), L-phenylalanine (≥98,5%), L-proline (≥98,5%), L-serine (≥98,5%), L-threonine (≥99%), L-tryptophan (≥99%), L-tyrosine (≥99%), L-valine (≥98,5%).

5. Biotin will only dissolve under basic conditions. Mix the substance with 40 mL $H_2O$ and stir, then add KOH pellets until the biotin is completely dissolved. Fill up to the final volume of 50 mL with $H_2O$.

6. Use the following antibiotics concentrations for transformant selection: 100 mg/L ampicillin; 50 mg/L kanamycin, 34 mg/L chloramphenicol, or 25 mg/L tetracycline.

7. Plasmid preparations should be essentially salt-free for electroporation in order to prevent arcing. Thus, use $H_2O$ for the elution step during plasmid isolation rather than buffer. To remove any salt, fill the bottom of a petri dish with $H_2O$. Use stainless steel forceps to float a small (∅ 13 mm) 0.025 μM pore size filter disc (Millipore) on the water surface. Carefully spot the desired volume of the plasmid solution onto the center of the filter disc. Dialyze for 15–20 min at room temperature. Close the lid of the petri dish during dialysis to prevent sample loss due to evaporation. Recover the dialyzed plasmid solution.

8. pQE-80L is good for the construction of N-terminal hexahistidine tag-fusions with the target protein. If another N-terminal tag is desired, or if the target protein shall be tagged at the C-terminus only, the single *Eco*RI site immediately downstream of the T5 promoter must be used for cloning. Note that the

restriction of pQE-80L with *Eco*RI removes the ribosome-binding site (56, 57). For proper translation initiation, the insert must contain an *Eco*RI site followed by a ribosome-binding site upstream of the coding sequence. The correct spacing between the ribosome-binding site and the translation start site is important (56). A C-terminal tag should be fused with the insert sequence, e.g., by PCR with appropriate primers, prior to insertion into pQE-80L. If you use an alternative tag, e.g., Strep-tag (58), consider a potential interference of the amino acid substitution with the tag function. Otherwise, choose an alternative tag if histidine analogs are incorporated into the target protein.

9. Alternatively, competent *E. coli* cells may be prepared and transformation performed as described in Subheading 3.3.
10. Store overnight cultures at 4°C for no longer than a week.
11. The baffled flask ensures proper aeration of the *E. coli* culture.
12. Repeatedly determine $OD_{600}$, during the lag phase perform an $OD_{600}$ measurement approximately every 30 min, during log phase every 15–20 min.
13. The dilution of the inoculum is 1:1,000; other dilutions may be used but the determined limiting canonical amino acid concentration applies only for the same dilution as an undefined amount of canonical amino acids is introduced with the LB medium of the inoculum.
14. In order to avoid pipetting errors due to very small volumes, prepare a 50 mM solution of the canonical amino acid in $H_2O$ and dilute to 5 and 0.5 mM with sterile $H_2O$. Based on our experience, the limiting canonical amino acid concentrations are usually in the micromolar range. The concentrations indicated here cover a relatively broad range in order get a rough idea of the limiting concentration of a specific canonical amino acid. For a more accurate determination, repeat the growth assay within a smaller concentration range.
15. As the cells are fragile, use large tips (1 mL or 5 mL) to resuspend and work on ice all the time.
16. For instance, apply a voltage of 1.5 kV, a resistance of 100 Ω and a capacitance of 25 μF when using the Gene Pulser apparatus (Bio-Rad Laboratories, Hercules, CA). It is important to immediately recover the pulsed cells with 1 mL LB medium still in the electroporation cuvette.
17. Use a baffled 2 L flask for proper aeration of the *E. coli* culture.
18. Autoclave the $H_2O$ in a baffled 2 L flask and add the other components after cooling. The medium should have room temperature or may be warmed to 37°C for inoculation. Cold medium severely delays cell growth.

19. It is important to use the same dilution of the inoculum as for assessing the limiting canonical amino acid concentration, otherwise, the cells will either not reach mid-log (inoculum too small) or will grow higher (too much inoculum) due to the unidentified contribution of canonical amino acids contained in the overnight LB culture.
20. With fast growing strains (starting from a 1:1,000 inoculum mid-log is reached within 6 h), analog incorporation can be performed in one working day. If the strain needs more than 6 h to reach mid-log, the main culture should be inoculated with the overnight culture late in the evening and induction should be performed early the next day in order to keep the starvation phase, i.e., the incubation of the cells in medium that is depleted for one canonical amino acid, as short as possible. The time between inoculation and induction should not exceed 12 h as the likelihood for the strain to lose the amino acid auxotrophy in depleted medium increases with time.
21. Optionally, sample 1 $OD_{600}$ of cells for later analysis by SDS-PAGE (59) before (noninduced culture) and at the end of the induction (induced culture).
22. The temperature during induction in the presence of the NCAA should be 30°C or lower. Target protein expression at 37°C usually does not lead to substantially increased protein amounts yet puts more stress on the expression strain. As well, the induction period should not be prolonged unnecessarily, e.g., overnight, as this can lead to the degradation of the target protein and higher amounts of only partially labeled target protein (see Fig. 1b).

## Acknowledgments

The author would like to thank Katrin Biermann for recording the growth curve and the determination of the limiting tyrosine concentration shown in Fig. 2. This work was supported by a grant from the Ministry of Science, Research and the Arts of Baden-Württemberg (*Az*: 33-720.830-6-04) to B. Wiltschi.

### References

1. Hammer, K., Mijakovic, I., and Jensen, P. R. (2006) Synthetic promoter libraries – tuning of gene expression, *Trends Biotechnol* 24, 53–55.
2. Mijakovic, I., Petranovic, D., and Jensen, P. R. (2005) Tunable promoters in systems biology, *Curr Opin Biotechnol* 16, 329–335.
3. Braatsch, S., Helmark, S., Kranz, H., Koebmann, B., and Jensen, P. R. (2008) Escherichia coli strains with promoter libraries constructed by Red/ET recombination pave the way for transcriptional fine-tuning, *Biotechniques* 45, 335–337.

4. Marzi, S., Fechter, P., Chevalier, C., Romby, P., and Geissmann, T. (2008) RNA switches regulate initiation of translation in bacteria, *Biol Chem* 389, 585–598.
5. Hartig, J. S. (2007) Teaching bacteria new tricks – with RNA switches, *Angew Chem Int Ed Engl* 46, 7741–7743.
6. Zhang, J., Lau, M. W., and Ferré-D'Amaré, A. R. (2010) Ribozymes and riboswitches: modulation of RNA function by small molecules, *Biochemistry* 49, 9123–9131.
7. Malys, N., and McCarthy, J. (2011) Translation initiation: variations in the mechanism can be anticipated, *Cell Mol Life Sci* 68, 991–1003.
8. Filbin, M. E., and Kieft, J. S. (2009) Toward a structural understanding of IRES RNA function, *Curr Opin Struct Biol* 19, 267–276.
9. Wiltschi, B., and Budisa, N. (2006) Basic requirements for reprogramming intracellular protein translation, In *Molecular Interactions – Bringing Chemistry to Life: Proceedings of the International Beilstein Workshop* (Hicks, M. G., and Kettner, C., Eds.), pp 93–109, Beilstein-Institut zur Förderung der Chemischen Wissenschaften, May 15th – 19th, Bozen, Italy.
10. Young, T. S., and Schultz, P. G. (2010) Beyond the canonical 20 amino acids: expanding the genetic lexicon, *J Biol Chem* 285, 11039–11044.
11. Liu, C. C., and Schultz, P. G. (2010) Adding new chemistries to the genetic code, *Annu Rev Biochem* 79, 413–444.
12. Wang, Q., Parrish, A. R., and Wang, L. (2009) Expanding the genetic code for biological studies, *Chem Biol* 16, 323–336.
13. Hassan, A. Q. (2008) Site-specific incorporation of chemical probes into proteins for NMR, *ACS Chem Biol* 3, 524–526.
14. Lepthien, S., Merkel, L., and Budisa, N. (2010) In vivo double and triple labeling of proteins using synthetic amino acids, *Angew Chem Int Ed Engl* 49, 5446–5450.
15. Merkel, L., Schauer, M., Antranikian, G., and Budisa, N. (2010) Parallel incorporation of different fluorinated amino acids: on the way to "teflon" proteins, *Chembiochem* 11, 1505–1507.
16. Arnold, U. (2009) Incorporation of non-natural modules into proteins: structural features beyond the genetic code, *Biotechnol Lett* 31, 1129–1139.
17. Rodgers, K. J., and Shiozawa, N. (2008) Misincorporation of amino acid analogues into proteins by biosynthesis, *Int J Biochem Cell Biol* 40, 1452–1466.
18. Link, A. J., and Tirrell, D. A. (2005) Reassignment of sense codons in vivo, *Methods* 36, 291–298.
19. Connor, R. E., and Tirrell, D. A. (2007) Non-canonical amino acids in protein polymer design, *Polymer Rev* 47, 9–28.
20. Budisa, N. (2004) Prolegomena to future experimental efforts on genetic code engineering by expanding its amino acid repertoire, *Angew Chem Int Ed Engl* 43, 6426–6463.
21. Johnson, J. A., Lu, Y. Y., Van Deventer, J. A., and Tirrell, D. A. (2010) Residue-specific incorporation of non-canonical amino acids into proteins: recent developments and applications, *Curr Opin Chem Biol* 14, 774–780.
22. Hoesl, M. G., Acevedo-Rocha, C. G., Nehring, S., Royter, M., Wolschner, C., Wiltschi, B., Budisa, N., and Antranikian, G. (2011) Lipase congeners designed by genetic code engineering, *ChemCatChem* 3, 213–221.
23. Budisa, N. (2006) *Engineering the genetic code – expanding the amino acid repertoire for the design of novel proteins*, Wiley-VCH Verlag GmbH & Co KGaA, Weinheim.
24. Rosenthal, G. (1982) *Plant nonprotein amino and imino acids: biological, biochemical and toxicological properties*, Academic Press, New York.
25. Hunt, S. (1985) The non-protein amino acids, In *Chemistry and biochemistry of the amino acids* (Barrett, G., Ed.), pp 55–137, Chapman and Hall, London.
26. Shoulders, M. D., Satyshur, K. A., Forest, K. T., and Raines, R. T. (2010) Stereoelectronic and steric effects in side chains preorganize a protein main chain, *Proc Natl Acad Sci USA* 107, 559–564.
27. Budisa, N., Minks, C., Medrano, F. J., Lutz, J., Huber, R., and Moroder, L. (1998) Residue-specific bioincorporation of non-natural, biologically active amino acids into proteins as possible drug carriers: Structure and stability of the per-thiaproline mutant of annexin V, *Proc Natl Acad Sci USA* 95, 455–459.
28. Pal, P. P., Bae, J. H., Azim, K. M., Hess, P., Friedrich, R., Huber, R., Moroder, L., and Budisa, N. (2005) Structural and spectral response of Aequorea victoria green fluorescent proteins to chromophore fluorination, *Biochemistry* 44, 3663–3672.
29. Minks, C., Huber, R., Moroder, L., and Budisa, N. (1999) Atomic mutations at the single tryptophan residue of human recombinant annexin V: effects on structure, stability, and activity, *Biochemistry* 38, 10649–10659.
30. Hoesl, M. G., Larregola, M., Cui, H., and Budisa, N. (2010) Azatryptophans as tools to study polarity requirements for folding of green fluorescent protein, *J Pept Sci* 16, 589–595.
31. Kurschus, F. C., Pal, P. P., Bäumler, P., Jenne, D. E., Wiltschi, B., and Budisa, N. (2009) Gold

fluorescent annexin A5 as a novel apoptosis detection tool, *Cytometry A* 75A, 626–633.

32. Budisa, N., and Pal, P. P. (2004) Designing novel spectral classes of proteins with a tryptophan-expanded genetic code, *Biol Chem* 385, 893–904.
33. Bae, J. H., Rubini, M., Jung, G., Wiegand, G., Seifert, M. H. J., Azim, M. K., Kim, J.-S., Zumbusch, A., Holak, T. A., Moroder, L., Huber, R., and Budisa, N. (2003) Expansion of the genetic code enables design of a novel "gold" class of green fluorescent proteins, *J Mol Biol* 328, 1071–1081.
34. Beatty, K. E., Fisk, J. D., Smart, B. P., Lu, Y. Y., Szychowski, J., Hangauer, M. J., Baskin, J. M., Bertozzi, C. R., and Tirrell, D. A. (2010) Live-cell imaging of cellular proteins by a strain-promoted azide-alkyne cycloaddition, *Chembiochem* 11, 2092–2095.
35. Dieterich, D. C., Hodas, J. J. L., Gouzer, G., Shadrin, I. Y., Ngo, J. T., Triller, A., Tirrell, D. A., and Schuman, E. M. (2010) In situ visualization and dynamics of newly synthesized proteins in rat hippocampal neurons, *Nat Neurosci* 13, 897–905.
36. Beatty, K. E., and Tirrell, D. A. (2008) Two-color labeling of temporally defined protein populations in mammalian cells, *Bioorg Med Chem Lett* 18, 5995–5999.
37. Yoo, T. H., Link, A. J., and Tirrell, D. A. (2007) Evolution of a fluorinated green fluorescent protein, *Proc Natl Acad Sci USA* 104, 13887–13890.
38. Lepthien, S., Hoesl, M. G., Merkel, L., and Budisa, N. (2008) Azatryptophans endow proteins with intrinsic blue fluorescence, *Proc Natl Acad Sci USA* 105, 16095–16100.
39. Wolschner, C., Giese, A., Kretzschmar, H. A., Huber, R., Moroder, L., and Budisa, N. (2009) Design of anti- and pro-aggregation variants to assess the effects of methionine oxidation in human prion protein, *Proc Natl Acad Sci USA* 106, 7756–7761.
40. Budisa, N., Rubini, M., Bae, J. H., Weyher, E., Wenger, W., Golbik, R., Huber, R., and Moroder, L. (2002) Global replacement of tryptophan with aminotryptophans generates non-invasive protein-based optical pH sensors, *Angew Chem Int Ed Engl* 41, 4066–4069.
41. Steiner, T., Hess, P., Bae, J. H., Wiltschi, B., Moroder, L., and Budisa, N. (2008) Synthetic biology of proteins: tuning GFPs folding and stability with fluoroproline, *PLoS ONE* 3, e1680.
42. Tang, Y., and Tirrell, D. A. (2001) Biosynthesis of a highly stable coiled-coil protein containing hexafluoroleucine in an engineered bacterial host, *J Am Chem Soc* 123, 11089–11090.
43. Montclare, J. K., Son, S., Clark, G. A., Kumar, K., and Tirrell, D. A. (2009) Biosynthesis and stability of coiled-coil peptides containing (2 S,4R)-5,5,5-trifluoroleucine and (2 S,4 S)-5, 5,5-trifluoroleucine, *Chembiochem* 10, 84–86.
44. Budisa, N., Wenger, W., and Wiltschi, B. (2010) Residue-specific global fluorination of Candida antarctica lipase B in Pichia pastoris, *Mol Biosyst* 6, 1630–1639.
45. Cirino, P. C., Tang, Y., Takahashi, K., Tirrell, D. A., and Arnold, F. H. (2003) Global incorporation of norleucine in place of methionine in cytochrome P450 BM-3 heme domain increases peroxygenase activity, *Biotechnol Bioeng* 83, 729–734.
46. Beiboer, S. H. W., Berg, B. V. D., Dekker, N., Cox, R. C., and Verheij, H. M. (1996) Incorporation of an unnatural amino acid in the active site of porcine pancreatic phospholipase A2. Substitution of histidine by 1,2,4-triazole-3-alanine yields an enzyme with high activity at acidic pH, *Protein Eng* 9, 345–352.
47. Parsons, J. F., Xiao, G., Gilliland, G. L., and Armstrong, R. N. (1998) Enzymes harboring unnatural amino acids: mechanistic and structural analysis of the enhanced catalytic activity of a glutathione transferase containing 5-fluorotryptophan, *Biochemistry* 37, 6286–6294.
48. Anfinsen, C. B., and Corley, L. G. (1969) An active variant of staphylococcal nuclease containing norleucine in place of methionine, *J Biol Chem* 244, 5149–5152.
49. Liu, X., Silks, L. A., Liu, C., Ollivault-Shiflett, M., Huang, X., Li, J., Luo, G., Hou, Y.-M., Liu, J., and Shen, J. (2009) Incorporation of tellurocysteine into glutathione transferase generates high glutathione peroxidase efficiency, *Angew Chem Int Ed Engl* 48, 2020–2023.
50. Fowden, L., Lewis, D., and Tristram, H. (1967) Toxic amino acids: their action as antimetabolites, In *Adv Enzymol Relat Areas Mol Biol* (Nord, F., Ed.), pp 89–163, John Wiley & Sons, Inc.
51. Speicher, D. W. (2002) Mass spectrometry, *Curr Protoc Protein Sci* Chapter 16, Unit 16.11.
52. Studier, W. F., Rosenberg, A. H., Dunn, J. J., and Dubendorff, J. W. (1990) Use of T7 RNA polymerase to direct expression of cloned genes, In *Methods Enzymol* (David, V. G., Ed.), pp 60–89, Academic Press.
53. Studier, F. W., and Moffatt, B. A. (1986) Use of bacteriophage T7 RNA polymerase to direct selective high-level expression of cloned genes, *J Mol Biol* 189, 113–130.

54. Rosenberg, A. H., Lade, B. N., Dao-shan, C., Lin, S.-W., Dunn, J. J., and Studier, F. W. (1987) Vectors for selective expression of cloned DNAs by T7 RNA polymerase, *Gene* 56, 125–135.
55. Ayyadurai, N., Neelamegam, R., Nagasundarapandian, S., Edwardraja, S., Park, H., Lee, S., Yoo, T., Yoon, H., and Lee, S.-G. (2009) Importance of expression system in the production of unnatural recombinant proteins in Escherichia coli, *Biotechnol. Bioprocess Eng.* 14, 257–265.
56. Kozak, M. (1983) Comparison of initiation of protein synthesis in procaryotes, eucaryotes, and organelles, *Microbiol Rev* 47, 1–45.
57. Shine, J., and Dalgarno, L. (1974) The 3′-terminal sequence of Escherichia coli 16 S ribosomal RNA: complementarity to nonsense triplets and ribosome binding sites, *Proc Natl Acad Sci USA* 71, 1342–1346.
58. Schmidt, T. G. M., and Skerra, A. (2007) The Strep-tag system for one-step purification and high-affinity detection or capturing of proteins, *Nat Protoc* 2, 1528–1535.
59. Laemmli, U. K. (1970) Cleavage of structural proteins during the assembly of the head of bacteriophage T4, *Nature* 227, 680–685.
60. Sophianopoulou, V., and Diallinas, G. (1995) Amino acid transporters of lower eukaryotes: regulation, structure and topogenesis, *FEMS Microbiol Rev* 16, 53–75.
61. Budisa, N., Steipe, B., Demange, P., Eckerskorn, C., Kellermann, J., and Huber, R. (1995) High-level biosynthetic substitution of methionine in proteins by its analogs 2-aminohexanoic acid, selenomethionine, telluromethionine and ethionine in Escherichia coli, *Eur J Biochem* 230, 788–796.

# Part III

## Optimizing the Host Systems

# Chapter 14

# Using Transcription Machinery Engineering to Elicit Complex Cellular Phenotypes

**Amanda M. Lanza and Hal S. Alper**

## Abstract

Cellular hosts are widely used for the production of chemical compounds, including pharmaceutics, fuels, and specialty chemicals. However, common metabolic engineering techniques are limited in their capacity to elicit multigenic, complex phenotypes. These phenotypes can include non-pathway-based traits, such as tolerance and productivity. Global transcription machinery engineering (gTME) is a generic methodology for eliciting these complex cellular phenotypes. In gTME, dominant mutant alleles of a transcription-related protein are screened for their ability to reprogram cellular metabolism and regulation, resulting in a unique and desired phenotype. gTME has been successfully applied to both prokaryotic and eukaryotic systems, resulting in improved environmental tolerances, metabolite production, and substrate utilization. The underlying principle involves creating mutant libraries of transcription factors, screening for a desired phenotype, and iterating the process in a directed evolution fashion. The successes of this approach and details for its implementation and application are described here.

**Key words:** Complex phenotype, Transcription machinery, gTME, Engineered phenotype, Sigma factor, TATA-binding protein, Metabolic engineering

## 1. Introduction

Many prokaryotic and eukaryotic cellular systems are attractive hosts for the sustainable production of chemicals, fuels, and pharmaceuticals. Classical metabolic engineering methodologies are well-suited for improving pathway fluxes and increasing product yields. However, these tools are quite limited in their ability to elicit complex cellular phenotypes. Examples of complex phenotypes include chemical tolerances, faster growth rates, morphology, and higher bioconversion rates. Complex phenotypes differ from metabolic phenotypes in part due to the number of genes regulating these traits. Metabolic phenotypes are often controlled by a handful

Wilfried Weber and Martin Fussenegger (eds.), *Synthetic Gene Networks: Methods and Protocols*, Methods in Molecular Biology, vol. 813, DOI 10.1007/978-1-61779-412-4_14, © Springer Science+Business Media, LLC 2012

of genes, whereas no singular gene or pathway is responsible for complex, multigenic traits (1, 2). The improvement of complex cellular phenotypes has been a long standing goal of the food and biotechnology industry well before the advent of recombinant DNA technology (3) and continues to be actively researched (4).

Global transcription machinery engineering (gTME) is a generic methodology for engineering complex phenotypes. The major premise of this approach is that introducing dominant mutant alleles of generic transcription-related proteins can reprogram gene networks and cellular metabolism. By linking this mutant protein expression with phenotype selection and screening, it is possible to identify mutants eliciting novel, complex cellular phenotypes. However, since the linkage between mutant protein sequence and function is difficult to predict de novo, this method is typically reduced to practice by creating mutant libraries of transcription factors and selecting for the ones that improve a desired phenotype. Since its development in 2006, this method has been implemented in *Escherichia coli*, *Saccharomyces cerevisiae*, and *Lactobacillus plantarum* and used to isolate strains with improved complex and metabolic phenotypes, including high ethanol tolerance (5, 6), overproduction of lycopene (6), and improved xylose fermentation (7). Thus, gTME is a useful tool for strain development that can be applied on its own or in conjunction with other cellular engineering techniques. The description of this method below begins by comparing gTME to alternative methods, and then provides the theoretical framework for the approach, followed by highlights of recent successes. This extended introduction is followed by a detailed description of the gTME method for both a generic phenotype and the specific example of improved ethanol tolerance in yeast.

### 1.1. Alternative Methods for Engineering Complex Cellular Phenotypes

Several rational and combinatorial methods exist for strain development. Most traditional, classical approaches utilize a mutation–selection strategy. The most common chemical mutagenic agent is *N*-methyl-N′-nitro-*N*-nitrosoguanidine (NTG), although treatment with radiation and UV rays are good alternatives. Chemical mutagenesis is limited, however, in its ability to engineer complex phenotypes. It is a slow and laborious process, making screening of large libraries hard to implement (4). Additionally, it has been shown that the phenotypic diversity achieved by NTG mutagenesis is much lower than that of gTME (8). Despite limitations of chemical mutagenesis, it has been widely and successfully used. A major application of this method is in antibiotic production, where strains producing penicillin in excess of 50 g/L have been isolated (4). Nevertheless, one particular challenge of this approach for complex phenotypes is the extreme library size needed to screen the low-probability event of multiple genes being impacted.

More recently, genomics-inspired approaches have been used to access complex phenotypes. Approaches, such as parallel gene trait mapping (9), multi-SCale Analysis of Library Enrichments (SCALE) (10, 11), shotgun genomics (12, 13), transposon mutagenesis (14–17), and genome shuffling (18, 19), have all been developed to map genotype to phenotype. These approaches are quite powerful, as they select for traceable changes to a cell that may be easily identified and transferred. As an example, SCALE can be used to quickly and efficiently screen and analyze genomic libraries in surrogate hosts (20). These libraries contain different size DNA fragments, or scales, cloned into expression vectors and are grown competitively under selective conditions associated with the desired phenotypes. This method was initially validated in *E. coli* to identify genes associated with Pine-sol resistance, a complex tolerance phenotype (9). This approach has recently been modified and updated in a methodology termed "Trackable multiplex recombineering" (TRMR) (21), which uses synthetic cassettes and homologous recombination to identify up- and down-regulation targets across the genome. One particular complex phenotype investigated using TRMR is cellulosic hydrolysate tolerance. Finally, the use of high-throughput gene complementation (through both metagenomics and bioprospecting) has been explored as a method for importing function found in one cell into a desired host of interest. This method is most typically used for isolating novel enzymes (22). Collectively, these methods still attempt to link complex phenotypes to a single fragment of genomic DNA (containing only handfuls of genes). Thus, a more global approach may be required for complex cellular phenotypes.

### 1.2. gTME Theoretical Framework

gTME is an alternative approach for imparting the multigenic changes necessary to elicit complex phenotypes. Transcriptional profiling of cells expressing mutant transcription factors reveals that these factors can elicit the differential expression of hundreds of genes (5) compared to wild-type cells. Thus, the phenotypes enabled by gTME are not limited to a single-gene or single-pathway like traditional approaches discussed above. Moreover, this broad-action increases the likelihood of engineering complex phenotypes resulting from the concerted manipulation of several genes.

To accomplish this reprogramming, the method of gTME can be broken down into four basic steps with two optional steps depending on the end goal and the desired level of phenotype improvement (Fig. 1):

- First, a target gene involved in a key aspect of transcription activity is selected. Suitable targets include generic transcription factors known to interact directly or indirectly with hundreds of genes.

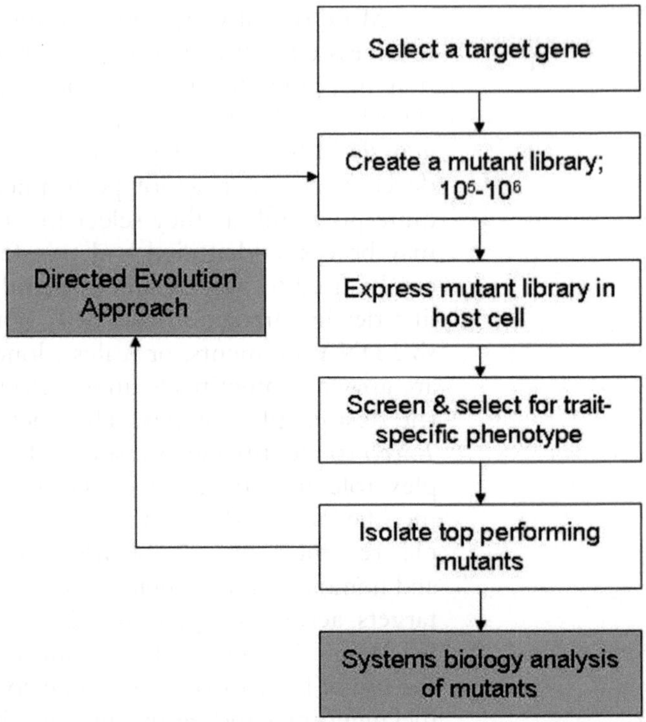


Fig. 1. General methodology for gTME. The general framework for the gTME approach is depicted. Shaded boxes highlight optional steps depending on the goal and scope of the project.

- Second, a mutant library (often comprising roughly $10^5$–$10^6$ different mutant versions of this selected protein) is created.
- Third, the mutated gene library is expressed in the host cell. Co-expression of the mutant and endogenous, chromosomal version creates a screen for dominant mutant phenotypes. In this regard, the wild-type version allows for the maintenance of crucial cellular functions, whereas the mutant version can impart phenotype-specific cellular reprogramming.
- Fourth, mutant proteins imparting the desired reprogramming and thus phenotype of interest can be identified using trait-specific screening and selection. Due to the use of large libraries, a high-throughput screen is almost always necessary.
- Fifth, the process (steps 2–4) can be reiterated using a directed evolution approach since the phenotype of interest is linked to a single mutant protein of interest.
- Sixth, it is possible to use a systems biology approach for understanding the underlying reprogramming and mechanism. This step is not necessary if the end goal is simply to obtain an improved strain.

One limitation of gTME is that it can only access latent cellular potential, not introduce de novo function to a cell. This limitation arises from the mode of action, namely altering gene expression not protein function. Specifically, while mutant factors result in gene expression changes across the cell, the functionality of those genes is fixed. Thus, unlike chemical mutagenesis, there is no potential to evolve other enzymes by strictly using a gTME approach. This is a major difference between gTME and other methods for engineering phenotype like chemical mutagenesis, metabolic engineering and metagenomics. However, these methods can be combined to effectively modify existing cellular mechanisms and introduce new functionality through heterologous DNA.

### 1.3. Examples of Applications and Future Directions of gTME

Transcription machinery engineering has been applied to several prokaryotes and to the eukaryote yeast. In *E. coli* and *L. plantarum*, sigma factors including $\sigma^{70}$ (encoded by *rpoD*) and $\sigma^{38}$ (encoded by *rpoS*) as well as the RNA polymerase itself, encoded by *rpoA*, (23) have been targets for gTME (6, 8, 24), whereas in yeast strains the focus has been on *taf25* and *spt15* (5, 7). Most of the mutant transcription factors identified from these varied screens contained either a small number of mutations (on the order of 1–5) or had large modifications (such as significantly truncated proteins).

In bacterial systems, the *rpoD* gene encoding the main sigma-70 factor has been a target of choice. In one study, a single mutant library of *rpoD* was developed and enabled the screening of three improved phenotypes: improved tolerance to ethanol, lycopene overproduction, and multiple, simultaneous phenotypes (6). In the first case, an *E. coli* mutant with a 6 h doubling time in 60 g/L of ethanol was identified. Development of a lycopene overproducing mutant through gTME was compared to single and multigene knockouts. A single round of gTME performed better than a single gene knockout, and combining the two methods resulted in strains able to produce more than 7 mg/L of lycopene. Tolerance to ethanol and SDS was selected to explore the ability of gTME to impart multiple, simultaneous phenotypes. Independent selection for an ethanol and an SDS mutant, followed by co-expression of both mutants, was found to be superior to either sequential or simultaneous selection strategies (6). A separate study in *L. plantarum* isolated mutants with increased tolerance to low pH and high lactic acid concentration, both of which are industrially relevant conditions (8). In the case of lactic acid tolerance, a single amino acid substitution was made to *rpoD*, whereas low pH tolerance was a result of four amino acid substitutions and a truncation of the protein (8). A third study looked at improved hyaluronic acid (HA) production in recombinant *E. coli* by targeting both *rpoD* and *rpoS*. The top performing mutant accumulated 560 mg/L of HA (24).

The majority of gTME work in yeast has focused on the TATA-binding protein encoded by the *SPT15* gene in *S. cerevisiae*. The first study looked at two mutant libraries for *spt15* and *taf25* for improved ethanol tolerance in high glucose fermentations (5). The *spt15* mutant outperformed the *taf25* mutant, giving a 12-fold improvement in optical density over a control strain when grown in 100 g/L of glucose and 6% ethanol by volume. Subsequent transcriptional profiling identified major gene targets of the mutant *spt15-300*, which when knocked out resulted in a loss of capacity for the ethanol tolerant phenotype (5).

*Spt15* mutant libraries for ethanol tolerance were also constructed and tested in a yeast diploid and Kyokai 7, an industrial strain for sake production (unpublished). Previously identified mutants for laboratory strains were not effective in Kyokai 7, whereas a new mutant identified by screening in the industrial strain has negligible impact on *S. cerevisiae*. This indicates the strain-specific behavior of *spt15* mutants, and therefore the importance of screening mutants in an environment relevant to the desired phenotype.

A third study screened an *spt15* mutant library for xylose utilization in *S. cerevisiae* (7). The control strain was able to grow on glucose but not able to utilize xylose. A single strain was isolated that was able to grow modestly on 50–150 g/L of xylose as well as mixed sugar cultures. However, the mutant strain had decreased ethanol yield compared to the control strain, even when grown on the preferred glucose carbon source, indicating further strain improvements are necessary before this could be applied to an industrial biomass process (7).

Despite its infancy, the technique of gTME has demonstrated an ability to elicit complex cellular phenotypes in both prokaryotes and simple eukaryotes. These phenotypes often result from dramatic reprogramming of innate gene expression, resulting in significant shifts in metabolism and ultimately cellular performance. More importantly, these multiple, simultaneous changes are being conducted by a single, generic target. Thus, this approach transforms complex phenotype elicitation into the directed evolution of a single protein. This methodology could also be employed to identify genes involved in a particular pathway or phenotype that is not well understood. gTME represents a significant improvement to more traditional methods, such as chemical mutagenesis or single-gene-based metabolic engineering. Using the method of gTME in conjunction with techniques like metabolic engineering, directed evolution, and cell adaptation could be a powerful combination facilitating the isolation of complex, cellular phenotypes.

## 2. Materials

### 2.1. Selection of Gene Target

1. Genome sequence or (at minimum), sequence of target gene.
2. Genomic DNA or cDNA from the host organism. The Promega Wizard Genomic DNA Purification Kit works well for most cell types.
3. Standard PCR reagents and gene-specific primers.

***2.2. Construction of Mutant Library***

1. Broad-host expression vector compatible with both host organism and *E. coli*, containing a promoter of the appropriate strength.
2. Restriction enzymes and T4 DNA ligase for basic cloning.
3. Sequencing primers and mutagenesis primers.
4. Error-prone PCR (epPCR) kit or reagents.
5. Column cleanup kit for DNA fragments.
6. Competent *E. coli* cells.
7. Large petri dishes (150 × 10 mm) and LB-agar supplemented with an appropriate antibiotic.
8. Sterile plate scrapers.
9. 30% glycerol by volume, sterile filtered.
10. LB media and culture tubes.
11. Plasmid extraction kit.
12. Device for measuring optical density.
13. Incubator for 37°C growth of *E. coli* cells.

***2.3. Mutant Expression in Host Strain***

1. Competent cells for the host organism and transformation reagents.
2. Large petri dishes (150 × 10 mm) and media supplemented with agar and the appropriate selection media for the host organism.
3. Sterile plate scrapers.
4. Incubator for growth of host cells.

***2.4. Phenotype Selection Strategies***

1. Selection conditions and proper flasks/tubes.
2. Incubator.
3. Plasmid recovery kit for the host organism.

***2.5. A Directed Evolution Approach***

1. Same reagents as Subheading 2.2.

***2.6. Analysis of Selected Mutants***

1. Sequencing primers.
2. Sequence alignment software, such as ClustalW.
3. Whole cell RNA extraction kit.
4. Global microarrays specific to host organism.

## 3. Methods

As described above, the basic gTME paradigm can be described by four main steps and two optional steps. These steps include: (1)

selecting a target of interest, (2) creating a mutant library of the target, (3) expressing the mutant library in a host strain, (4) selecting for the phenotype of interest, (5) reiterating the process using directed evolution, and (6) evaluating the mutant using a systems biology approach. The detailed procedure for carrying out gTME depends significantly on the desired phenotype and cellular host. Thus, to provide an adequate description of this method, each of the six steps is described in two ways in this section. First, this method is described generically. Second, in order to more clearly illustrate the methodology, the specific example of improved ethanol tolerance in yeast was selected as a case study (5).

### 3.1. Selection of Gene Target

#### 3.1.1. Generic Methodology

1. Selection of a suitable gene target is the first step in gTME. In order to be most effective at generically impacting metabolism, global regulators of basic transcription have traditionally been selected. As a starting point, it is useful to consider a high-confidence target, or a gene in which dominant mutations have previously been shown to exist. To date, transcription factors associated with the basic RNA polymerase system, such as *rpoD* and *rpoS*, as well as the polymerase itself, *rpoA*, have been targets in prokaryotic systems (6, 8, 23, 24). In yeast, *taf25* and *spt15* have been selected as targets (5, 7). Additional potential targets in yeast include the three RNA polymerases and approximately 75 general transcription factors (5). In order to obtain optimal results, it may be necessary to select and test multiple targets, as the ideal target may be phenotype specific (see Note 1).
2. The sequence of the target gene should be obtained from sequence databases or de novo sequencing. For many model organisms, transcription factors and associated proteins have been sequenced and characterized. This information greatly facilitates target selection. However, in other microbes of interest, these genes may not have been studied and genome sequence annotation may be limited. In these cases, it may be necessary to select a target gene by comparing to better understood organisms using sequence homology.

#### 3.1.2. Ethanol Tolerance Example

1. Two known transcription factors, *TAF25* and *SPT15*, were selected as targets.
2. The genome sequence of BY4741 is known and sequences for both genes were found using the Saccharomyces Genome Database.

### 3.2. Construction of a Mutant Library

#### 3.2.1. Generic Methodology

1. First, a suitable expression vector must be found to clone the target gene selected in the steps above. The expression vector should contain an origin of replication, selectable marker, and promoter region for the host organism (described further in step 2 below), as well as an antibiotic marker and origin of replication for *E. coli* (or another selected host for routine plasmid propagation).

2. A suitable promoter used to express the mutant transcription factor must be selected. The strength of the promoter can have an impact on gTME selection and screening, as well as the magnitude of the dominant phenotype, and should be considered as part of vector selection. It has previously been shown that a relatively strong constitutive promoter can be used with low copy number plasmids (see Note 2). Choice of promoter strength also depends on the ploidy of the strain.
3. Primers should be designed to amplify the gene of interest from the genome. Restriction enzyme sites should be appended to both primers to allow for cloning into the selected expression plasmid.
4. Genomic DNA is extracted for the desired host organism. The target gene is amplified from genomic DNA using a polymerase chain reaction (see Note 3) and the primers designed above. This product should be cloned directly behind the vector's promoter using basic molecular biology techniques.
5. This completed plasmid should be sequence confirmed to ensure that the gene was cloned in-frame and does not contain any sequence errors. This plasmid serves as the control vector for experiments, as well as starting point for library construction.
6. A set of mutagenesis primers should be designed to amplify and mutate the target gene. To ensure the maximum amount of mutation, these primers are designed to have nearly complete homology to the vector (see Note 4).
7. A mutant library using epPCR should be constructed using the control vector as a template. This mutant library can be constructed using various error prone PCR techniques including: mutant polymerases, nucleotide analogues, or increased Mg and Mn concentrations (25). Typically, nine 50 μL reactions are performed to obtain enough pooled DNA.

   epPCR by mutant polymerases is the most common method for producing mutated copies of the target gene. An epPCR reaction is assembled identically to standard PCR, except it uses a polymerase with an increased likelihood for incorporating mismatched base pairs. Mutation frequency can be controlled by changing the template concentration, where less initial template increases the mutation frequency (see Note 5).

   epPCR by nucleotide analogues is another option that uses pyrimidine analogues that are recognized by polymerases and incorporated in place of dTTP and dCTP (26). Individual base pair substitution rates range from $1^{-2}$ to $4.4^{-2}$, whereas overall mutation rates were found to be as high as $1.9^{-1}$.

   epPCR by increased concentrations of $MgCl_2$ and the addition of $MnCl_2$ have both been shown to increase the mutation rate of a standard PCR reaction with the use of traditional enzymes, such as *Taq* (27). The presence of either compound helps stabilize mismatched base pairs. $MgCl_2$ will typically be added at a concentration up to 7 mM (a 4.7-fold increase from a typical reaction). $MnCl_2$ is not typically present in PCR reactions and can be added at a concentration of 0.5 mM just prior to thermocycling, as it can precipitate.

8. Pooled DNA from the epPCR step should be purified using standard molecular biology techniques. This mixture should be digested with the appropriate restriction enzymes (see Note 6).
9. The fragments and the vector can be ligated (usually overnight), then transformed into a bacterial host, such as DH10β or DH5α using standard techniques. Typically, around 10–15 standard transformation reactions are performed using 2.5 μL of the ligation mixture for each transformation.
10. Cells are then plated onto large LB-agar plates (see Note 7) supplemented with the appropriate antibiotic and grown overnight at 37°C.
11. Colonies are counted from the plates after around 20 h of incubation to determine the library size (see Note 8).
12. Colonies are then pooled using sterile media and a sterile cell scraper to create a liquid library of the cells.
13. Portions of this liquid library are stored as glycerol stocks (in 15% glycerol by volume).
14. The remainder of this library is harvested for plasmids using methods such as Qiagen Mini-prep spin kit. Often, around 20 samples from the library are harvested for plasmids.
15. Final plasmids are pooled and DNA concentration is measured using a spectrophotometer.

#### 3.2.2. Ethanol Tolerance Example

1. The p416-TEF-mut2 plasmid was selected as an expression vector. This supports bacterial, as well as yeast replication. The prokaryotic and eukaryotic selection markers are ampicillin and uracil, respectively. This is a lower-copy number CEN-based plasmid.
2. The TEF-mut2 promoter was selected to drive mutant gene expression. This is a constitutive promoter found to give 7% expression compared to the native TEF promoter, from which it was derived (28).
3. Primers were designed with NheI and SalI restriction sites. For TAF25: TCGAGTGCTAGCAAAATGGATTTTGAGGAAGA TTACGATand CTAGCGGTCGACCTAACGATAAAAGTC TGGGCGACCT, for SPT15: TCGAGTGCTAGCAAAAT GGCCGATGAGGAACGTTTAAAGG and CTAGCGGTC GACTCACATTTTTCTAAATTCACTTAGCACA.

4. Yeast genomic DNA was extracted using the Promega Wizard Kit and a standard haploid laboratory strain of yeast, BY4741. The TAF25 and SPT15 genes were amplified from gDNA using *Taq* polymerase and digested with NheI and SalI.
5. The resulting plasmids were sequence verified to ensure the correct sequences of TAF25 and SPT15.
6. Generic mutagenesis primers were designed based on the p416 vector and NheI and SalI restriction sites. GCATAGCAATCT AATCCAAGTTTTCTAGAATG and ATAACTAATTACATG ACTCGAGGTCGACTTA were chosen as the two mutagenesis primers.
7. Mutagenesis was performed using the GeneMorph II Random Mutagenesis Kit from Strategene. Varied template concentrations (using the p416-TEF-mut2-TAF25 or SPT15-based plasmids) were selected to enable low, medium, and high mutation rates.
8. The fragments were purified using Qiagen PCR cleanup kit, and digested overnight at 37°C using NheI and SalI. The vector was digested with XbaI and SalI overnight.
9. The fragments were ligated with the vector overnight at 16°C using T4 DNA ligase. The mixture was transformed into competent DH5α.
10. The transformation mixture was plated on LB-agar plates supplemented with 100 μg/mL of ampicillin and grown overnight at 37°C.
11. The colonies were counted and library size was determined to be $10^5$.
12. Colonies were scraped off plates to create a liquid library.
13. A portion of the liquid library was stored in a 15% glycerol mixture at −80°C.
14. Plasmids were extracted from the library using a Qiagen Miniprep spin kit.
15. The plasmids were pooled and the concentrations were around 300 ng/μL.

### 3.3. Mutant Expression in Host Strain

#### 3.3.1. Generic Methodology

1. Extracted plasmids are library-transformed into the host strain using a method suitable for the organism.
2. The transformation mixture is plated on a solid media containing a selectable marker specific to the host strain.
3. Colonies are then scraped off the plates using sterile media to create a liquid library of the cells; glycerol stocks can be made at this point.

#### 3.3.2. Ethanol Tolerance Example

1. Plasmids containing mutant libraries for TAF25 and SPT15 were transformed into competent BY4741 using the standard

Geitz's lithium acetate protocol. A total of 1 μg of plasmid was used for each transformation mixture.

2. The transformation mixture was plated on drop-out media (a total of forty-eight 150 × 10 mm plates were used) lacking uracil. The plates were incubated for 2–3 days at 30°C.
3. Colonies were scraped from the plates into a liquid media to proceed directly with phenotypic selection.

### 3.4. Phenotype Selection Strategies

#### 3.4.1. Generic Methodology

gTME requires a large library size in order to effectively isolate mutants conferring a desired phenotype. Because of this library size, a high-throughput screen for the isolation of desired mutants is essential. The screening method and the specific conditions vary depending on the phenotype that is being isolated (see Note 9). Common phenotypes studied to date include tolerance to environmental conditions, metabolite production and multigenic phenotypes. Screening for each of these traits should be uniquely approached and may require multiple rounds of library construction and refinement before identifying mutations conferring the desired phenotype.

1. Screening for improved tolerance (see Note 10) involves identifying an environmental condition for which tolerance in the host organism is desired.

   Expose the mutant library to gradually harsher conditions, selecting against those mutants unable to grow in the environment. Typically, starting values at or above the minimum inhibitory concentration of wild-type cells are used for rapid selection.

   As cells from the library grow, this initial selection phase can be repeated by diluting and growing the surviving culture under the same environmental condition.

   The environmental condition can be made increasingly harsher, thereby reducing the surviving population.

   After several rounds (typically 5–10), this liquid culture can then be plated on solid media for the selection of individual clones.

   Individually selected clones can then be tested again for growth under the environmental conditions (typically, 20–50 clones are isolated from plates for initial testing).

   Finally, the plasmid DNA should be extracted, retransformed and the resulting clones again tested for growth under the environmental condition (see Note 11).

2. Screening for metabolite production (see Note 12) involves developing a high-throughput screen to detect the compound of interest. The most common screens are colorimetric, enzymatic, or a direct measurement of the compound.

Colorimetric screening, when possible, is ideal because it is fast and requires minimal labor. Candidates can be screened by eye or using a spectrophotometer (see Note 13).

Enzymatic screens are another option if the metabolite of interest interacts with enzymes whose products or cofactors result in a detectable photometric shift (see Note 14).

For those metabolites that cannot be screened by either a color change or enzymatic method, it may be necessary to directly measure metabolite concentration. If the metabolite is secreted, stationary phase culture can be pelleted and tested using an instrument like a YSI biochemical analyzer, HPLC, or GC-MS. Additionally, the product of interest could be stained for or indirectly linked to a fluorescent protein (like GFP) and detected using FACS.

Select individual colonies from the library and grow in 96-well plates (or on solid media, depending on the screen).

Screen each clone for metabolite production and compare performance to the control strain.

Isolate top performing clones and repeat the screen using 5 mL culture tubes to confirm earlier results.

For those clones exhibiting sufficient metabolite production, extract the plasmid DNA, retransform it and once more test the clone for metabolite production (see Note 11).

3. Selection of multigenic phenotypes; when dealing with more than one phenotype, the order of selection is important (see Note 15).

   Starting from the full mutant library, develop and carry out a selection strategy for the first phenotype.

   Identify and isolate top performing clones.

   Repeat the screening to confirm earlier results.

   Extract the plasmid DNA, retransform it, and conduct a final test of the phenotype.

   Starting from the full mutant library, develop and carry out a selection strategy for the second, and any subsequent, phenotypes.

   Co-express combinations of the top performing mutants for each phenotype in the host organism.

   Evaluate these combinations for performance under both conditions of interest.

#### 3.4.2. Ethanol Tolerance Example

Yeast strains with an increased tolerance to ethanol were selected using a tolerance screen.

Increased ethanol tolerance was selected as the desired phenotype.

Selection was initially performed in YSC-URA media supplemented with 100 g/L of glucose and 5% ethanol by volume.

30 mL of culture in 30 × 115 mm closed top tubes were grown vertically at 30°C.

The *taf25* library was subcultured four times under these conditions.

The *spt15* library was subcultured twice under the initial condition and then twice at 120 g/L glucose and 6% ethanol.

The mixture was plated on YSC-URA and 20 surviving colonies selected for both *taf25* and *spt15* mutations.

These selected clones were grown in overnight cultures and assayed for growth rate in 60 g/L glucose and 5% ethanol. Improvement in growth performance was determined by OD, as compared to the control.

Plasmids were extracted from the clones showing improved growth. These plasmids were retransformed and the ethanol-tolerant phenotype once more validated. The *spt15-300* mutant (containing the F177S, Y195H, and K218R mutations) was found to impart the most significantly improved phenotype.

### 3.5. A Directed Evolution Approach to gTME

#### 3.5.1. Generic Methodology

Directed evolution is a protein engineering algorithm by which the fitness of a protein can be enhanced through iterative mutagenesis and selection. Such an approach can be applied to gTME for increased library diversity and improved phenotypes. Unlike traditional approaches, the directed evolution algorithm is applicable to gTME since the phenotype improvement is linked to a mutant protein. Directed evolution can be performed in multiple ways. First, the top performing mutants isolated in Subheading 3.4 Phenotype selection strategies can be remutagenized (following the methods of Subheading 3.2) and then selection can be repeated. A previous study merging directed evolution and gTME showed that after two rounds of mutagenesis, the fold improvement in phenotype was incremental (6). Alternatively, it is possible to isolate a subset of the top performing mutants and perform gene shuffling to create a new, diverse library (see Note 16). For the case of iterative mutagenesis, the directed evolution method would entail:

1. Isolate a top performing mutant following Subheading 3.4. This mutant serves as the starting point for the construction of a new library. Proceed as described in Subheading 3.2, starting with step 7 and using the mutant target gene as the template for mutagenesis.
2. Express the new mutant library in a host strain (see Subheading 3.3).
3. Select for an improved phenotype (see Subheading 3.4). During the selection phase, the new mutant library should be compared both to the wild-type, endogenous target gene, as well as the mutant target gene isolated after initial library construction.

4. This process can be iterated more than once to obtain further improved phenotypes.

*3.5.2. Ethanol Tolerance Example*

For the case of improved ethanol tolerance in yeast by mutant *spt15*, a directed evolution approach was not used. However, this approach could have been performed by using *spt15-300* as a template for creating a mutagenesis library.

### 3.6. Analysis of Selected Mutants

*3.6.1. Generic Methodology*

1. Sequence analysis of the target gene
   After isolating a mutant cell line conferring a desired phenotype, it may be of interest to determine the mutation(s) associated with those changes. Identifying the locations and types of mutations that contribute to a specific phenotype can provide useful information about the functionality of transcription related proteins.
   The plasmid carrying the mutant target gene is recovered from the cell line.
   The target gene is sequenced using standard technology and primers both upstream and downstream of the gene.
   The sequence is compared with the endogenous target gene to identify mutations. Previous studies have found mutations ranging from single amino acid substitutions to domain truncations (see Note 17).
2. Transcription profiling
   Mutant transcription factors are selected for their ability to indirectly regulate transcription of hundreds of genes; therefore, a desired phenotype is likely the result of unique interactions between the target gene and other genomic loci. An analysis of the cell's transcriptome, compared to the control cell line, can identify both up and down-regulated genes.
   The control and mutant cell lines are grown under desired media conditions until mid-log phase.
   Whole cell RNA is extracted. The extraction method varies depending on the cell type.
   The whole cell RNA is subjected to a global microarray analysis (see Note 18).
   The genes most impacted by the mutant transcription factor can further be probed using traditional knockouts and over-expression.

*3.6.2. Ethanol Tolerance Example*

1. Sequence analysis of *spt15-300* ethanol tolerant mutant
   The *spt15-300* plasmid was extracted using the Zymoprep yeast plasmid miniprep kit and transformed into *E. coli* DH5α. This plasmid was isolated and sequenced using forward primer TCACTCAGTAGAACGGGAGC and reverse primer AATA GGGACCTAGACTTCAG.
   The sequence of *spt15-300* and the wild-type gene were compared using ClustalW. *Spt15-300* was found to contain

three point mutations, each resulting in an amino acid residue change.

2. Transcription profiling of *spt15-300* ethanol tolerant mutant

   The control strain and the *spt15-300* mutant were grown in YSC-URA medium supplemented with 100 g/L of glucose and 5% ethanol to an OD of 0.4–0.5.

   Whole cell RNA was extracted using the Ambion RiboPure Yeast RNA extraction kit.

   Microarray services were contracted through Ambion using Affymetrix yeast 2.0 arrays. Arrays were run in triplicate with biological replicates.

   Over 100 genes were found to be diversely expressed in the presence of the *spt15* mutant and increased ethanol concentration. One hundred eleven genes were found to be up-regulated, with only 21 down-regulated. Twelve of the most highly expressed genes in the mutant were individually deleted. These deletions were then found to result in a loss of mutant capacity (5), indicating their individual importance in ethanol tolerance as well as confirming that such a phenotype relies on concerted expression of multiple genes.

## 4. Notes

1. Previous studies have shown that different gene targets elicit different phenotypic effects. For example, both *taf25* and *spt15* mutant libraries were screened for improved ethanol tolerance, but the *spt15* mutants outperformed the *taf25* mutants, demonstrating that different members of the transcription machinery have differential influence over phenotypic responses (5). A similar effect was seen with *rpoD* and *rpoS* libraries (24). Thus, for a given phenotype, it may be necessary to select and test multiple target genes before finding one that has the desired impact on cellular metabolism.
2. Many gTME studies have successfully used low copy plasmids coupled together with relatively high constitutive promoters (5, 6, 24). An alternative option is to use the native promoter for the target gene; however, expression may not be constitutive and may be subject to regulation. In a higher ploidy strain, promoter strength must be increased to provide a more profound dominant phenotype (unpublished).
3. While *Taq* polymerase is the most common enzyme choice, many other polymerases are available and optimized for different conditions, including long amplicons, high-fidelity, and high GC-content templates.

4. The last nucleotides of the 5′ primer should be ATG and the last nucleotides of the 3′ primer should be TAA (or another stop codon). This will allow all possible residues within the target gene to be subjected to mutation in the next step.
5. Stratagene's GeneMorph II Random Mutagenesis kit recommends 560 ng for a low mutation rate, 280 ng for medium and 28 ng for high (8). A typical error rate for the low mutation is 0–4.5 mutations per kilobase, 4.5–9 for medium and, 9–16 for the high mutation rate.
6. The mutant library can be digested with DpnI to remove copies of the circular template DNA, reducing the likelihood that the control plasmid is present in the transformation mixture and thus increasing the diversity of the library.
7. Because of the large library size, it is ideal to use 150 × 10 mm plates for library transformation mixtures. This will ensure a better distribution of transformants across the plate, allow for antibiotic selection and facilitate selection of individual colonies. Typically around 300–500 μL of transformation mixture is plated on each plate.
8. A typical desirable library size is $10^5$–$10^6$ (5, 6, 8).
9. In general, high-throughput selections are sensitive to environmental conditions, including media composition, temperature, and selection stringencies and there exists a dependency between isolated mutants and the conditions under which they are identified (unpublished). This phenomenon is especially important when applying gTME to industrial strains. Decisions, such as minimal or complex media and shaking rate, for example, should be made in advance and maintained throughout selection to ensure that isolated mutants perform consistently.
10. Screening for improved tolerance to environmental conditions is perhaps the most straightforward of selection strategies because undesired mutants are unable to compete and therefore die off. However, the specifics of the screen depend on the phenotype. For example, when identifying ethanol tolerant mutants, an initial concentration of 50 g/L of ethanol was used (6). In another study attempting to identify mutants tolerant to low pH, the library was initially exposed to either pH of 4.60 or 3.85 (8).
11. This final step is important for confirmation that the improved phenotypic condition is a result of the mutated transcription-related gene and not other, genomic mutations that could have occurred to a particular cell.
12. Metabolite production can be a more difficult phenotype to screen for because the concentration of the desired compound must be directly or indirectly measured, and it is not possible to select against colonies lacking the desired phenotype.

13. A colorimetric screen was used to detect improved lycopene production in yeast. Colonies were initially screened for a reddish hue (6).
14. An enzymatic assay was successfully used to screen for glucoronic acid production. The enzyme, uronate dehydrogenase, converts glucuronate to glucarate, resulting in an accumulation of NADH, detectable at 340 nm (29), whereas NAD+ is not.
15. In a study looking for mutants tolerant to both ethanol and SDS, four distinct search strategies were examined; isolation of ethanol tolerant mutants followed by SDS tolerance, isolation of SDS tolerant mutants followed by ethanol tolerance, simultaneous selection of ethanol and SDS, or independent selection of ethanol and SDS mutants followed by co-expression of these mutants. This final search strategy was found to impart the most significant phenotype overall (6). A similar effect was found when isolating mutants adapted to high lactic acid and low pH, although the two phenotypes were not additive (8).
16. Gene shuffling is an alternative to directed evolution in which portions of various genes are recombined to increase genetic diversity (18, 19).
17. A variety of different mutations have been shown to confer unique phenotypes. For example, a single, nonsynonymous mutation to *rpoD* was found to significantly increase growth of bacteria in high lactic acid concentrations (8). In other cases, three or more amino acid substitutions were found to confer improved phenotypes (5, 8). In several cases, top performing mutant sigma factors were found to be a result of a truncation (6, 24). These truncations resulted in a loss of conserved regions and because sigma factors are involved in many cellular processes, it is hypothesized that loss of these regions can generate significant metabolic improvements. Alternatively, the majority of the point mutations in an *rpoD* mutant conferring high HA yield occurred in the nonconserved region of the gene (24).
18. Microarray studies can be contracted through a service, such as Asuragen or Cogenics, and should be run in triplicate and include biological references for statistical confidence. Microarray analysis will not only identify large sets of genes associated with a particular phenotype, but it can be used to identify a small subset of genes most closely linked to the phenotype (i.e., largest change in expression compared to the control).

## References


1. Stephanopoulos, G., Alper, H., and Moxley, J. (2004) Exploiting biological complexity for strain improvement through systems biology, *Nat Biotechnol 22*, 1261–1267.
2. Tyo, K. E., Alper, H. S., and Stephanopoulos, G. N. (2007) Expanding the metabolic engineering toolbox: more options to engineer cells, *Trends Biotechnol 25*, 132–137.
3. Parekh, S., Vinci, V. A., and Strobel, R. J. (2000) Improvement of microbial strains and fermentation processes, *Appl Microbiol Biotechnol 54*, 287–301.
4. Patnaik, R. (2008) Engineering complex phenotypes in industrial strains, *Biotechnol Prog 24*, 38–47.
5. Alper, H., Moxley, J., Nevoigt, E., Fink, G. R., and Stephanopoulos, G. (2006) Engineering yeast transcription machinery for improved ethanol tolerance and production, *Science 314*, 1565–1568.
6. Alper, H., and Stephanopoulos, G. (2007) Global transcription machinery engineering: A new approach for improving cellular phenotype, *Metabolic Engineering 9*, 258–267.
7. Liu, H., Yan, M., Lai, C., Xu, L., and Ouyang, P. (2010) gTME for improved xylose fermentation of Saccharomyces cerevisiae, *Appl Biochem Biotechnol 160*, 574–582.
8. Klein-Marcuschamer, D., and Stephanopoulos, G. (2008) Assessing the potential of mutational strategies to elicit new phenotypes in industrial strains, *Proc Natl Acad Sci USA 105*, 2319–2324.
9. Gill, R. T., Wildt, S., Yang, Y. T., Ziesman, S., and Stephanopoulos, G. (2002) Genome-wide screening for trait conferring genes using DNA microarrays, *Proc Natl Acad Sci USA 99*, 7033–7038.
10. Gall, S., Lynch, M. D., Sandoval, N. R., and Gill, R. T. (2008) Parallel mapping of genotypes to phenotypes contributing to overall biological fitness, *Metab Eng 10*, 382–393.
11. Warnecke, T. E., Lynch, M. D., Karimpour-Fard, A., Sandoval, N., and Gill, R. T. (2008) A genomics approach to improve the analysis and design of strain selections, *Metab Eng 10*, 154–165.
12. Jin, Y. S., and Stephanopoulos, G. (2007) Multi-dimensional gene target search for improving lycopene biosynthesis in Escherichia coli, *Metab Eng 9*, 337–347.
13. Kang, M. J., Lee, Y. M., Yoon, S. H., Kim, J. H., Ock, S. W., Jung, K. H., Shin, Y. C., Keasling, J. D., and Kim, S. W. (2005) Identification of genes affecting lycopene accumulation in Escherichia coli using a shot-gun method, *Biotechnol Bioeng 91*, 636–642.
14. Hemmi, H., Ohnuma, S., Nagaoka, K., and Nishino, T. (1998) Identification of genes affecting lycopene formation in Escherichia coli transformed with carotenoid biosynthetic genes: candidates for early genes in isoprenoid biosynthesis, *J Biochem 123*, 1088–1096.
15. Alper, H., Miyaoku, K., and Stephanopoulos, G. (2005) Construction of lycopene-overproducing E. coli strains by combining systematic and combinatorial gene knockout targets, *Nat Biotechnol 23*, 612–616.
16. Winterberg, K. M., Luecke, J., Bruegl, A. S., and Reznikoff, W. S. (2005) Phenotypic screening of Escherichia coli K-12 Tn5 insertion libraries, using whole-genome oligonucleotide microarrays, *Appl Environ Microbiol 71*, 451–459.
17. Badarinarayana, V., Estep, P. W., 3rd, Shendure, J., Edwards, J., Tavazoie, S., Lam, F., and Church, G. M. (2001) Selection analyses of insertional mutants using subgenic-resolution arrays, *Nat Biotechnol 19*, 1060–1065.
18. Patnaik, R., Louie, S., Gavrilovic, V., Perry, K., Stemmer, W. P., Ryan, C. M., and del Cardayre, S. (2002) Genome shuffling of Lactobacillus for improved acid tolerance, *Nat Biotechnol 20*, 707–712.
19. Wang, Y., Li, Y., Pei, X., Yu, L., and Feng, Y. (2007) Genome-shuffling improved acid tolerance and L-lactic acid volumetric productivity in Lactobacillus rhamnosus, *J Biotechnol 129*, 510–515.
20. Lynch, M. D., Warnecke, T., and Gill, R. T. (2007) SCALEs: multiscale analysis of library enrichment, *Nat Methods 4*, 87–93.
21. Warner, J. R., Reeder, P. J., Karimpour-Fard, A., Woodruff, L. B., and Gill, R. T.(2010) Rapid profiling of a microbial genome using mixtures of barcoded oligonucleotides, *Nat Biotechnol* 28 8:856–862.
22. Schmeisser, C., Steele, H., and Streit, W. R. (2007) Metagenomics, biotechnology with non-culturable microbes, *Appl Microbiol Biotechnol 75*, 955–962.
23. Klein-Marcuschamer, D., Santos, C. N., Yu, H., and Stephanopoulos, G. (2009) Mutagenesis of the bacterial RNA polymerase alpha subunit for improvement of complex phenotypes, *Appl Environ Microbiol 75*, 2705–2711.
24. Yu, H., Tyo, K., Alper, H., Klein-Marcuschamer, D., and Stephanopoulos, G. (2008) A high-throughput screen for hyaluronic acid accumulation in recombinant Escherichia coli transformed by libraries of engineered sigma factors, *Biotechnol Bioeng 101*, 788–796.
25. Arnold, F. H., and Georgiou, G. (2003) *Directed evolution library creation : methods and protocols*, Humana Press, Totowa, N.J.

26. Zaccolo, M., Williams, D. M., Brown, D. M., and Gherardi, E. (1996) An approach to random mutagenesis of DNA using mixtures of triphosphate derivatives of nucleoside analogues, *J Mol Biol 255*, 589–603.
27. Cadwell, R. C., and Joyce, G. F. (1992) Randomization of genes by PCR mutagenesis, *PCR Methods Appl 2*, 28–33.
28. Alper, H., Fischer, C., Nevoigt, E., and Stephanopoulos, G. (2005) Tuning genetic control through promoter engineering, *Proc Natl Acad Sci USA 102*, 12678–12683.
29. Yoon, S. H., Moon, T. S., Iranpour, P., Lanza, A. M., and Prather, K. J. (2009) Cloning and characterization of uronate dehydrogenases from two pseudomonads and Agrobacterium tumefaciens strain C58, *J Bacteriol 191*, 1565–1573.

# Chapter 15

# Streamlining of a *Pseudomonas putida* Genome Using a Combinatorial Deletion Method Based on Minitransposon Insertion and the Flp-*FRT* Recombination System

**Audrey Leprince, Danielle Janus, Víctor de Lorenzo, and Vitor Martins dos Santos**

## Abstract

Here, we document a technique to reduce the size of the genome of *Pseudomonas putida* by using a combinatorial mini-Tn*5*-targeted Flp-*FRT* recombination system. This method combines random insertions with the site-specific Flp-*FRT* recombination system to generate successive random deletions in a single strain in which parts of the genome are excised via the action of the cognate flippase. For this purpose, we have generated two mini-Tn*5* transposon mutant libraries with single and double integrations of either mini-Tn*5 KpF* alone or mini-Tn*5 KpF* in parallel with mini-Tn*5 TF*, respectively. These mini-Tn*5* transposons carry different selectable markers and each has an *FRT* (Flippase Recognition Target) site. Mapping of the position of both mini-Tn*5* transposons in the chromosome of *P. putida* was conducted by Arbitrary Primed-PCR (AP-PCR). Subsequent sequencing of the PCR fragments led to the identification of the coordinates of the transposons and the orientation of both *FRT* sites. Under specific laboratory conditions, both *FRT* sites were recognized by the flippase, and the deletion of a nonessential intervening genomic segment along with the transposon backbones occurred without inheritance of any marker genes.

**Key words:** Combinatorial deletion, Flp-*FRT* recombination system, *Pseudomonas putida*, Mini-Tn*5* transposon, Reduced genome, Streamlining

## 1. Introduction

The "top-down" approach of Synthetic Biology seeks to understand and reprogram existing biological systems by their modification. One approach involves reduction of an existing genome for the sake of simplicity and predictability, as it has been successfully

Wilfried Weber and Martin Fussenegger (eds.), *Synthetic Gene Networks: Methods and Protocols*, Methods in Molecular Biology, vol. 813, DOI 10.1007/978-1-61779-412-4_15, © Springer Science+Business Media, LLC 2012

applied to several organisms such as *Escherichia coli*, *Mycoplasma genitalium*, *Haemophilus influenza*, and *Bacillus subtilis* (1–5). The resulting microbe should be more predictable and controllable in comparison to the wild-type strain because genome reductions may decrease the complexity of regulatory and metabolic networks and finally lead to a strain that is easily manageable. Therefore, several techniques have been developed to engineer strains from different organisms in a high-throughput manner by deleting large sections of the microbe's genome and thereby removing genes that are nonessential under specified laboratory conditions. For *E. coli*, a combination of transposon mutagenesis and a site-specific recombination system such as Cre/*loxP* led to the creation of random deletions (6–9). In this chapter, we focus on *Pseudomonas putida*, a soil bacterium that has the natural ability to degrade organic solvents, mainly aromatic compounds such as *p*-hydroxybenzoate, protocatechuate, aldehydes, and acids through the different metabolic pathways into high-value compounds and is thus of great biotechnological relevance (10, 11). Therefore, the generation of *P. putida* strains that are easier to manipulate would be a great benefit for industrial applications.

### *1.1. Rationale of the Random Deletion System*

The successful application of a combinatorial deletion method based on minitransposon insertions to *P. putida* relies on the known properties of the Flp recombinase (Flp) of the yeast *Saccharomyces cerevisiae* and the 34-bp long *FRT* sequences that are used for the Flp-mediated recombination (12–17).

In a first step, two modified mini-Tn*5* transposons, mini-Tn*5 KpF* and mini-Tn*5 TF*, were generated to guarantee random insertion of the *FRT* sites into the genome of *P. putida*. Mini-Tn*5 KpF* is based on the vector pBAM1 and carries a kanamycin (Km) resistance cassette, the *pyrF* operon, and one *FRT* site. The second mini-Tn*5* transposon, mini-Tn*5 TF*, is based on the plasmid pJMT6 and carries a tellurite (Tel) resistance cassette and one *FRT* site (18). All the elements that need to be transferred to the chromosome by transposition have to be located between the outer- and inner-end transposase recognition sites (OE and IE) (19). These mini-Tn*5* derived transposons were transferred to the *P. putida* chromosome via two independent conjugation experiments thereby generating two different kinds of insertion mutant libraries, one having only the mini-Tn*5 KpF* inserted in the chromosome (Single Minitransposon (SMT) mutant strains) and the other one having both mini-Tn*5* transposons integrated (Two Minitransposon (TMT) mutant strains). The successful insertion of each mini-Tn*5* derivative as a single copy was verified by Southern blot hybridization with probes specific for the Km and the Tel resistance cassettes, as appropriate. To precisely map the position of both mini-Tn*5* transposons in all TMT mutant strains, AP-PCR was conducted for each mutant and the resulting PCR fragments were sequenced, subsequently revealing the coordinates of the minitransposons in the chromosome and the orientation of both *FRT* sites to each other (20).

Depending on the position of the minitransposons and the orientation of the *FRT* sites, *P. putida* TMT mutants were chosen for the final deletion step. For this purpose, the plasmid pBBFLP, which carries the Flp recombinase, was transferred to the mutant by triparental mating (21). The expression of the Flp recombinase finally led to the loss of the genomic fragment, which could be verified by its phenotype. As further confirmation, specific primers were chosen to amplify the scar that was left behind. Subsequent sequencing analysis of the PCR fragments revealed the excision of the genomic fragment that was situated between both *FRT* sites. To further increase the percentage of genomic deletions, the resulting strain can be used for the same random gene deletion process starting again with the integration of the mini-Tn*5* transposons.

## 2. Materials

### 2.1. Bacterial Strains and Plasmids

1. Genome streamlining is carried out in the *P. putida* strain TEC1 (Rifampicin resistant and uracil auxotroph mutant of strain KT2440). *E. coli* CC118λ*pir* is used for the replication of mini-Tn*5* derivative carrying plasmids and *E. coli* HB101 participates in the triparental mating as helper strain via the presence of plasmid RK600.
2. Plasmid pBAM1, which is an improved version of pUT Km1 but still contains the same single *Not*I restriction site within the minitransposon, carries the kanamycin resistance cassette between the O- and I-ends recognized by the transposase (Tnp). pBAM1 is used to construct the mini-Tn*5* *KpF* derivative by cloning the *pyrF* operon and one *FRT* site into the *Not*I site. This minitransposon is randomly inserted into the chromosome of *P. putida* TEC1 to generate *P. putida* SMT mutants.
3. Plasmid pJMT6, which is a pUT derivative carrying the tellurite resistance cassette, is used to construct the mini-Tn*5* *TF* derivative by cloning one *FRT* site into the *Not*I site. This minitransposon is designed to be randomly inserted into the chromosome of *P. putida* SMT to generate TMT mutants.
4. Plasmid patt*FRT* carries one *FRT* site and one Multiple Cloning Site (MCS), $Ap^R$.
5. pGEM®-T Easy vector (Promega, Madison, USA).
6. Plasmid pBBFLP ($Tc^R$) carries the Flp recombinase, which is used in the Flp-*FRT* site-specific recognition system of the yeast *S. cerevisiae*. The recombinase recognizes two *FRT* sites

either in the same orientation, which leads to the deletion of the fragment situated between both *FRTs* and their subsequent recombination, or in opposite direction, which leads to the inversion of the genomic fragment situated in between (22).

#### 2.2. Bacterial Strain Cultures

1. Luria Bertani (LB) medium (23) for *E. coli* strains. Overnight culture at 37°C.
2. LB supplemented with 0.002% uracil (ura) and 0.2% citrate (cit) for culture of *P. putida* strains, overnight cultures at 30°C.
3. M9 minimal medium; 12.8 g/l $Na_2HPO_4 \times 7H_2O$, 3 g/l $KH_2PO_4$, 0.5 g/l NaCl, 1 g/l $NH_4Cl$, 2 mM $MgSO_4$, citrate, and uracil. Sterilization of the magnesium sulfate by filtration. The medium is used for the selection of conditional deleted mutants and counterselection of *E. coli*.
4. LB or M9 plates supplemented with 1.5% bacto agar or Select agar, respectively.
5. LB or M9 media supplemented with 50 μg/ml piperacillin (Pip), 50 μg/ml kanamycin (Km), 40 μg/ml potassium tellurite (Tel), 50 μg/ml chloramphenicol (Cm), 15 μg/ml tetracycline (Tc), 8 μg/ml nalidixic acid (Na, to counterselect *E. coli*), and 5% sucrose (w/v, suc) for *E. coli* and *P. putida* strains when required. Tetracycline is dissolved in methanol; all other chemicals are soluble in water. All solutions are conserved at –20°C.
6. 300 μg/ml 5-fluoorotic acid (FOA, Zymo Research) dissolved in 0.5 M Tris–HCl pH 8.0 and stored at –20°C in small aliquots.

#### 2.3. Kits and Laboratory Equipment

1. Sterile Falcon tubes for cell culture.
2. Genomic DNA extraction kit (Wizard® Genomic DNA Purification, Promega).
3. Plasmid extraction (PureYield™ Plasmid Miniprep System, Promega).
4. Spectrophotometer.
5. PCR cleanup (QIAquick PCR purification Kit, Qiagen) and gel extraction (QIAquick Gel Extraction Kit, Qiagen) kits.
6. Taq DNA Polymerase (Qiagen) and High Fidelity PCR Enzyme Mix (Fermentas), supplied with the corresponding buffers.
7. Gel electrophoresis apparatus and equipment.

#### 2.4. Conjugation

1. 0.45-μm filter (mixed cellulose filter, Millipore).
2. LB Agar.
3. Petri dishes.

4. 10 mM $MgSO_4$ solution, sterilized by filtration.
5. M9 minimal medium (Molecular Cloning manual (23)).

#### 2.5. Southern Blot Analyses

1. Southern blot hybridization equipment: nylon membrane, Amersham ECL Direct Nucleic Acid Labeling and Detection Systems kit (GE Healthcare; Freiburg, Germany).
2. Hybridization tube and hybridizer.
3. Luminescent Image Analyzer LAS-3000 (Fujifilm).

## 3. Methods

The general procedure for random genomic deletion is presented in Fig. 1. The first round of deletion is presented in detail, which uses *P. putida* TEC1 as wild-type strain and which leads to the generation of *P. putida* $\Delta_1$ mutant.

#### 3.1. Construction of Two Mini-Tn5 Derivatives for Insertion in P. putida TEC1

The different derived minitransposons are designed to be easily inserted in the *P. putida* TEC1 genome (uracil auxotroph but resistant to fluoroorotic acid) and to be recognized by the Flp recombinase at the end of the process. The generation of the two mini-Tn*5 KpF* and *TF* derivatives is described in this section.

1. Mini-Tn*5 KpF*. Amplify the *pyrF* operon from *P. putida* TEC1 using primers *pyrF*1F (5′-cg*GGATCC*GTCGGGCCGTG-GGCAACG-3′) and *pyrF*2R (5′-cccgAAGCT TTTACCCAC GGATCTCCGC-3′), which add *Bam*HI and *Hin*dIII restriction sites at the extremities.
2. Clone the 876 bp fragment into the corresponding sites of the plasmid patt*FRT* downstream of the *FRT* site by ligation using T4 DNA ligase (Fermentas). The ligation mixture is transferred to *E. coli* DH5 for replication.
3. After plasmid extraction, amplify the <*FRT*::*pyrF*> fragment by PCR using primers Fp 1 F ( *Not*I) (5′-ataagaat*GCGGCCGC*GT-TGACAAAGGGAATCAGGG-3′) and Fp 2R (5′-ataagaat-*GCGGCCGC*AAGCTTTTAC-3′), which added *Not*I sites to both extremities, and purify it from the gel after electrophoresis.
4. Clone the purified fragment into the *Not*I restricted pBAM1 plasmid by overnight ligation at 4°C to generate plasmid pBAM/*KpF*.
5. Selection of the appropriate orientation for the <*FRT*::*pyrF*> fragment by restriction analyses.
6. Mini-Tn*5 TF*. Amplify by PCR an *FRT* fragment using primers: F 10 F (5′-ATTCCGGGC ATTGCTGTTGAC-3′) and F 11R (5′-GGCCTGGCACCTATAATTGAACC-3′). Purify the PCR

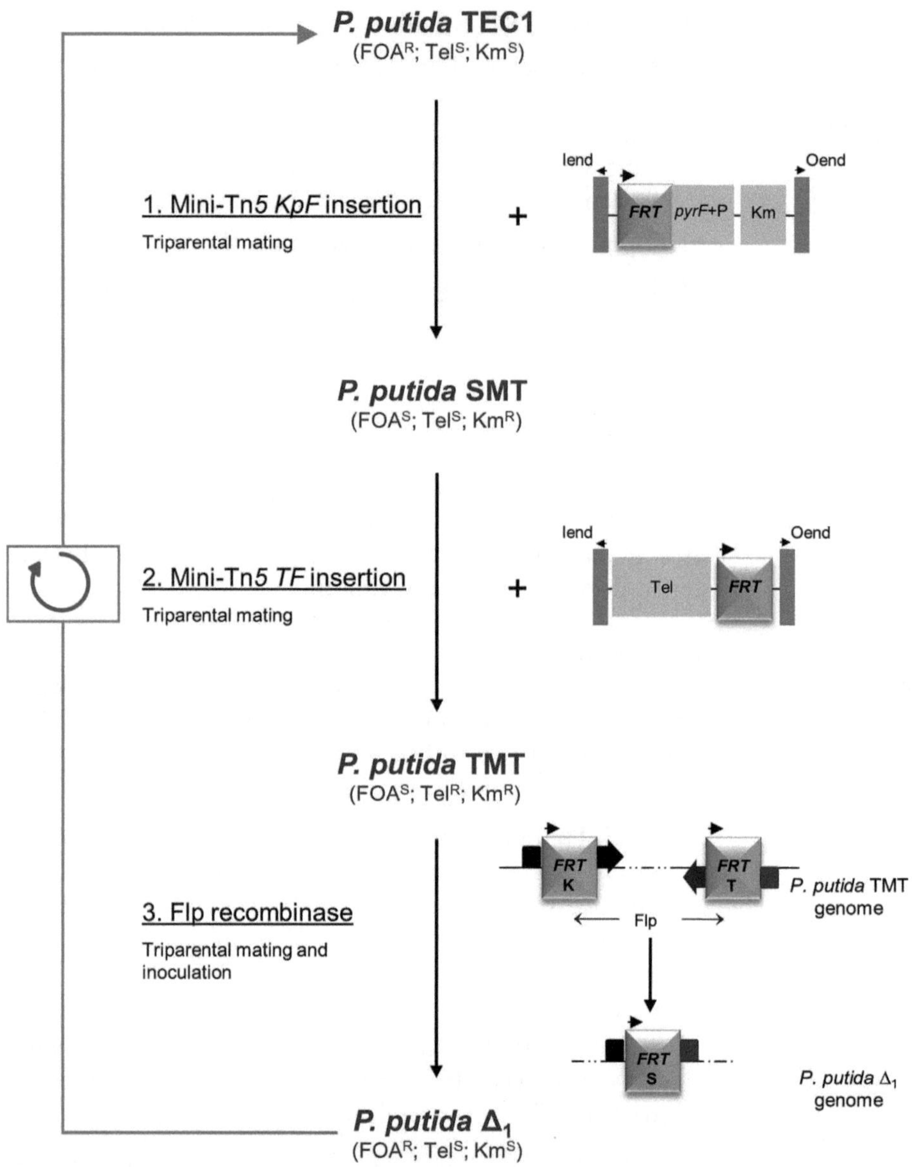


Fig. 1. General scheme of the random deletion process using *P. putida* TEC1 as wild-type strain. The first step allows the insertion by triparental mating of mini-Tn*5 KpF* in the genome of TEC1 and the generation of SMT mutants. The mini-Tn*5 KpF* is represented on the *top right side*. The orientation of the *FRT* site is indicated by a *black arrow*. *Gray boxes* symbolize the different elements of composition of the mini-Tn*5*: Km, the kanamycin resistance cassette; *pyrF*+P, the *pyrF* gene under its own promoter (P); *FRT*, the *FRT* site; O- and lend, the ends of the mini-Tn*5*. The second step allows the insertion by triparental mating of mini-Tn*5 TF* in the genome of selected SMT mutants and the generation. The mini-Tn*5 TF* is described on the *middle right side*. The *gray box* Tel represents the tellurite resistance cassette. The third and last step leads to genomic deletion by insertion of pBBFLP and expression of the Flp recombinase. The configuration of both mini-Tn*5 KpF* and *TF* derivatives in a *P. putida* TMT mutant is represented by *two gray boxes* ***FRT*** **K** and ***FRT*** **T**, respectively. The reduced genome of *P. putida* $\Delta_1$ is shown by a *single gray box* ***FRT*** **s**, which represents the scar formed after deletion and recombination between both *FRT* sites. The whole process is repeated by using for the second deletion *P. putida* $\Delta_1$ as wild-type strain.

product and clone it into pGEM®-T Easy vector. *Not*I restriction of the pGEM <*FRT*> delivers the *FRT* fragment framed by two *Not*I sites. *Not*I restriction of pJMT6 allows overnight ligation with the *FRT* at 4°C. The ligated products generate pUT/*TF*.

### 3.2. Establishment of Single- and Double-Inserted Mutants, *P. putida* SMT and TMT

Random transpositions of the mini-Tn*5* derivatives into the genome of *P. putida* TEC1 and generation of the SMT and TMT mutant strains are described in this section and are further verified for single and double insertion events by Southern blot analyses

#### 3.2.1. Generation of a Putative *P. putida* SMT Mutant Library

1. Inoculate separately in 3 ml LB, supplemented with the appropriate antibiotics, *P. putida* TEC1, *E. coli* CC118λ*pir* (pBAM/*KpF*), and *E. coli* HB101 (pRK600). Incubate overnight under continuous shaking (180 rpm) at the appropriate temperature.
2. Measure the optical density (OD) at 600 nm. Use a ratio of 1:1:2 of donor/helper/acceptor strains.
3. Centrifuge the strains separately at 18,000 × *g* for 4 min at 4°C. Discard the supernatant and resuspend each strain in 1 ml LB.
4. Centrifuge the strains at 18,000 × *g* for 3 min at 4°C. Discard the supernatant from each tube.
5. Resuspend the *E. coli* pBAM/*KpF* pellet (donor) in 200 μl LB. Resuspend the *E. coli* HB101 pellet (helper) with the previous mixture. Finally, mix the resuspension with *P. putida* TEC1 pellet. Apply these 200 μl, containing the three strains, to a single sterile filter (see Note 1), previously placed on an LB plate. Incubate the plate at 30°C (see Note 2) for 7–9 h.
6. After incubation resuspend the filter in 2 ml 10 mM $MgSO_4$. Vortex until the drop is detached from the filter and dissolved in the solution.
7. Plate 50 μl of the nondiluted mixture and 50 μl of each 1/10 and 1/100 diluted mixture on M9 + cit + ura + Km and incubate overnight at 30°C.
8. Prepare M9 plates containing only uracil and citrate and plates supplemented in addition with either Km, FOA, or Pip. Select 200 colonies from the conjugation and streak each of them on all four plates (see Note 3).
9. Select the putative mutants, which grow on M9 + cit + ura plates and exhibit the following phenotype: $Km^R$, $FOA^S$, and $Pip^S$.

   A library of SMT mutants, which carry a single *FRT* site in the genome, is created by collecting single colonies exhibiting this phenotype.

#### 3.2.2. Establishment of Putative *P. putida* TMT Mutants

Genomic deletions are monitored by the presence and orientation of two *FRT* sites recognized by the Flp recombinase. The insertion of the second *FRT* is carried out by the transposition of mini-Tn*5* *TF* into the genome.

1. Pool ten putative *P. putida* SMT mutants into one single LB culture (20 ml in 100 ml flask) supplemented with citrate, uracil, and kanamycin. Incubate the acceptor strains under continuous shaking overnight at 30°C. Inoculate separately *E. coli* CC118λ*pir* (pUT/*TF*) and *E. coli* HB101 (pRK600) overnight at 37°C.
2. Measure the $OD_{600nm}$. Use a ratio of 1:1:2 of donor/helper/acceptor strains to repeat the same procedure as described in Subheading 3.2, item 1 (3–6).
3. Plate 50 μl of the nondiluted mixture and 50 μl of each 1/10 and 1/100 diluted mixture on M9 + cit + ura + Km + Tel. Incubate overnight at 30°C.
4. Prepare M9 plates containing only citrate and uracil and plates supplemented in addition with Km, Tel, FOA, or Pip. Select 200 colonies from the conjugation and streak each of them on all five plates.
5. Select the putative mutants, which grow on M9 + cit + ura plates and exhibit the following phenotype: $Km^R$, $Tel^R$, $FOA^S$, and $Pip^S$.

   A library of 600 putative *P. putida* TMT mutants are separately grown overnight at 30°C in M9 medium supplemented with citrate, uracil, kanamycin, and tellurite to conserve them with glycerol at −80°C (see Note 4).

#### 3.2.3. Confirmation of the Mutants by Southern Blot Hybridization Experiments

In order to verify the insertion as a single copy of each mini-Tn*5* derivative in the genome of the putative mutants, genomic DNA (gDNA) is extracted for each of them and used as material for the Southern blot hybridization.

1. Extract and purify the gDNA with the Genomic DNA extraction kit. Measure the concentration with the NanoDrop at 260 nm. Restrict two times 4 μg of the purified DNA by incubating the DNA with *Eco*RV (see Note 5) for 2 h at 37°C.
2. Meanwhile prepare the probes by PCR amplification of the Km cassette from plasmid pBAM/*KpF* using the primers: Km-pBAM F (5′-TATTCAGCGTGAAACGAGCTGTAG-3′) and Km-pBAM R (5′-CGGATTATCAATGCCATATTTCTG-3′) and amplification of the Tel cassette from pUT/*TF* using the primers: Tel fw (5′-GAAGCAGGCGAGAAACTGAC-3′) and Tel rv (5′-TTGATGAGCGTCGTCTGAAC-3′). Extract and purify the probes from a 0.8% agarose gel.
3. In order to hybridize separately the samples with the Km and Tel probes, load two 0.8% agarose gels with the same restricted

DNA samples as well as the positive (plasmid pBAM/*KpF* or pUT/*TF*) and negative (restricted genomic DNA from *P. putida* TEC1) controls and the probes (one per gel) (see Note 6). Run the gels at 100 V for 40 min and stain them afterward for 15 min in a solution of ethidium-bromide (0.03% (v/v)). Take a picture of the gels for further comparison with the blots.

4. Rinse the gels in a HCl solution (250 mM, depurination step) for 11–12 min under gentle agitation, until bromophenol blue from the loading dye turns yellow and clean briefly the gels with distilled water. Pour a solution of NaCl (1.5 M) and NaOH (0.5 M) onto the gels and denature for 25 min after the color turns back to blue. Rinse briefly with distilled water. Finally, proceed with the neutralization step by pouring a solution of NaCl (1.5 M in Tris–HCl, pH 7.5) for 30 min.
5. Transfer the agarose-embedded fragmented DNA to two positively charged nylon membranes. Allow the DNA to migrate overnight at room temperature (see Note 7).
6. Label the probes following the provider's instructions.
7. Place each nylon membrane on the wall of a hybridization tube and add the hybridization buffer containing the labeled probe, one probe per tube (see Note 8). Incubate under continuous rotation overnight at 42°C.
8. Wash the membranes and add the detection reagents as described in the provider's instructions.
9. Detect the chemiluminescence with the Luminescent Image Analyzer LAS-3000.

The samples revealing a single band on the blots, which indicates a single transposition event for each mini-Tn*5* derivative, are kept for further experiments.

### 3.3. First Genomic Deletion and Generation of *P. putida* $\Delta_1$ Mutant

#### 3.3.1. Determination of the Position of Two Mini-Tn5 Derivatives in a *P. putida* TMT Mutant

After random transposition of both mini-Tn*5* derivatives and verification of the unique presence of each of them by Southern blot analyses, AP-PCR is carried out with the genomic DNA of every TMT mutant to obtain the coordinates of both inserted minitransposon derivatives. For each mutant, two amplification reactions are performed corresponding to the mapping of mini-Tn*5* *KpF* and mini-Tn*5* *TF*. An AP-PCR is composed of two consecutive rounds of amplification (see Note 9) (24):

1. First round: add buffer (final concentration: 1×), dNTPs (final concentration: 0.2 mM), dimethyl sulfoxide (DMSO, final concentration: 2.0%) (see Note 10), $MgCl_2$ (final concentration: 3 mM), DNA template (between 200 and 600 ng), forward and reverse primers (final concentration: 0.2 μM), and Taq DNA polymerase (1.5 U) to a 50 μl-reaction. Use one random primer, which binds to the genomic DNA, ARB1 (5′-GGCCACGCGTCGACTAGTACNNNNNNNNNN GATAT-3′) and one specific primer situated close and toward

one end of the mini-Tn*5*, either APPCR-Km1 (5′-CTACAGCTCGTTTCACGCTGAATA-3′) or kilA rev (5′-ACGCTTTGTTCTTCCATTCG-3′), which corresponds to the mapping of mini-Tn*5 KpF* and mini-Tn*5 TF*, respectively. Perform the amplification reaction according to the following program: 5 min at 95°C; 6 cycles of 30 s at 95°C, 60 s at 30°C, and 90 s at 72°C; 30 cycles of 30 s at 95°C, 60 s at 52°C, and 90 s at 72°C; and 4 min at 72°C.

2. Second round: use 2 μl of the PCR products from the first round as DNA template. Add the different components to a 100 μl-reaction. Use the following set of primers: ARB2 (5′-GGCCACGCGTCGACTAGTAC-3′), whose sequence is known and matches the defined sequence of random primer ARB1, and APPCR-Km3 (5′-CTTGTGCAATGTAACA TCAGAG-3′) or Iend Rev2 (5′-CTGTCTCTTATACA CATC-3′), which correspond to the extremity of mini-Tn*5 KpF* or mini-Tn*5 TF*, respectively. Perform the second amplification reaction according to the following program: 1 min at 95°C; 30 cycles of 30 s at 95°C, 30 s at 56°C, and 90 s at 72°C; and 4 min at 72°C.
3. Clean the PCR products with the QIAquick PCR Purification kit and send them for sequencing with the external primer (either APPCR-Km3 or Iend Rev2) (see Note 11).

The results are analyzed by applying the sequences in the Basic Local Alignment Search Tool for nucleotides (blastn) from NCBI, using *P. putida* KT2440 as the reference genome. A series of coordinates that deliver information regarding the relative position of both mini-Tn*5* derivatives as well as the direction of the two *FRT* sites is associated with the TMT mutants' library for each mutant. This information allows targeting the potentially successful candidates for genomic deletion, as both *FRT* sites need to be in the same orientation (21).

#### 3.3.2. First Genomic Deletion and Generation of a *P. putida* $\Delta_1$ Mutant Strain

Depending on the position of the minitransposons and the orientation of the two *FRT* sites, we first choose the *P. putida* TMT 407 mutant for the deletion step (see Note 12). This one comprises the insertion of pBBFLP vector via triparental mating and the expression of the Flp recombinase leading to the loss of the genomic fragment. The method is described hereafter:

1. Inoculate overnight the chosen *P. putida* TMT, *E. coli* HB101 (pRK600), and *E. coli* CC118λ*pir* (pBBFLP) at the appropriate temperature under continuous shaking.
2. Measure the $OD_{600nm}$ of each strain and use a ratio of 2:1:1, respectively.
3. Proceed as described in Subheading 3.2, item 1 (3–6).
4. Plate 50, 100, and 200 μl of the nondiluted mixture on LB + cit + ura + Na + Tc. Incubate overnight at 30°C.

5. Resuspend the plate exhibiting single colonies with LB medium (1 ml) supplemented with citrate, uracil, and tetracycline and dilute into 50 ml of the same medium. Incubate overnight at 30°C under continuous shaking.
6. Plate 50 μl of the nondiluted culture and 50 μl of a 1/10 and a 1/100 diluted culture on LB + cit + ura + Na + suc. Incubate at 30°C.
7. Screen single colonies for genomic deletion by streaking them on four different LB plates. Supplement the first plate only with citrate and uracil to collect biomass. Supplement separately the others with citrate and uracil to allow growth for the mutant and in addition with Km, Tel, or FOA. Incubate the plates at 30°C.

A putative deleted *P. putida* $\Delta_1$-407 mutant is selected from the LB plate supplemented with uracil and citrate when the following phenotype is observed: $Km^S$, $Tel^S$ and $FOA^R$. The confirmation of the deletion is carried out by PCR and sequencing as described in the following paragraph.

##### 3.3.3. Confirmation of the First Genomic Deletion

The mapping of mini-Tn*5 KpF* and mini-Tn*5 TF* in the genome of *P. putida* TMT 407 delivered precise information regarding the disrupted genes, which are indicated as gene A and gene B, respectively. Gene A is located 200 ORFs upstream of gene B. Specific primers (A fw and B rv) are designed with the Primer3-web (v. 0.4.0) interface of the Primer3 software, which correspond to 100–300 bp-long sequences situated upstream of mini-Tn*5 KpF* and downstream of mini-Tn*5 TF*, respectively. Four colonies, which represent the same putative deleted mutant, are selected for the amplification of the new scar that is theoretically left after deletion and recombination between both *FRT* sites. The PCR products are directly subjected to sequencing as follows:

1. Extract and purify the genomic DNA from the four colonies with the Genomic DNA extraction kit.
2. Amplify the scar by PCR using primers A fw and B rv. Use 300–500 ng genomic DNA and add buffer, dNTPs, primers, DMSO, $MgCl_2$, and Taq polymerase, in the same concentrations as previously mentioned, to a 100 μl-reaction.
3. Visualize the PCR products on a 0.8% agarose gel and purify them with the QIAquick PCR purification Kit (see Note 13). Send the purified PCR product for direct sequencing with each primer.

The sequences are aligned by Blast with the genome of *P. putida* KT2440 to match with the disrupted genomic fragments. The verification of the sequence of the scar itself, which is composed of two mini-Tn*5* ends and one *FRT* site, is allowed by alignment with both mini-Tn*5* derivatives. The four colonies are confirmed to lack the same 200 ORFs-genomic fragment; therefore, they all correspond to *P. putida* $\Delta_1$-407 mutant. Moreover, no mutations are found in the sequence of the scar.

### 3.4. Successive Deletion in an Already Reduced Strain

#### 3.4.1. Repetition and Improvement of the Deletion Method

Once the method is established and confirmed for the first genomic deletion, the process is repeated using *P. putida* $\Delta_1$-407 as the new wild type. The purpose is to increase the percentage of genomic deletions in the same strain following a random process. The same method for the triparental mating (see Subheading 3.2, item 1 (1–7)) is applied to the deleted mutant to insert the mini-Tn*5* *KpF*.

1. Resuspend the single colonies obtained after triparental mating with 1 ml LB supplemented with citrate, nalidixic acid, and kanamycin. Dilute into 50 ml of the same medium and grow at 30°C to mid-exponential phase.
2. Use the culture directly as the acceptor strain for the triparental mating involving *E. coli* HB101 (pRK600) and *E. coli* CC118λ*pir* (pUT/*TF*, donor strain) following the steps 2 and 3 described in Subheading 3.2, item 2.
3. Resuspend the single colonies with 1 ml of LB medium supplemented with citrate, nalidixic acid, kanamycin, and tellurite and diluted in 50 ml. Grow the culture at 30°C under continuous shaking until mid-exponential phase.
4. Use the culture directly as the acceptor strain for the triparental mating involving *E. coli* HB101 (pRK600) and *E. coli* CC118λ*pir* (pBBFLP, donor strain) following the steps 2–7 described in Subheading 3.3.

This last step allows the selection of *P. putida* $\Delta_1$-407 TMT mutants, which correspond to the first deleted mutant carrying two mini-Tn*5* derivatives and three *FRT* sites in the genome. Owing to the loss of both antibiotic resistance cassettes after the first deletion, it is easy to assess their presence after the second excision process. The colonies that exhibit one of the following phenotypes: ($Km^S$, $Tel^R$) or ($Km^R$, $Tel^S$) or ($Km^S$, $Tel^S$), $Pip^S$ and $FOA^R$ are selected for further experiments.

#### 3.4.2. Verification and Mapping of the Second Genomic Deletion

In order to increase the number of potential *P. putida* $\Delta_1$-407 TMT candidates no preselection is carried out between each step of mini-Tn*5*'s insertion and final deletion. However, this time-saving protocol does not allow the user to map the intermediary positions of both mini-Tn*5* derivatives. An in silico analysis predicts 24 different combinations of the three *FRT* sites before deletion by varying their relative position and their orientation. In order to identify the deleted region, it is important to first reveal the phenotype of the mutant (obtained in the last paragraph), which indicates the removal of one or both mini-Tn*5* derivatives. Second, considering the simultaneous presence of three *FRT* sites in the chromosome of the $\Delta_1$ TMT mutant, it is helpful to assess the role of the *FRT* from the scar that might also have been recognized by the flippase during the second round of deletion. For this purpose, the scar is amplified by PCR using the same primers and protocol as in Subheading 3.3, item 3. The size of the PCR products is verified on agarose gel. In the present case study, the selected colony highlights the following phenotype: $Km^S$, $Tel^R$, $Pip^S$, and $FOA^R$, involving the loss of the kanamycin resistance cassette. The agarose gel did not reveal any amplification of the scar. These two results indicate the interaction between the Flp recombinase, the *FRT* site from the scar and the *FRT* site in the mini-Tn*5 KpF*. Thus, the prediction of the "predeletion" configuration can be reduced to only four cases (See Fig. 2). Furthermore, the new scar that has been formed (***FRT*** **n** in Fig. 2) appears to have the same sequence independently of the case. Based on this information, the mapping of the deleted region is carried out as follows using different sets of primers to carry out three parallel AP-PCRs:

1. Confirm the downstream part of the new scar by AP-PCR using the two following sets of primers: X1/ARB1 (1st round) and X2/ARB2 (2nd round) (Fig. 3a) and the same protocol as given in Subheading 3.3 item 1 (1–3) (see Note 14).
2. Amplify by AP-PCR the upstream part of the new scar by using the following sets of primers: Y1/ARB1 and Y2/ARB2 (Fig. 3b) and the same protocol as given in Subheading 3.3 item 1 (1–3).
3. Amplify by AP-PCR the upstream and downstream parts of the mini-Tn*5 TF* present in the genome after the second deletion, as described above. Use the following set of primers: Z1/ARB1 and Z2/ARB2 (Fig. 3c) and the same protocol as previously described for AP-PCR. The corresponding part of the genome matching with the sequencing result indicates, meanwhile, the position and the orientation of the mini-Tn*5 TF*.

The sequencing results of step 1 and 2 provide the necessary information regarding the part of the genome that was excised by action of the Flp recombinase. In the present example, the second excision enlarged the deletion size of the first due to the interaction of the flippase with the *FRT* site of the first scar and the *FRT* site of the second mini-Tn*5 KpF* (see Fig. 3). The total size of the deletion after the two rounds of experiments reached 4.7% of the chromosome (see Note 15).

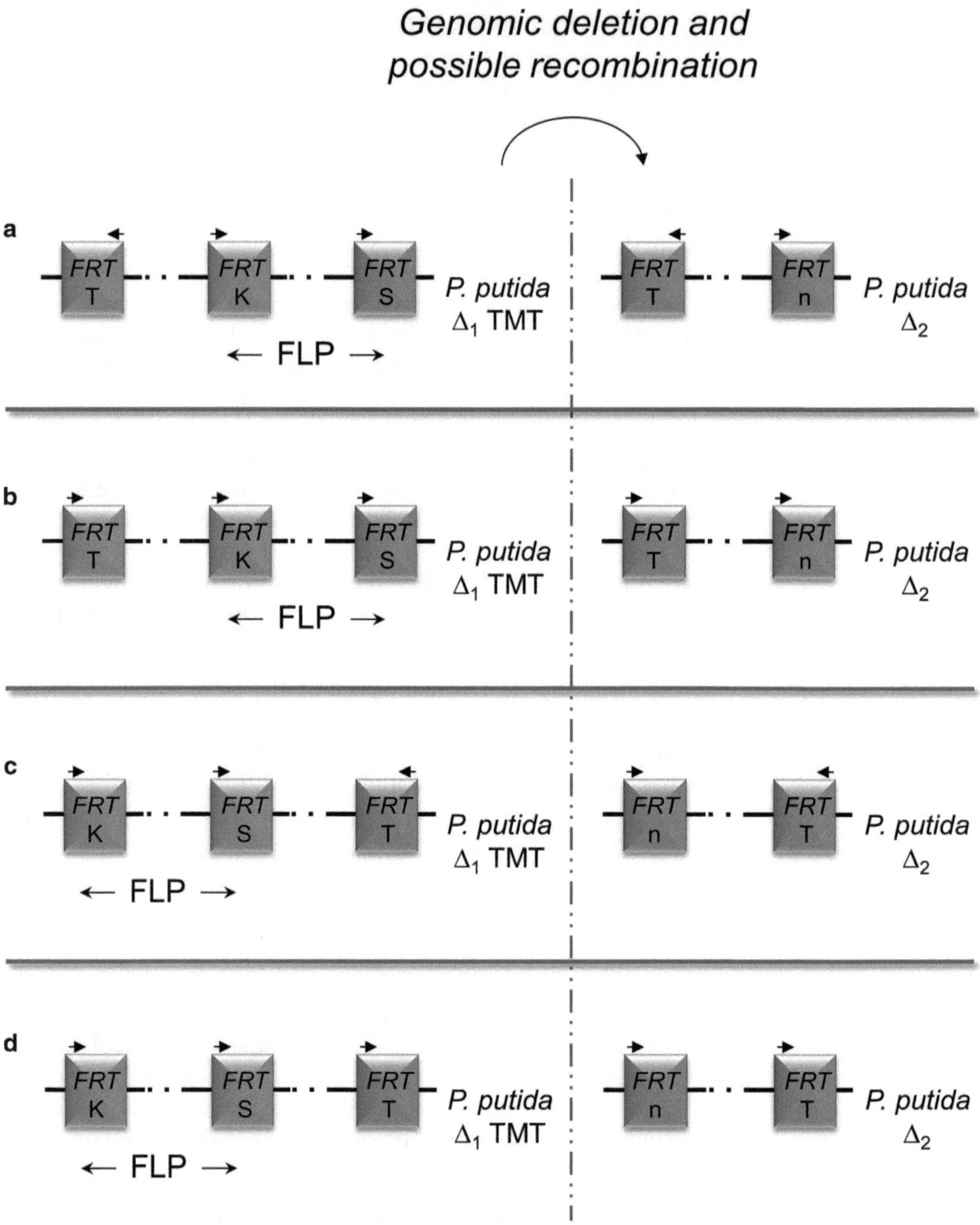


Fig. 2. Possible configurations of the three *FRT* sites in the genome of *P. putida* $\Delta_1$ TMT mutant leading to the reduced *P. putida* $\Delta_2$ strain, which exhibits the following phenotype: $Km^S$, $Tel^R$. Each mini-Tn*5* derivative is represented by a *box*. Its orientation is determined by the direction of the *FRT* site (marked as a *black arrow*). Mini-Tn*5 KpF* and mini-Tn*5 TF* are indicated by the letters K and T, respectively. The scar (S) formed after the first deletion/recombination is represented by a *box* and the orientation of its *FRT* site is highlighted by a *black arrow*. In all cases, the Flp recombinase (FLP) initiates the second deletion and the recombination between the *FRT* sites from mini-Tn*5 KpF* and from the scar (*double arrow*). In each case, the new scar (n) formed after recombination is the same and is symbolized on the *right side of the figure*. The part downstream of the scar after deletion remains unmodified after the second deletion. (**a** and **b**) The direction of the *FRT* in the mini-Tn*5 TF*, negative and positive, respectively, differentiates these two cases: (**a**) the recombination case shown is the only possibility, (**b**) the recombination case shown is the only one leading to the apparent phenotype of the mutant. (**c** and **d**) The scar is situated between both mini-Tn*5* derivatives. The *FRT* site from mini-Tn*5 TF* is oriented negatively or positively, respectively.

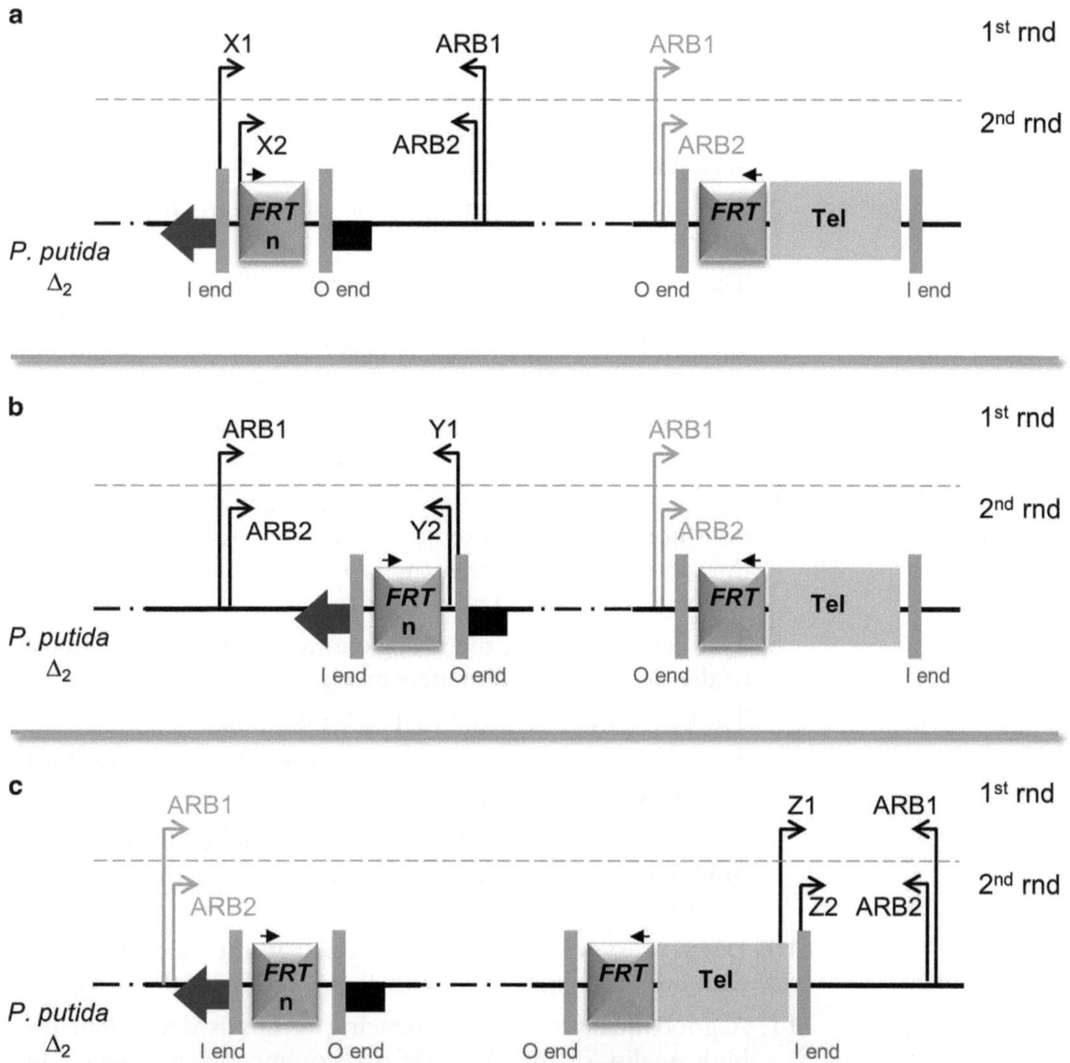


Fig. 3. Mapping of the second deletion for a *P. putida* $\Delta_2$ mutant. Two *FRT* sites are present in the genome: from the second scar (framed by I- and O end) and from the mini-Tn*5 TF* (framed by O end and the Tel cassette). (**a**) AP-PCR is performed using the two sets of primers (in *black*): X1/ARB1 and X2/ARB2 to control the presence of the downstream gene (*black box*), which is conserved from the first deletion. The primers ARB1 and ARB2 in *gray* indicate the different possibilities of annealing in the genome (also in case **b** and **c**). (**b**) AP-PCR is performed using the two sets of primers (*in black*): Y1/ARB1 and Y2/ARB2 to sequence the fragment situated upstream of the newly formed scar (***FRT* n**, *dark gray box*). (**c**) AP-PCR is performed using the two sets of primers (*in black*): Z1/ARB1 and Z2/ARB2 to highlight the position and orientation of the mini-Tn*5 TF*, which did not participate in the second deletion.

## 4. Notes

1. Leave the plate in a sterile environment for 10–30 min for drying the drop slightly on the filter.
2. The incubation temperature corresponds to the one of the acceptor strain.
3. There is always a possibility for the mini-Tn*5* carrying plasmid to fully integrate in the chromosome. In order to verify the absence of the backbone of the vector, the exconjugants are screened for their sensitivity to piperacillin.
4. To facilitate the conservation process, *P. putida* SMT and TMT mutants can be placed in 96-well plates at −80°C.
5. Choose the restriction enzyme in such a way that the genome does not contain the site in a significantly recurrent way. Take into account the fact that the enzyme should not recognize a site within the mini-Tn*5* derivative.
6. Take care of diluting the probe enough to reduce the intensity of the signal on the blot after revelation.
7. The time can be reduced to 4–5 h minimum.
8. Take care to mix first the probe with the hybridization buffer before touching the blots.
9. Primers ARB1 and ARB2 were directly applied from the work of O'Toole and colleagues, already tested for the genome of *P. putida* strains.
10. DMSO serves as an enhancing agent in the case of genomes with high GC content such as *P. putida* KT2440 (61.5%).
11. A good quality of the sequencing results is dependent on a high quality of the DNA. It is recommended to finally elute the PCR products with distilled water. Use 40 μl and leave it with closed cap at room temperature for 3–4 min. Centrifuge and measure the DNA concentration with NanoDrop. Send the appropriate concentration of DNA and primers according to the sequencing service.
12. The action of the Flp recombinase leads to the deletion of the fragment situated between two *FRT* sites only when these sites are positioned in the same direction. If this requirement is fulfilled, it is interesting to estimate the size of the fragment and the nature of the genes to delete. This latest information indicates a lower or higher probability of success in the deletion process.
13. If unspecific products appear to be amplified (more than one band on the gel), extract and purify the band corresponding to the size predicted in silico.

14. Adapt the annealing temperature in function of the two primers used for the PCR. It should be in a range of maximum 5°C below the melting temperature of both primers.
15. The high-throughput method for random genomic deletion offers the possibility to obtain excisions of nonessential fragments from the genome under specific laboratory conditions in a faster and more reliable way. It avoids intermediary steps of mapping one mini-Tn5 derivative after the other for a high number of SMT and TMT mutants and focuses directly on the mapping of the deletion obtained after the different steps. However, it happens that in function of the possible configurations found for the three *FRT* sites in *P. putida* $\Delta_1$ TMT mutant the AP-PCR needs to be repeated a certain number of times requiring optimization of the sets of primers and of the different cycles. It is, therefore, conceivable to carry out the second round of deletion by repeating each step from the first round, meaning using intermediary mappings.

## References


1. Kolisnychenko, V., Plunkett, G., 3rd, Herring, C. D., *et al.* (2002) Engineering a reduced Escherichia coli genome, *Genome Res* *12*, 640–647.
2. Hashimoto, M., Ichimura, T., Mizoguchi, H., *et al.* (2005) Cell size and nucleoid organization of engineered Escherichia coli cells with a reduced genome, *Mol Microbiol* *55*, 137–149.
3. Hutchison III, C. A., Peterson, S. N., Gill, S. R., *et al.* (1999) Global transposon mutagenesis and a minimal mycoplasma genome, *Science* *286*, 2165–2169.
4. Pósfai, G., Plunkett Iii, G., Fehér, T., *et al.* (2006) Emergent properties of reduced-genome Escherichia coli, *Science* *312*, 1044–1046.
5. Kobayashi, K., Ehrlich, S. D., Albertini, A., *et al.* (2003) Essential Bacillus subtilis genes, *Proc Natl Acad Sci U S A* *100*, 4678–4683.
6. Abremski, K., Hoess, R., and Sternberg, N. (1983) Studies on the properties of P1 site-specific recombination: evidence for topologically unlinked products following recombination, *Cell* *32*, 1301–1311.
7. Hoess, R. H., and Abremski, K. (1985) Mechanism of strand cleavage and exchange in the Cre-lox site-specific recombination system, *J Mol Biol* *181*, 351–362.
8. Sternberg, N., and Hamilton, D. (1981) Bacteriophage P1 site-specific recombination. I. Recombination between loxP sites, *J Mol Biol* *150*, 467–486.
9. Sternberg, N., Hamilton, D., and Hoess, R. (1981) Bacteriophage P1 site-specific recombination. II. Recombination between loxP and the bacterial chromosome, *J Mol Biol* *150*, 487–507.
10. Jimenez, J. I., Minambres, B., Garcia, J. L., *et al.* (2002) Genomic analysis of the aromatic catabolic pathways from Pseudomonas putida KT2440, *Environ Microbiol* *4*, 824–841.
11. Wackett, L. P. (2003) Pseudomonas putida – a versatile biocatalyst, *Nat Biotechnol* *21*, 136–138.
12. Sadowski, P. D. (1995) The Flp recombinase of the 2-microns plasmid of Saccharomyces cerevisiae, *Prog Nucleic Acid Res Mol Biol* *51*, 53–91.
13. Chen, Y., and Rice, P. A. (2003) New insight into site-specific recombination from Flp recombinase-DNA structures, *Annu Rev Biophys Biomol Struct* *32*, 135–159.
14. Huang, L. C., Wood, E. A., and Cox, M. M. (1997) Convenient and reversible site-specific targeting of exogenous DNA into a bacterial chromosome by use of the FLP recombinase: the FLIRT system, *J Bacteriol* *179*, 6076–6083.
15. Hoang, T. T., Karkhoff-Schweizer, R. R., Kutchma, A. J., *et al.* (1998) A broad-host-range Flp-FRT recombination system for site-specific excision of chromosomally-located DNA sequences: application for isolation of unmarked Pseudomonas aeruginosa mutants, *Gene* *212*, 77–86.
16. Storici, F., Coglievina, M., and Bruschi, C. V. (1999) A 2-microm DNA-based marker

recycling system for multiple gene disruption in the yeast Saccharomyces cerevisiae, *Yeast* **15**, 271–283.

17. Kopke, K., Hoff, B., and Kuck, U. (2010) Application of the yeast FLP/FRT recombinase system in filamentous fungi: Marker recycling and construction of knockout strains devoid of heterologous genes, *Appl Environ Microbiol* **76**, 4664–4674.
18. Sanchez-Romero, J. M., Diaz-Orejas, R., and De Lorenzo, V. (1998) Resistance to tellurite as a selection marker for genetic manipulations of Pseudomonas strains, *Appl Environ Microbiol* **64**, 4040–4046.
19. Reznikoff, W. S. (2003) Tn5 as a model for understanding DNA transposition, *Mol Microbiol* **47**, 1199–1206.
20. Leprince, A., De Lorenzo, V., Völler, P., *et al.* (submitted) Deletion of large genomic segments of Pseudomonas putida through recombination of pairs of FRT sites randomly inserted with specialized mini-Tn5 transposons.
21. Delas Heras, A., Carreno, C. A., and De Lorenzo, V. (2008) Stable implantation of orthogonal sensor circuits in Gram-negative bacteria for environmental release, *Environ Microbiol* **10**, 3305–3316.
22. Schweizer, H. P. (2003) Applications of the Saccharomyces cerevisiae Flp-FRT system in bacterial genetics, *J Mol Microbiol Biotechnol* **5**, 67–77.
23. Sambrook, J., Maniatis, T., and Fritsch, T. (1989) *Molecular cloning: a laboratory manual*, Cold Spring Harbor Laboratory, Cold Spring Harbor, N. Y.
24. O'Toole, G. A., Pratt, L. A., Watnick, P. I., *et al.* (1999) Genetic approaches to study of biofilms, *Methods Enzymol* **310**, 91–109.

# Chapter 16

# Transposon-Based and Plasmid-Based Genetic Tools for Editing Genomes of Gram-Negative Bacteria

**Esteban Martínez-García and Víctor de Lorenzo**

## Abstract

A good part of the contemporary synthetic biology agenda aims at reprogramming microorganisms to enhance existing functions and/or perform new tasks. Moreover, the functioning of complex regulatory networks, or even a single gene, is revealed only when perturbations are entered in the corresponding dynamic systems and the outcome monitored. These endeavors rely on the availability of genetic tools to successfully modify *á la carte* the chromosome of target bacteria. Key aspects to this end include the removal of undesired genomic segments, systems for the production of directed mutants and allelic replacements, random mutant libraries to discover new functions, and means to stably implant larger genetic networks into the genome of specific hosts. The list of gram-negative species that are appealing for such genetic refactoring operations is growingly expanding. However, the repertoire of available molecular techniques to do so is very limited beyond *Escherichia coli*. In this chapter, utilization of novel tools is described (exemplified in two plasmids systems: pBAM1 and pEMG) tailored for facilitating chromosomal engineering procedures in a wide variety of gram-negative microorganisms.

**Key words:** Synthetic biology, Insertional mutagenesis, Genetic stability, Mini-transposons, Chromosomal deletions, Minimal genome

## 1. Introduction

The way that goes from genetically engineered microorganisms to synthetic, or at least heavily refactored counterparts involves the stepwise replacement of growing portions or their naturally occurring genomes by rationally designed and chemically manufactured DNA (1). Although the current ability to synthesize long genomic segments is now in the range of 1 Mb (2), the contemporary level of knowledge does not allow assembling new activities or genetic circuits involving more than 20–30 kb of engineered DNA. It is thus likely that still for some time most synthetic biology endeavors

Wilfried Weber and Martin Fussenegger (eds.), *Synthetic Gene Networks: Methods and Protocols*,
Methods in Molecular Biology, vol. 813, DOI 10.1007/978-1-61779-412-4_16, © Springer Science+Business Media, LLC 2012

of this sort will focus on handling a relatively short range of DNA sizes, whether for deletions from existing chromosomes, for genomic replacements of alleles by designed variants, or by straight implantation of new sequences. Numerous genetic tools exist in *E. coli* to this end (3, 4) but the state of affairs for other biotechnologically relevant gram-negative bacteria, such as *Pseudomonas putida* is far less satisfactory. In this chapter, two strategies are described for implementing a large number of genetic manipulations in the chromosome of a large variety of gram-negative microorganisms. For this purpose, *P. putida* was used as the target bacterium, and the constructs named pBAM1 and pEMG (Fig. 1) adopted to give details of the underlying concepts and their practical application. As explained below, these plasmids are tailored for either implantation/insertion of heterologous DNA segments in the genome of the targeted strain as well as for directed mutagenesis or deletion or prespecified chromosomal regions. Even though the procedures are different for each plasmid system, their utilization share a good deal of the biological materials listed in the corresponding section below.

The organization and properties of pBAM1 (Fig. 1a) make it suitable to be used either for creating saturated random transposon

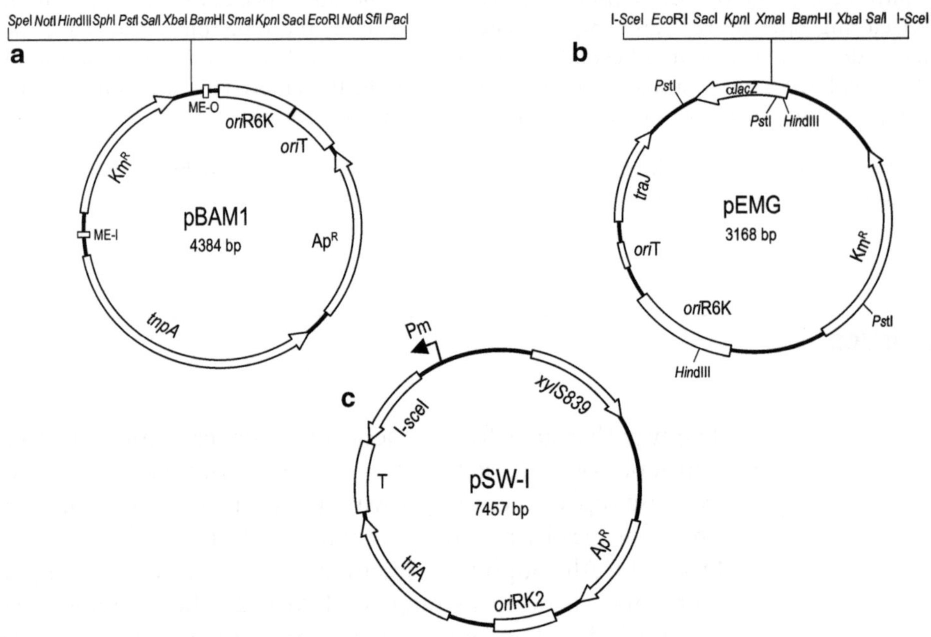


Fig. 1. Plasmids maps. (**a**) pBAM1. Functional elements of the plasmid include relevant restriction sites, antibiotic markers (Ap ampicillin, Km kanamycin), transposase (*tnpA*), origin of replication (R6K), the origin of transfer region (*oriT*), mosaic element 0 (ME-0), and mosaic element I (ME-I), as shown (8). (**b**) pEMG. Note functional elements, in particular multiple cloning site (MCS) flanked by I-SceI sites and merged in α*lacZ* sequence (9). (**c**) pSW-I (I-SceI), Broad host range plasmid for conditional expression of I-SceI nuclease (11). The *Pm* promoter is activated by 3-methylbenzoate because of the action of a variant of the XylS regulator, which tightly controls transcription (12).

mutant libraries or for stably introducing gene networks or functional cassettes into the genome of a specific bacterial host. Both of these properties are due to the special characteristics of mini-transposons (5, 6). The pBAM1 plasmid is composed of four functional blocks (Fig. 1a). The first segment correspond to the plasmid selectable marker, ampicillin. Next, an R6K origin of replication that makes its maintenance dependent of the trans supply of the p protein (*pir* gene; (7)). Thus, pBAM1 must be replicated in specialized *E. coli* strains, which expresses the pi protein from a lysogenic phage, such as *E. coli* CC118*λpir* (5). A Tn*5* transposase borne by the same plasmid (*tnpA*) recognizes the end sequences of the mini-Tn*5* transposon module (ME-I and ME-O) and catalyzes the random motion of the mini-Tn*5* cassette the target genome (4). All of these features have been individually optimized, cured of the most common restriction sites present within its sequence, and then assembled and chemically synthesized de novo (8). The pBAM1 frequencies of transposon insertions when applied to *P. putida* are in the range of $10^{-3}$ when the plasmid is delivered to the recipient by mating (*see* procedure below), or $10^{-7}$ when electroporation is used as an alternative method of suicide donation (8). On the other hand, the pEMG plasmid (Fig. 1b) is used to generate directed scar-less deletions, as well as allelic replacements in the genome (9). This genetic system is a recreation of the method developed by (3) for the same purpose in *E. coli.* The procedure is based on the homologous recombination forced by the appearance of double-strand breaks (DSB) in the genome of the target bacterium upon cleavage in vivo by I-SceI, a homing endonuclease from *Saccharomyces cerevisiae* that recognizes an 18-bp DNA sequence (10). The I-SceI recognition sequence is not present in any of the microbial genomes sequenced so far, as revealed by blast search against 1,379 completed genomes (http://www.ncbi.nlm.nih.gov/sutils/genom_table.cgi). In this way, integration of pEMG into the chromosome of choice endows flanking I-SceI target sequences to the extremes of α*lacZ*-pUC19-based polylinker. Intracellular expression of the I-SceI enzyme in live bacteria is brought about in vivo by the cognate pSW-I (I-SceI) plasmid hereafter named pSW-I (Fig. 1c; 11). Transient expression of the nuclease is tightly controlled in pSW-I by means of the 3-methylbenzoate-inducible promoter Pm (12). The steps of the deletion strategy process include (1) cloning regions homologous to those flanking the desired deletion/replacement into pEMG, (2) cointegrating the resulting plasmid into the genome of the target host, (3) introduction of pSW-I into cells bearing the cointegrate, (4) induction of the DSBs, (5) selection of the deleted/replaced strain and (5) pSW-I curation. Details for all these procedures with pBAM1 and pEMG follow.

## 2. Materials

### 2.1. Strains

See Table 1.

### 2.2. Media and Reagents

1. Luria Bertani (LB): 10 g/l tryptone, 5.0 g/l yeast extract, 5 g/l NaCl, dissolve in $H_2O$ and autoclave. For LB-agar plates, the same adding 1.5% agar.
2. Minimal medium (M9; see Note 1). 10× stock solution of M9: weight 42.5 g $Na_2HPO_4 \cdot 2H_2O$, 15 g $KH_2PO_4$, 2.5 g NaCl and 5 g $NH_4Cl$ and dissolve in 500 ml of $H_2O$. Other stocks: 1 M $MgSO_4$, 20% citrate (as selective carbon source for *Pseudomonas*), 1.6% agar solution. Prepare the four stocks and autoclave them separately. Dilute components in sterile water to final concentrations 1× M9 salts, 2 mM $MgSO_4$, 0.2% citrate (where required, 1.4% agar).
3. Ampicillin (Ap): 150 mg/ml in $H_2O$, filter sterilize. Store at −20°C. For *E. coli* cells use it at a final concentration of 150 μg/ml; for *P. putida* 500 μg/ml.
4. Kanamycin (Km): 50 mg/ml in $H_2O$, filter sterilize. Store at −20°C. Use it at a final concentration of 50 μg/ml.
5. Chloramphenicol (Cm): 30 mg/ml in ethanol. Store at −20°C. Use it at a final concentration of 30 μg/ml.
6. 5-Bromo-4-chloro-3-indolyl-β-D-galactopyranoside (Xgal): 2% in dimethylformamide. Use it at a final concentration of 40 μg/ml. Store at −20°C, protect from light.

**Table 1**

| Strain (plasmid) | Description/relevant characteristics | References |
|---|---|---|
| *E. coli* strains | | |
| CC118λ*pir* (pBAM1) | Δ(*ara-leu*), *araD*, Δ*lac*X174, *galE*, *galK*, *phoA*, *thi1*, *rpsE*, *rpoB*, *argE (Am)*, *recA1*, lysogenic λ*pir*. pBAM1 recipient strain | (8, 20) |
| DH5α (pSW-I) | *supE44*, Δ*lacU169*, (φ*80 lacZDM15*), *hsdR17*, (*rk⁻mk⁺*), *recA1*, *endA1*, *thi1*, *gyrA*, *relA*. pSW-I recipient strain | (8, 21) |
| DH5αλ*pir* (pEMG) | DH5α, lysogenic λ*pir*. pEMG recipient strain | (8) |
| HB101(pRK600) | $Sm^r$, *hsdR⁻M⁺*, *pro*, *leu*, *thi*, *recA*. Host of pRK600, mating helper strain | (22) |
| *Pseudomonas putida* strains | | |
| KT2440 | Prototrophic, wild-type strain | (14) |
| KT2442 | KT2440, spontaneous rifampycin resistant | (23) |
| MAD1 | KT2442, $Tel^R$, chromosomal insertion of mini-*Tn*5 [*xylR*, *Pu-lacZ*] | (24) |

7. Isopropyl β-D-1-thiogalactopyranoside (IPTG): 1 M in water. Use it at a final concentration 1 mM. Store at −20°C.
8. Sucrose: 300 mM in $H_2O$. Autoclave and keep at room temperature.
9. 3-methylbenzoic acid (Fluka #89890): 0.5 M in water, slowly add concentrated (5 M) NaOH until completely dissolved, and filter sterilize. Use it at a final concentration of 15 mM. Store at room temperature.

### 2.3. DNA Techniques

1. For plasmid preparations one can use, among others, the Wizard Plus SV Minipreps kit (Promega).
2. PCR-amplified DNA and agarose DNA extractions are commonly purified with the NucleoSpin Extract II kit (MN).
3. In DNA ligations, the Quick Ligation™ kit (NEB) is employed.
4. PCR: For general PCR reactions, prepare first a PCR reaction mix in an Eppendorf tube. The values given below are calculated for a final volume of 25 μl per tube:
   (a) 5 μl buffer 5×.
   (b) 1.5 μl $MgCl_2$ 25 mM.
   (c) 1 μl of upstream primer 5 μM.
   (d) 1 μl of downstream primer 5 μM.
   (e) 0.5 μl dNTPs 10 mM.
   (f) 0.3 μl DMSO (specially recommend for organism with a high GC genome content).
   (g) 0.2 μl Taq Polymerase.

   Then, add 15.5 μl of sterile $H_2O$ into each of the PCR reaction tubes. Pick fresh single colonies directly from a plate and transfer it directly to the PCR reaction tube containing the 15.5 μl of $H_2O$. Vortex the PCR reaction mixture and distribute 9.5 μl into PCR tubes. Set up the PCR machine with the appropriate Tm (depending on primer composition) and extension time.

### 2.4. PCR Primers

#### 2.4.1. pBAM1 Mini-transposon-Related Primers

ME-I Primers (8)

- ME-I-extR: 5′-CTCGTTTCACGCTGAATATGGCTC-3′. PCR round-1.
- ME-I-intR: 5′-CAGTTTTATTGTTCATGATGATATA-3′. PCR round-2 and sequence.

ME-O Primers (8)

- ME-O-extF: 5′-CGGTTTACAAGCATAACTAGTGCGGC-3′. PCR round-1.
- ME-O-intF: 5′-AGAGGATCCCCGGGTACCGAGCTCG-3′. PCR round-2 and sequence.

Arbitrary primers (13; see Note 2)

- ARB6: 5′-GGCACGCGTCGACTAGTACNNNNNNNNNNACGCC-3′. PCR round-2.
- ARB2: 5′-GGCACGCGTCGACTAGTAC-3′. PCR round-2.

*2.4.2. Primers for the Deletion/Replacement System*

- pSW-F: 5′-GGACGCTTCGCTGAAAACTA-3′. Presence/absence of pSW-I plasmid.
- pSW-R: 5′-AACGTCGTGACTGGGAAAAC-3′. Presence/absence of pSW-I plasmid.

## 3. Methods

### *3.1. pBAM1 Transposon Mutagenesis*

The pBAM1 plasmid (8) can be delivered into recipient cells by either mating or electroporation, at user's choice. Each of the two plasmids introduction method is explained below. The following protocol is optimized for the gram-negative soil bacterium *P. putida* (14, 15) but can surely be employed with minor modifications in other bacteria.

*3.1.1. Preparing the Mating Mixture*

1. Aerobically grow overnight cultures of:
   (a) Donor cells: *E. coli* CC118λ*pir* (pBAM1) in LB with ampicillin 150 μg/ml at 37°C.
   (b) Helper cells: *E. coli* HB101 (pRK600) in LB with chloramphenicol 30 μg/ml at 37°C.
   (c) Recipient cells: *P. putida* cells in LB at 30°C.
2. Distribute 1,000 μl of each of the overnight cultures in separate 1.5 ml Eppendorf tubes.
3. Centrifuge the three Eppendorf tubes at 10,000×*g* for 2 min at room temperature in a tabletop centrifuge.
4. Dispose of supernatants and add 1,000 μl of 10 mM $MgSO_4$ to each Eppendorf tube and resuspend cells.
5. Adjust the $OD_{600}$ of the three cultures to 0.03 (~$3 \times 10^7$ cells) in a 1.5 ml Eppendorf tube.
6. Mix the three strains in a 1:1:1 ratio (e.g.: 100 μl of each strain) in an empty test-tube and complete with 4.7 ml of 10 mM $MgSO_4$.
7. Vortex the mixture and pass through a Millipore filter disk (0.45-μm pore-size, 13 mm diameter) using a 10 ml sterile syringe.
8. Place the filter onto an LB agar plate and incubate for a maximum of 5 h at 30°C (in this case, the optimum temperature for recipient cells).

9. With the help of flamed sterile forceps take the filter from the LB plate and introduce it into a test-tube with 5.0 ml of 10 mM $MgSO_4$ and vortex to resuspend cells.
10. Plate appropriate dilutions (see Note 3) onto M9 citrate with Km 50 μg/ml (selective media for exconjugant *P. putida* cells) onto 140 mm diameter Petri dishes. Also plate separately onto M9 citrate with Km 50 μg/ml as negative controls, *E. coli* CC118λ*pir* (pBAM1), *E. coli* HB101 (pRK600), and the recipient strain (Fig. 2).

*3.1.2. Preparing P. putida Electrocompetent Cells*

Protocol adapted from ref. 16.

1. Grow overnight a 20 ml LB culture of *P. putida* in a 100 ml flask at 30°C with shaking.
2. Distribute the whole culture in a 50 ml sterile Falcon and centrifuge at 5,000×*g* at room temperature for 10 min.
3. Remove supernatant and resuspend the pellet gently in 10 ml of 300 mM sucrose and centrifuge as in step 2.

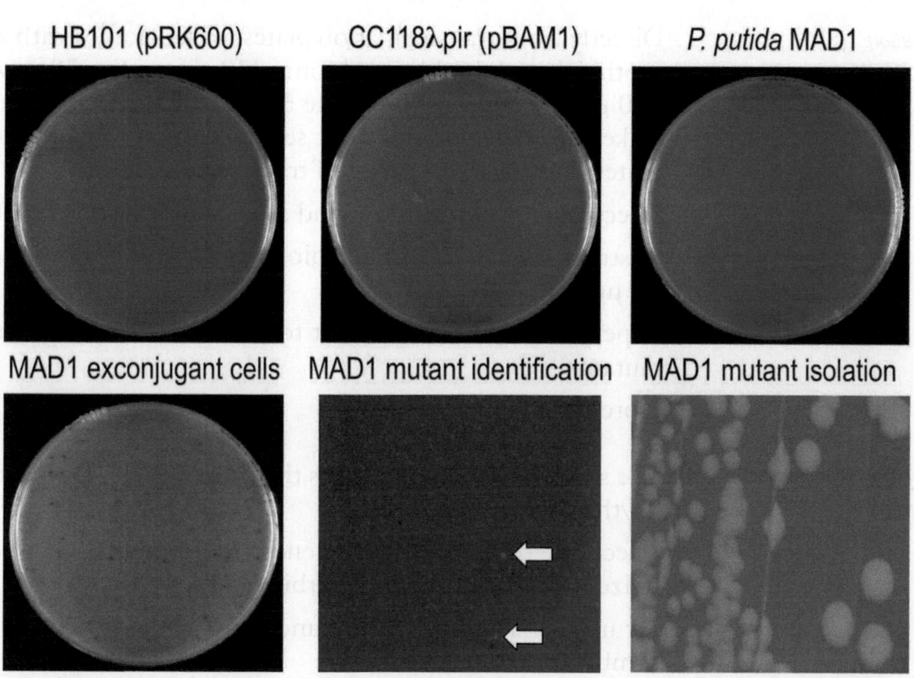


Fig. 2. Representative stages of a pBAM1-based transposition experiment. This picture corresponds to a pBAM1 transposition into *P. putida* MAD1 strain (24). The MAD1 strain bears in its genome an *xylR*/*Pu-lacZ* fusion that makes the strain to respond to *m*-xylene, thereby turning blue in M9 citrate + Km 50 μg/ml + Xgal medium + *m*-xylene vapors. Thus, transposition of the mobile element into any of these genes will render white cells. This setup pictures the typical output of a phenotypical transposition screening. In the upper part, the negative controls of the experiment are shown. These include *E. coli* HB101 (pRK600), *E. coli* CC118λ*pir* (pBAM1), and *P. putida* MAD1 cells plated onto M9 citrate + Km 50 μg/ml + Xgal and exposed to *m*-xylene vapors. The first picture on the lower part pictured a plate onto which 200 μl from the mating mix was spread. The second lower picture is a zoom in of the exconjugant plate, where two white clones can be observed. After several isolation passes, white clones are picked (*third lower picture*) which are later subjected to arbitrary PCR for determining the specific transposon insertion point.

4. Discard supernatant, add 1 ml of 300 mM sucrose, resuspend the pellet and transfer it to a 1.5 ml Eppendorf tube.
5. Centrifuge the culture at 8,000 × *g* for 2 min at room temperature.
6. Dispose of supernatant and add 500 μl of 300 mM sucrose, resuspend, and distribute in 100 μl aliquots in 1.5 ml Eppendorf tubes (normally the cell concentration should be around ≈$6 \times 10^{10}$ cells/ml).

#### 3.1.3. Electro-transformation

1. Add 100 ng of pBAM1 plasmid to a 100 μl aliquot of recipient cells, mix gently, and transfer to a 2 mm gap width electroporation cuvette (BioRad Gene Pulser® Cuvette) and proceed to electroporate.
2. Quickly, add 1.0 ml of LB and incubate 1 h at 30°C with shaking.
3. Plate 100 μl dilutions onto M9 citrate plus Km 50 μg/ml (using 140 mm diameter Petri dish). If made in smaller plates (90 mm diameter), add a proportionately lesser amount of cells.

#### 3.1.4. Isolation of Transposon Mutants

1. Directly from transformation plates, pick colonies with a sterile toothpick and double patch onto M9 citrate Km 50 μg/ml Ap 500 μg/ml and on M9 citrate 50 μg/ml. This procedure is to make sure that one does not select recipient cells that cointegrated the plasmid instead of transposed mutants.
2. Select kanamycin-resistant and ampicillin-sensitive colonies.
3. Restreak isolated single colonies in M9 citrate plus Km 50 μg/ml.
4. Repeat step 3 two times, just to be sure of working with pure mutants.
5. Store the mutants at −80°C.

#### 3.1.5. Mapping Transposon Insertion Sites

1. Use single mutant colonies as the source of the DNA template for the PCR round-1.
2. Select one of the transposon ends (ME-I or ME-O) to characterize its insertion point by arbitrary PCR (17).
3. Set up the arbitrary PCR round-1 using the following primer combination:
   (a) For ME-I end: ARB6/ME-I-extR.
   (b) For ME-O end: ARB6/ME-O-extF.
4. Transfer a fresh single colony from a plate directly into the PCR reaction tube.
5. The arbitrary PCR round-1 conditions are: 5 min at 95°C, 6 cycles of 30 s at 95°C, 30 s at 30°C, and 1 min and 30 s at 72°C; 30 cycles of 30 s at 95°C, 30 s at 45°C, and 1 min and 30 s at 72°C (17).

6. Use 1 μl of the PCR round-1 as the template for arbitrary PCR round-2.
7. Set up the arbitrary PCR round-2 (This time adjust the quantities of PCR reagents for a 50 μl final volume) using the following primers:
   (a) For ME-I end: ARB2/ME-I-intR.
   (b) For ME-O end: ARB2/ME-O-intF.
8. The conditions for the arbitrary PCR round-2 are: 1 min at 95°C, 30 cycles of 30 s at 95°C, 30 s at 52°C, and 1 min and 30 s at 72°C, followed by an extra extension period of 4 min at 72°C (17).
9. Clean up the PCR product obtained in the second round (see Note 4) using a commercial column kit and send to sequence it with the one of the primers:
   (a) For ME-I end: ME-I-intR.
   (b) For ME-O end: ME-O-intF.
10. After analyzing the sequencing result one should be able to identify, at least, 30-bp of the mini-transposon end. An example of the complete transposon mapping process is represented in Figs. 2 and 3. In this case, mutants were sought that produced a white phenotype on X-gal medium, derived from a previously engineered *lacZ*+ derivative of *P. putida* (Table 1).

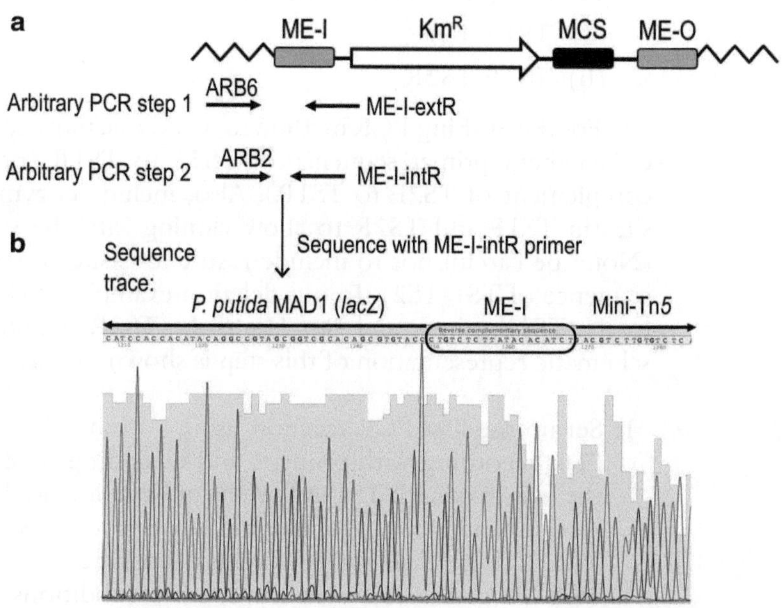


Fig. 3. Example of the principal steps involved in the identification of the mini-transposon landing spot. (**a**) A picture of the mini-Tn*5* module of pBAM1 with the position of the ME-I primers together with the arbitrary primers used in the process. (**b**) A portion of a characteristic sequence chromatogram that included marked the position of ME-I, the mini-Tn*5* segment and the sequence portion corresponding to the *lacZ* gene of the *P. putida* MAD1 white clone obtained from the transposition experiment in this strain.

11. Once the site of insertion is determined, discard the sequence that belongs to the transposon and analyze the rest with a blast tool (http://blast.ncbi.nlm.nih.gov/Blast.cgi) to identify the specific genomic coordinates of the integration point -or look for similar sequences in other bacteria to figure out the inserted gene.

### 3.2. Deletion/Replacement System Protocol

For this deletion/replacement process, the I-SceI expressing plasmid (pSW-I) can be introduced before (preferred option if multiple deletions within the same strain are desired) or after obtaining the cointegrate (if one plans a single deletion). Only the second option is described here. To facilitate a specific example of the deletion protocol one example of the procedure is detailed below. To this end, a chromosomal region of *P. putida* KT2440 between the genome coordinates 5680657-5690333 was chosen. This region comprises seven genes that resemble a type IV pili operon.

#### 3.2.1. Primers Design

General comment: This cloning step could be performed by two different methods: (1) PCR upstream and downstream regions and cloning each region separately into the plasmid or (2) by splicing by overlap extension (SOE)-PCR (18). A detailed description the SOEing PCR protocol follows below.

1. Identify targeting sequences, these localize upstream (TS1) and downstream (TS2) the region one plans to delete.
2. Design two pairs of primers that span each 500 bp of the targeted sequences (see Note 5).
   (a) TS1F-TS1R
   (b) TS2F-TS2R

For the SOEing PCR method to work, one must add the reverse complement primer sequence of TS2F to TS1R (or the reverse complement of TS2F to TS1R). Also, include enzyme restriction sites in TS1F and TS2R to allow cloning into the suicide vector (Note: be careful not to include restrictions sites present in DNA sequence of TS1-TS2). In our deletion example, *Eco*RI was added to the TS1F primer and *Bam*HI to the TS2R oligonucleotide. A schematic representation of this step is shown in Figs. 4a and 7a.

#### 3.2.2. SOEing PCR

1. Set up the first PCR reaction using a fresh colony as template (Tm according with primers, and extending time depends on the length of each TSs and on the polymerase used). It is highly recommended to use one of the error-proof polymerases available. Using *Pfu* DNA polymerase (Promega; see Note 6) and with TS of 500 bp each the specific conditions for our case were: 5 min at 95°C, 25 cycles of 30 s at 95°C, 30 s at 60°C, 1 min at 75°C, and an extra 5 min at 75°C.

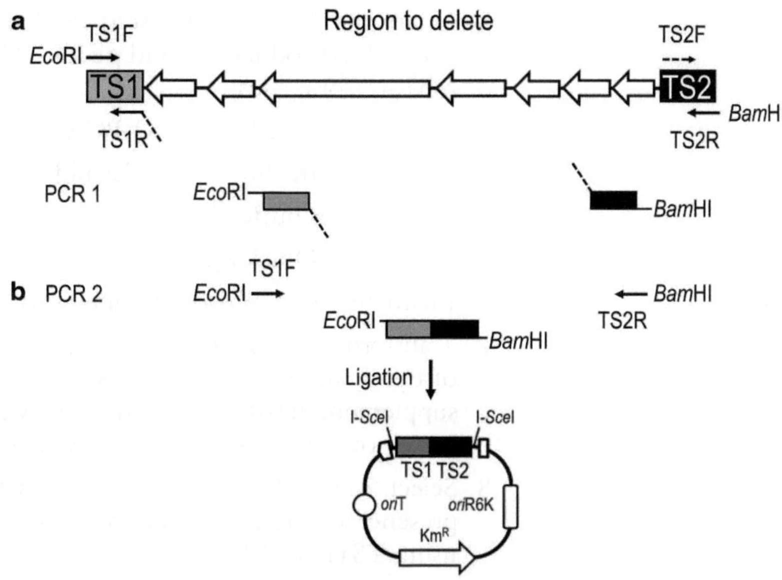


Fig. 4. (**a**) SOEing PCR scheme, with the two PCR rounds and the corresponding primers used. (**b**) A drawn of the TS1-TS2 cloning into pEMG to generate pEMG-TSs, where the position of the two I-SceI sites flanking the TS1-TS2 can be observed.

2. Clean up the PCR fragments with a commercial column kit (or similar) to remove unused primers (Fig. 7b).
3. Prepare the second PCR reaction using 1 μl from each of the reaction products of the first PCR as template. The set up of the PCR round 2 is 5 min at 95°C, 25 cycles of 30 s at 95°C, 30 s at 60°C, 2 min at 75°C, and an extra 5 min at 75°C.
4. Clean up the second PCR product using the column kit and resuspend in 33 μl of $H_2O$ (Fig. 7c).

*3.2.3. Cloning Targeting Sequences into pEMG*

1. Prepare a digestion of the pEMG plasmid with the appropriate enzymes (*Eco*RI and *Bam*HI in the example).
2. Purify from a 1% agarose gel the linear fragment (3.1 kb) of the pEMG plasmid.
3. Digest with the appropriate restriction enzymes the TS1-TS2 PCR fragment (*Eco*RI and *Bam*HI). Example of a TS1-TS2 digestion mix:
   (a) 33 μl of TS1-TS2 PCR
   (b) 4.5 μl Buffer 10×
   (c) 4.5 μl BSA
   (d) 1.5 μl enzyme 1
   (e) 1.5 μl enzyme 2
4. Inactivate the restriction enzymes e.g. by heating.

5. Ligate the TS1-TS2 digested PCR fragment into the pEMG plasmid to produce plasmid pEMG-TSs (Fig. 4b). The ligation mixture comprises:
   (a) 6 μl digested TS1-TS2 PCR fragment
   (b) 3 μl of the linearized plasmid
   (c) 10 μl of buffer 2×
   (d) 1 μl of Quick ligase
6. Incubate 5 min at room temperature.
7. Transform *E. coli* DH5αλpir competent cells (19) with an aliquot of the ligation mixture (see Note 7) and plate onto LB plates supplemented with Km 50 μg/ml, Xgal, and IPTG to discriminate potential positive clones by α-complementation.
8. Select a few white colonies, restreak them, and check for the presence of the correct plasmid (pEMG-TSs) by colony PCR using TS1F-TS2R.
9. Select a few positive clones and send them to sequence with (M13-forward and M13-reverse) to confirm that the target sequences did not incorporate any error due to the PCR amplification process.

*3.2.4. pEMG-TSs Cointegration*

The delivery of the pEMG plasmid could be done either by mating or electroporation. Here, only the electroporation technique is discussed (for mating just follow the same procedure described before but using *E. coli* DH5α*λpir* harboring pEMG-Ts and finally plate onto M9 citrate Km 50 μg/ml). To prepare electrocompetent cells of *P. putida* KT2440 (hereafter KT2440) follow the same steps described earlier.

*3.2.5. Transformation of KT2440 Electrocompetent Cells*

1. Add 200 ng of the construct plasmid to a 100 μl aliquot of KT2440, mix gently and transfer to a 2 mm gap width electroporation cuvette (BioRad Gene Pulser® Cuvette) and proceed to electroporate.
2. Add 1 ml of LB and incubate 1 h at 30°C with shaking.
3. Plate everything onto LB plus Km 50 μg/ml.
4. Streak a few colonies and use colony PCR, with the TS1F and TS2R as the primer combination, to check for cointegration events (see Note 8). A picture of the cointegration process is shown in Fig. 5a.

*3.2.6. Delivery of pSW-I*

1. Prepare as described before electrocompetent cells of the cointegrated strains.
2. The pSW-I plasmid described here bears the ampicillin resistance gene, but the plasmid is also available upon request with other resistance cassettes, such as gentamycin (pSW-2) and kanamycin (pSW-3) as specified in (9).

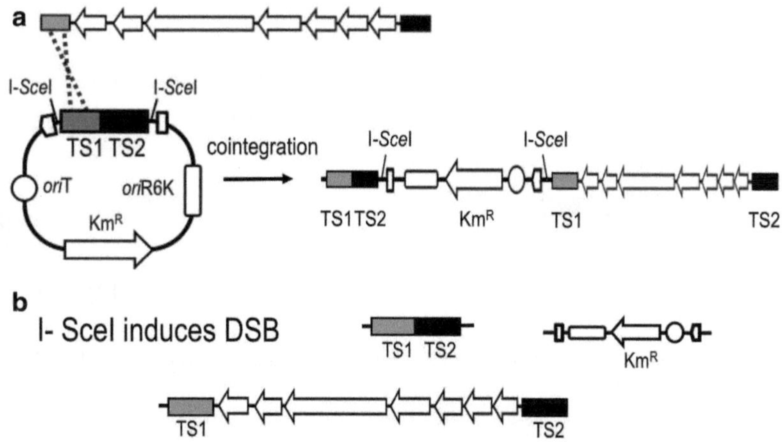


Fig. 5. (**a**) Cointegration step between the pEMG-TSs and the genome through their homology regions. (**b**) I-SceI induction and the formation of double-strand breaks (DSB).

3. Add 50 ng of pSW-I plasmid to a 100 μl aliquot of the cointegrate, mix gently, and transfer to a 2 mm gap width electroporation cuvette and proceed to electroporate.
4. Add 1 ml of LB and incubate 1 h at 30°C with shaking.
5. Plate several dilutions onto LB + Km 50 μg/ml + Ap 500 μg/ml (KT2440 is highly tolerant to ampicillin but such concentration allows plasmid selection).
6. Streak a few colonies and check the presence of the plasmid by PCR using the pSW-F and pSW-R primers (described in Subheading 2; see Note 9). The PCR conditions are 5 min at 95°C, 25 cycles of 30 s at 95°C, 30 s at 55°C, 30 s at 72°C, and with a final extra 5 min at 72°C. The product expected size if the pSW-I plasmid is present is 520 bp (see Note 10).

*3.2.7. Induction of the I-SceI Enzyme*

1. Grow a cointegrated clone that harbors the pSW-I plasmid with 15 mM 3-methylbenzoate (the compound that induce the expression of I-SceI) in a 5 ml LB tube (induction time ranges from 6 h to 16 h). See Fig. 5b.
2. Plate a few dilutions onto LB or LB plus Ap 500 μg/ml (if one wished to maintain the pSW-I plasmid for further deletions on the same strain).
3. Restreak colonies in LB and LB + Km 50 μg/ml to check the loss of the cointegrated plasmid.
4. Select a few kanamycin sensitive clones and differentiate by colony PCR between WT (Fig. 6a) and deleted clones (Fig. 6b). The proportion is usually close to 50% (Fig. 7d).
5. To entirely ensure that the deletion have occurred. It is highly recommended to design a primer pair (check-F and check-R in Fig. 7a) that hybridizes within a gene, or intergenic segment, of the to be deleted region. Always use the wild-type strain as the positive control for the PCR amplification (Fig. 7e).

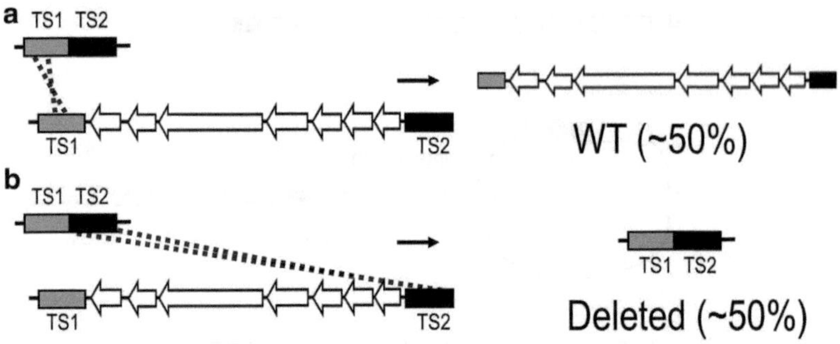


Fig. 6. (**a**) Recombination process between the TS1s to yield a wild-type strain. (**b**) Homologous recombination that renders a deleted/replaced strain.

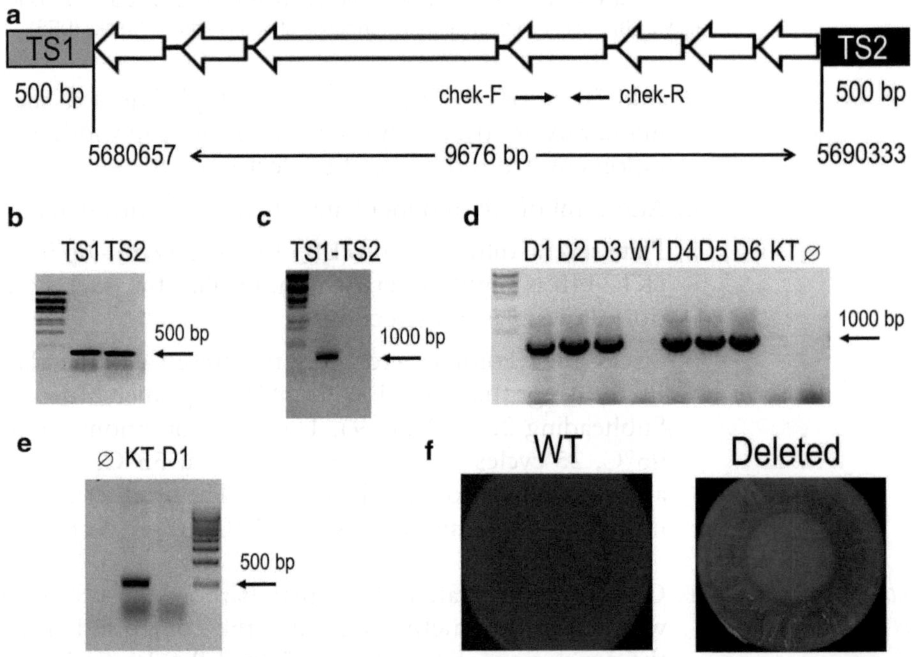


Fig. 7. Example of key steps in the deletion of a genomic region in *P. putida* KT2440. (**a**) Chromosomal organization of the deleted operon, including the upstream (TS1) and downstream (TS2) regions of homology, and the *P. putida* K2440 genomic coordinates of the region are shown. (**b**) Electrophoresis result of the SOEing PCR round 1, showing the purified DNA fragments of TS1 and TS2 (500 bp each). (**c**) Electrophoresis outcome of the SOEing PCR round 2, presenting the combined TS1-TS2 piece, a 1,000 bp product. (**d**) Electrophoresis showing the PCR products, using TS1F and TS2R as primer combination, to check whether the recombination process yielded deleted or wild-type cells. (**d**) Lanes D1-D3 correspond to deleted clones (amplification of a 1,000 bp band), whereas W1 is a revertant to wild-type. KT is *P. putida* KT2440 (wild-type control) and ∅ is a negative control (no DNA template). In this procedure, six out of seven clones analyzed were deleted strains and only one revert to a wild-type cell. (**e**) Confirmative result of the deletion, visualized after a PCR amplification using primers that hybridize within a gene inside the operon (primers check-F and check-R in figure). The wild-type strain produces a PCR-band while not the deleted strain, fact that confirms the removal of the operon. (**f**) Phenotypic characterization of the mutant strain. 2 μl of overnight cultures were spotted onto LB agar plates containing 40 mg/l of Congo red and 15 mg/l of Coomassie Brilliant Blue and let them grow at room temperature for several days. Note that the lack of the fimbrial genes endows the colony with a different morphology.

3.2.8. Curation of pSW-I

1. Pick a deleted clone and make several passages in LB without drugs. In order to do this, grow overnight in LB at 30°C with shaking, then transfer to a new tube with fresh LB and repeat the process for 3 days.
2. Plate appropriate dilutions onto LB agar plates to obtain separate colonies.
3. Restreak several colonies in LB and LB + Ap 500 μg/ml.
4. Select ampicillin sensitive colonies and double check the plasmid curation by colony PCR using the pSW-I primer pair (pSW-F and pSW-R) and the same PCR conditions as described before.
5. Keep a frozen stock in LB 20% glycerol at −80°C of the mutant strain.
6. Proceed to analyze phenotypic differences between the wild-type and mutant strains (Fig. 7e).

## 4. Notes

1. Instead of the M9 citrate medium used here for *Pseudomonas*, other bacterial species may require other selective media that allows only the growth of recipient exconjugants.
2. These arbitrary primers are optimal for the GC content of *Pseudomonas*.
3. When all of the protocol steps are followed, plating 200–300 μl (from the 5.0 ml mating mixture) onto 140 mm diameter Petri dishes is suggested.
4. Even if a smear (and not a clear product band) is seen after an electrophoresis of the second arbitrary PCR round, still send it to sequence.
5. The use of target sequences of 500-bp length each is recommended. However, longer homology sequences could be used (800-bp), but then more primers (e.g., TS1R and TS2F together with M13-forward and M13-reverse) will be needed to completely sequence the TS1-TS2 cloned in pEMG.
6. If no TS1s amplification product is obtained with *Pfu* polymerase, try with a regular *Taq* polymerase but with 20 cycles instead of 25, and send more clones to sequence.
7. Usually, using 8 μl from the ligation mixture renders a good number of colonies.
8. It is recommended to work with various cointegrated strains, separately or mixed as a pool of cointegrates.
9. Checking by PCR for the presence of pSW-I is not necessary and one can proceed without following this PCR step.

10. Once the pSW-I plasmid is in the host strain, proceed quickly to induce the I-SceI enzyme. Do not store plates at room temperature or even at 4°C for long because the pSW-I plasmid gets lost very easily.

## Acknowledgments

Authors are indebted to S.M. Wong for the kind gift of pSW (I-SceI) plasmid. Development of the genetic tools described here was funded by generous grants of the CONSOLIDER program of the Spanish Ministry of Science and Innovation, by EU contracts BACSIN and MICROME and by Funds of the Autonomous Community of Madrid (TMEMS). All plasmids and strains described here are available upon request.

### References

1. de Lorenzo V. (2010) Environmental biosafety in the age of Synthetic Biology: Do we really need a radical new approach? BioEssays 32: 926–931.
2. Gibson DG, Glass JI, Lartigue C et al (2010) Creation of a bacterial cell controlled by a chemically synthesized genome. Science 329: 52–56.
3. Posfai G, Kolisnychenko V, Bereczki Z et al (1999) Markerless gene replacement in *Escherichia coli* stimulated by a double-strand break in the chromosome. Nucleic Acids Res 27: 4409–4415.
4. Reznikoff WS. (2008) Transposon Tn*5*. Annu Rev Genet 42: 269–286.
5. Herrero M, de Lorenzo V, Timmis KN. (1990) Transposon vectors containing non-antibiotic resistance selection markers for cloning and stable chromosomal insertion of foreign genes in Gram-negative bacteria. J Bacteriol 172: 6557–6567.
6. de Lorenzo V, Herrero M, Jakubzik U et al (1990) Mini-Tn*5* transposon derivatives for insertion mutagenesis, promoter probing, and chromosomal insertion of cloned DNA in Gram-negative eubacteria. J Bacteriol 172: 6568–6572.
7. Kolter R, Inuzuka M, Helinski DR. (1978) Trans-complementation-dependent replication of a low molecular weight origin fragment from plasmid R6K. Cell 15: 1199–1208.
8. Martínez-García E, Calles B, Arévalo-Rodríguez M et al (2011) pBAM1: an all-synthetic genetic tool for analysis and construction of complex bacterial phenotypes. BMC Microbiol 11:38.
9. Martinez-Garcia E, and de Lorenzo V. (2011) Engineering multiple genomic deletions in Gram-negative bacteria: analysis of the multi-resistant antibiotic profile of Pseudomonas putida KT2440. Environ Microbiol. DOI:10.1111/j.1462-2920.2011.02538.x.
10. Colleaux L, D'Auriol L, Galibert F et al (1988) Recognition and cleavage site of the intron-encoded omega transposase. Proc Natl Acad Sci USA 85: 6022–6026.
11. Wong SM, Mekalanos JJ. (2000) Genetic footprinting with mariner-based transposition in *Pseudomonas aeruginosa*. Proc Natl Acad Sci USA 97: 10191–10196.
12. Blatny JM, Brautaset T, Winther-Larsen HC et al (1997) Improved broad-host-range RK2 vectors useful for high and low regulated gene expression levels in Gram-negative bacteria. Plasmid 38: 35–51.
13. Pratt LA, Kolter R. (1998) Genetic analysis of *Escherichia coli* biofilm formation: roles of flagella, motility, chemotaxis and type I pili. Mol Microbiol 30: 285–293.
14. Nelson KE, Weinel C, Paulsen IT et al (2002) Complete genome sequence and comparative analysis of the metabolically versatile *Pseudomonas putida* KT2440. Environ Microbiol 4: 799–808.
15. Dos Santos VA, Heim S, Moore ER et al (2004) Insights into the genomic basis of niche specificity of *Pseudomonas putida* KT2440. Environ Microbiol 6: 1264–1286.

16. Choi KH, Kumar A, Schweizer HP. (2006) A 10-min method for preparation of highly electrocompetent *Pseudomonas aeruginosa* cells: application for DNA fragment transfer between chromosomes and plasmid transformation. J Microbiol Methods 64: 391–397.
17. Das S, Noe JC, Paik S et al (2005) An improved arbitrary primed PCR method for rapid characterization of transposon insertion sites. J Microbiol Methods 63: 89–94.
18. Horton RM, Hunt HD, Ho SN et al (1989) Engineering hybrid genes without the use of restriction enzymes: gene splicing by overlap extension. Gene 77: 61–68.
19. Sambrook J, Maniatis T, Fritsch EF (1989) Molecular cloning a laboratory manual, Cold Spring Harbor Laboratory Press, Cold Spring Harbor, N.Y.
20. de Lorenzo V, Timmis KN (1994) Analysis and construction of stable phenotypes in Gram-negative bacteria with Tn*5*- and Tn*10*-derived minitransposons. Methods Enzymol 235: 386–405.
21. Hanahan D. (1983) Studies on transformation of *Escherichia coli* with plasmids. J Mol Biol 166: 557–580.
22. Sambrook J, Fritsch EF, Maniatis T (1989) Molecular cloning: A laboratory manual, Cold Spring Harbor. New York.
23. Regenhardt D, Heuer H, Heim S et al. (2002) Pedigree and taxonomic credentials of *Pseudomonas putida* strain KT2440. Environ Microbiol 4: 912–915.
24. Fernandez S, de Lorenzo V, Perez-Martin J. (1995) Activation of the transcriptional regulator XylR of *Pseudomonas putida* by release of repression between functional domains. Mol Microbiol 16: 205–213.

# Part IV

## Implementation of Synthetic Gene Networks in Different Host Systems

# Chapter 17

# Synthetic Networks: Oscillators and Toggle Switches for *Escherichia coli*


## Nicolas Perry and Alexander J. Ninfa



## Abstract

Bacterial synthetic gene networks are constructed by manipulating the regulation of genes inside a cell, with the purpose of eliciting novel regulatory behaviors. The methods for manipulating genes and gene regulation in *E. coli* are well established, making it the preferred host for basic studies of synthetic networks. We focus our work on constructing two kinds of synthetic gene networks: toggle switches (bistable systems) and oscillators. Toggle switches are capable of exhibiting two stable steady states of gene expression (OFF and ON) without stable intermediate states; the steady state reached by the system depends on the previous history of the system. Biological oscillators exhibit regular cycles in gene expression around an unstable steady state. Studying these two kinds of synthetic networks helps advance our understanding of natural bistable systems and oscillators, such as the circadian oscillators controlling gene expression in many types of cells, and the genetic systems controlling the cell cycle and differentiation in metazoans.



**Key words:** Genetic circuit, Genetic oscillator, Genetic switch, Genetic bistability, Recombinant DNA, Continuous culture automation


## 1. Introduction

The field of synthetic biology is a relatively new field, in which engineered genetic networks are used to produce novel cell behaviors (1, 2). However, the techniques that make construction of synthetic genetic networks possible are generally the same recombinant DNA methods that have been under development since the 1970s. In our lab, we use standard recombinant DNA techniques (e.g., ref. 3) as well as modern techniques for the genetic manipulation of *E. coli* (e.g., ref. 4) to produce synthetic gene networks. Each synthetic network is composed of one or more synthetic operons, which are linked to selectable markers to


Wilfried Weber and Martin Fussenegger (eds.), *Synthetic Gene Networks: Methods and Protocols*,
Methods in Molecular Biology, vol. 813, DOI 10.1007/978-1-61779-412-4_17, © Springer Science+Business Media, LLC 2012

form "modules" (5). A typical module will contain a promoter sequence, one or more structural genes encoding protein products, and a transcription terminator sequence; these are linked to an antibiotic resistance gene that is expressed independently and constitutively. For our work in developing novel regulatory networks, the promoter sequences are usually highly engineered to be repressed or activated by transcription factors that do not naturally regulate the associated structural genes (5, 6).

Synthetic biology genetic modules are typically constructed in plasmids. While some workers have chosen to directly use plasmids for the analysis of synthetic genetic networks in vivo (e.g., (7, 8)), we prefer to integrate the modules into the bacterial chromosome by homologous recombination. Integration of the modules into the chromosome increases the uniformity of gene expression in cell populations, since it fixes the copy number of the module in the cell; by comparison, plasmids that are not integrated into the chromosome may display a range of different copy numbers within cells. To bring about localization of modules to defined chromosomal locations, specially engineered plasmids, known as "landing-pad" plasmids, are used (9). These plasmids contain sequences homologous to chromosomal sequences that flank a multicloning site in the plasmid, where the synthetic module is inserted. The "homology regions" of the plasmid that flank the synthetic module allow recombination of the synthetic module onto the chromosome, where it comprises an insertion between the homology regions. To facilitate the recombination of modules onto the chromosome, cells are transformed with linearized DNA, and the RecBCD enzyme that typically degrades linear DNA inside the cell is inhibited (4, 9).

As mentioned before, we construct synthetic networks, comprising one or more modules, that are engineered to behave as toggle switches and oscillators. A few examples are shown in Fig. 1. One simple toggle switch consists of a single synthetic module that contains a structural gene for a transcriptional activator (the *glnG* product NRI that upon phosphorylation is the activator NRI ~ P), driven from a promoter that is activated by the same activator (Fig. 1a). The promoter of this module was engineered to render it repressed by the LacI repressor, which is expressed constitutively in the cells. The cells also contain the enzymatic activity required to bring about the phosphorylation of NRI to form the activator, NRI ~ P. The positive feedback of the activator on its own expression, and the antagonism between activator and repressor for control of the engineered promoter, result in bistability (5, 6). One manifestation of this bistability is hysteresis in the activity of the module; cells containing this device require a much higher concentration of inducer (IPTG) to turn ON the expression of the switch than they require to keep the system fully ON once it has been activated. A toggle switch that displayed bistability at an even greater range of conditions was obtained by adding a second

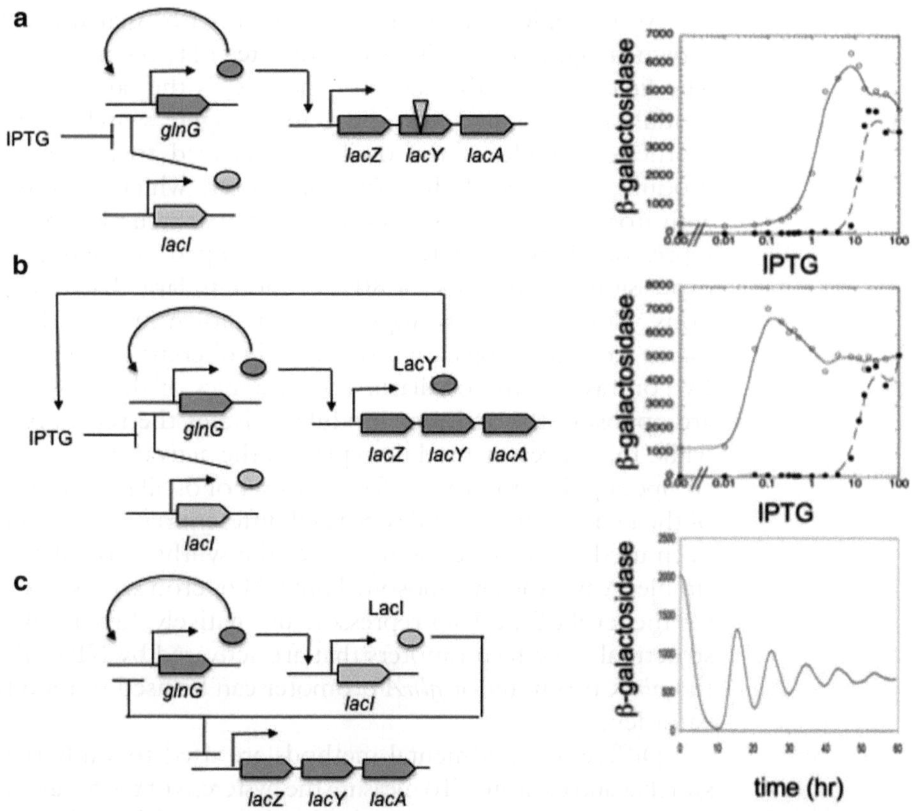


Fig. 1. Example synthetic genetic networks. (**a**) Toggle switch based on a single positive feedback loop. *Left*: schematic of the device. The *horizontal line* represents DNA, the *raised arrow* represents the site of transcription initiation, the *filled block arrows* represent structural genes, the *filled circles* represent the protein products of the structural gene, transcriptional activation is depicted by *arrows*, and transcriptional repression is indicated by *blunt-ended lines*. *Right*: typical results of a hysteresis experiment. *Filled circle*, naive cultures; *open circle*, preinduced cultures. (**b**) Toggle switch comprises two positive feedback loops and containing two synthetic modules. *Left*: schematic of the switch. Symbols are as in Panel (**a**). *Right*: results of a typical hysteresis experiment. (**c**) Genetic oscillator comprises two synthetic modules. *Left*: schematic of the device. *Right*: results of a typical oscillator experiment conducted using the automated continuous culture turbidostat instrument.

synthetic genetic module that provided a second positive feedback loop (Fig. 1b). In this system, the second synthetic module contained the structural genes for *lacZYA*, expressed from the *glnK* promoter, that is activated by activator. Thus, activator resulted in the expression of LacY (galactoside permease), which helped to internalize the IPTG inducer and keep cells induced at low IPTG concentrations. Comparison of the systems in Fig. 1a, b revealed that the two positive feedback loops (activator activates itself; activator activates LacY that brings in inducer to derepress activator expression) displayed strong synergy; this was explained as resulting from a dramatic increase in the sensitivity (apparent Hill coefficient or kinetic order) of the activator's effect on itself, which in turn increased the range of conditions at which the system was bistable and displayed hysteresis (9).

An example of a synthetic network with oscillator activity is shown in Fig. 1c (5). This system consisted of two synthetic modules. The first of these, the activator module, is the same as the single module incorporated into the genetic toggle switch. The second synthetic module of the oscillator, referred to as the repressor module, consists of the *glnK* promoter, which is activated by activator, driving the expression of the structural gene for LacI repressor. This oscillator produces damped oscillation of gene expression that persists for 80 h or more in large bacterial populations. Note that in the toggle switch shown in Fig. 1b, the two major feedback loops are positive and coherent with one another. By contrast, in the oscillator the dynamics of these two modules are opposite: the activator module activates the repressor module while the repressor module represses the activator module. These competing dynamics allow for the onset of oscillatory transcription of the genes in the modules. Several different reporter systems have been used to measure the output of the synthetic oscillator. In the simplest case, the chromosomal *lacZYA* operon serves as a reporter for the level of the LacI repressor; alternatively, fusions of the *lacZ* structural gene to promoters that are activated by NRI ~ P, such as the *glnK* promoter or *glnA* promoter can be used to monitor activator level.

Different experimental methods are used to study the toggle switches and oscillator. To measure the hysteresis of the toggle switches, overnight cultures are grown in the absence of the inducer, or with a saturating concentration of the inducer (0.4 mM IPTG). These overnight cultures and then diluted approx. million-fold into fresh media containing different concentrations of IPTG. Typically, 12 different concentrations of IPTG are examined using cells from the induced overnight culture and from noninduced overnight cultures, for a total of 24 flasks. These cultures are incubated until they reach early log phase, typically an optical density (OD) of about 0.4, after which the expression of the *lacZ* reporter is assayed (Fig. 1).

In the case of the synthetic genetic oscillators, constant monitoring and sampling of the growing culture is required. This is partly due to the oscillator affecting the growth of cells such that the rate of growth varies depending on the phase of the oscillator. Another complicating factor is that the period of oscillation is typically 9–12 h, and four or five oscillations are observed before the amplitude of oscillation decays to an unmeasurable amplitude. Thus, the measurement of oscillator function requires a continuous culture for about 80 h, during which culture turbidity is held constant and the *lacZ* reporter is assayed at regular intervals. For our initial experiments with this system (5), the cultures were grown in a continuous culture device that was monitored manually for the 80 h period, with occasional alteration of the flow rate of the nutrient pumps to maintain a fairly constant culture turbidity.

The difficulty of these experiments led to the development of a fully automated system, in which a computer-controlled turbidostat instrument maintains the culture at constant turbidity and periodically samples the culture (9). The samples are stored within a refrigerator at 4 C, where the *lacZ*-encoded β-galactosidase levels are stable for about 2 days. This automated system allows us to perform multiple experiments and long-duration experiments without the need for continuous monitoring, sampling, and assaying; once an experiment is started the investigator need only assay the stored samples once every 2 days.

## 2. Materials

### 2.1. Plasmids Vectors for Construction of Modules

As noted above, we use "landing-pad" plasmids that allow the facile recombination of synthetic modules onto the *E. coli* chromosome, at defined locations. The general features of these plasmids are shown in Fig. 2. Landing pad plasmids for integration of modules within the *rbs* (ribose utilization) operon or within the *glnKamtB* operon have been described (5), and a general scheme for developing landing-pad vectors has been presented (9). In our recent work, we have utilized the methods described in ref. 9 to produce additional landing pads that localize synthetic modules to the *leu* operon and *eda* operon (unpublished). An important feature of these landing pad vectors is that each contains a selectable marker within the portion of the plasmid that will become incorporated into the bacterial chromosome (Fig. 2), such that the synthetic module shows tight genetic linkage with the selectable marker.

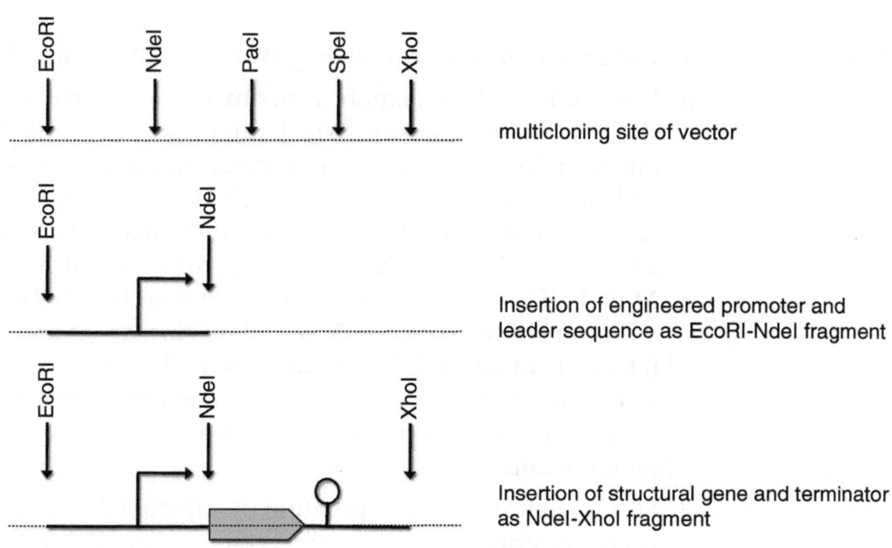


Fig. 2. Strategy for assembly of a synthetic operon within a multicloning site.

These selectable markers are used as a handle to move the synthetic module from strain to strain in the process of building up a synthetic genetic network. This strain building process typically involves the standard method of $P1_{vir}$-mediated generalized transduction (10). The selectable markers in currently available landing pads are as follows: *rbs* landing pad, gentamycin resistance; *glnK* landing pad, chloramphenicol resistance; *leu* landing pad, either chloramphenicol or kanamycin resistance; *eda* landing pad, either chloramphenicol or kanamycin resistance.

### 2.2. Recombinant DNA and PCR Materials

1. Restriction enzymes, DNA ligase, and reaction buffers (New England Biolabs).
2. Primers for PCR amplification (Invitrogen).
3. TAQ polymerase platinum and its reaction buffer (Invitrogen).
4. dNTP mix (10 mM ea) (Invitrogen).
5. Mastercycler TM Eppendorf Thermal cycler (Germany), for performing PCR reactions.
6. Gel electrophoresis for the analysis of recombinant DNA and PCR procedures uses agarose (Fisher Scientific) and TAE buffer (National Diagnostics), and gels are stained with Ethidium Bromide (Sigma) and visualized by long-wave UV transillumination.
7. PCR purification kit and gel-extraction kits (Qiagen).

### 2.3. Bacterial Transformation

1. Chemical transformation uses a standard calcium-chloride based method (11), using reagent-grade materials.
2. Electroporation is performed using a MicroPulser TM electroporator (BioRad).

### 2.4. Growth Media

1. Bacterial strains are routinely grown in LB Medium (3).
2. Hysteresis and oscillation experiments are performed with W-Salts Minimal Medium, which can be prepared with different nitrogen sources (such as ammonia, glutamine, or arginine) and different carbon sources (such as glucose, succinate, or casamino acids). Twofold concentrated W-salts medium contains: 21 g/L $K_2HPO_4$, 9 g/L $KH_2PO_4$, and 1.3 mL/L of 1 M $MgSO_4$. The final medium contains vitamin B1 (0.004% w/v), a carbon source (0.4% w/v of either glucose or succinate), a nitrogen source (0.2% w/v glutamine, 0.2% w/v ammonium sulfate, or 0.5% w/v casamino acids, or some combination of these), and any amino acids needed to satisfy auxotrophic requirements, at 0.004% (w/v).
3. SOC medium for electroporation of cells contains: 20 g/L w/v bacto-tryptone, 5 g/L bacto-yeast extract, 0.5 g/L NaCl, 0.186 g/L KCl, 0.952 g/L $MgCl_2$, and 3.6 g/L glucose.

4. Antibiotics are used at the following concentrations: gentamycin (Gibco), 5–15 μg/mL, typically requires determination of optimal concentration for each preparation of the drug; kanamycin (Sigma-Aldrich), 50 μg/mL; ampicillin (Sigma-Aldrich), 50 μg/mL; chloramphenicol (Sigma-Aldrich) 34 μg/mL.

### 2.5. Beta-galactosidase Activity Assays

1. Z buffer: 16.1 g/L $Na_2HPO_4$, 5.5 g/L $NaH_2PO_4$, 0.74 g/L KCl, 0.246 g/L $MgSO_4$.
2. Assay reactions contain 1.6 mL Z buffer, 40 μL of chloroform (reagent grade), 40 μL of 0.1% SDS (Sigma-Aldrich), and 0.3 mL of bacterial culture.
3. Assay substrate is 4 mg/mL *o*-Nitrophenyl-β-D-galactoside (ONPG, Sigma-Aldrich) in 0.1 M sodium phosphate buffer, pH 7.0.
4. Assay stop solution is 1 M sodium carbonate.

### 2.6. Computer-Controlled Continuous-Culture Turbidostat with Automated Sampling

The computer-controlled turbidostat system to run continuous cultures has been described in detail (9). This instrument is composed of components acquired from FIAlab Instruments (Bellevue, WA), but manufactured by different companies:

1. Two four-channel peristaltic pumps, VS Series from Alitea (Sweden).
2. One two-channel peristaltic pump, S Series frim Alitea.
3. Two Cheminert TM diverter valves from Valco Instruments Co. Inc. (Houston, TX).
4. One spectrometer, SD2000 from Ocean Optics (Dunedin, FL).
5. One autosampler, Aim3200 from AimLab (Australia).
6. All instruments are controlled with software written in Labview TM programming language, from National Instruments (Austin, TX), running on a standard PC utilizing a windows operating system.

## 3. Methods

### 3.1. Construction of Synthetic Modules

Synthetic modules are built up by a combination of standard gene cloning and PCR methods. As an example, we consider the linker region of the *leu* and *eda* landing pad vectors, which contains unique restriction sites for EcoRI, NdeI, PacI, SpeI, and XhoI (Fig. 2). A typical approach for building a synthetic operon would be to clone an engineered promoter and leader sequence region, obtained by PCR, between the EcoRI and NdeI restriction sites. This would be accomplished by using as a downstream primer a long oligonucleotide that contains an NdeI site, the sequence of the transcribed leader (bearing the translational regulatory signals), any operators that need to be added to a natural promoter, and at least 17 bp of the natural promoter to be manipulated. We typically add operators immediately downstream from the site of transcription initiation, similar to the location of the *lacO1* operator within the *lacZYA* operon (12). As an upstream primer for PCR of the promoter–leader sequence fragment, we use an oligonucleotide primer that contains an EcoRI site, and necessary upstream sequences to be added, such as distal operators or enhancers (9), followed by at least 17 bp of the natural promoter sequence. As our work typically utilizes enhancer-driven bacterial promoters that extend over considerable distances, PCR of genomic DNA is used to produce the promoter–leader sequence fragment. If a small fragment could contain all of the desired regulatory signals, a convenient strategy would be to simply have the entire EcoRI-NdeI fragment synthesized, eliminating the need for a PCR step. The NdeI site is used to delineate the boundary between the promoter–leader sequence and the structural genes of the synthetic operon because this restriction site contains an ATG sequence that can be perfectly aligned with the initiation codon of the (first) structural gene (9). PCR is used to produce a fragment that has an NdeI site exactly at the start codon of the structural gene, the structural gene sequence, a transcriptional terminator sequence downstream from the structural gene, and either a PacI, SpeI, or XhoI unique restriction site downstream from the terminator. For this PCR, the long downstream primer contains the desired restriction site, the transcriptional terminator sequence (9), and a portion of the structural gene, while the upstream primer contains the NdeI site and the beginning of the structural gene. The synthetic operon is then completed by cloning this PCR fragment into the plasmid already bearing the promoter–leader sequence fragment.

1. Restriction enzyme cleavage reactions are performed following the manufacturer's directions (New England Biolabs).
2. PCR reactions follow the instructions for the Platinum Taq Polymerasze enzyme (Invitrogen), except that we typically use an annealing temperature of 45°C instead of the recommended 55°C. In many cases, the amplification program is successful even if the conditions are not optimal; in obdurate cases, the times and temperatures for the three steps of PCR must be empirically optimized. PCR fragments are digested with the appropriate restriction enzymes, and then purified using a Qiagen PCR clean-up kit, following the manufacturer's directions.
3. Substrates for DNA ligation experiments are gel-purified from agarose gels, using the Qiagen gel extraction kit, according to the manufacturer's instructions.
4. DNA ligation utilizing Quick ligation kit is performed according to the manufacturer's instructions (New England Biolabs).

5. The results of PCR, restriction cleavage, and ligation reactions are monitored on agarose gels, as described (13). Miniprep of plasmid DNA is as described (3), using the alkaline lysis method.

### 3.2. Recombination of Synthetic Modules onto the Bacterial Chromosome and Building Strains with Multiple Modules

1. The landing-pad vector containing the synthetic module is cleaved with a restriction enzyme within the vector portion of the plasmid. In most cases, the BspHI restriction enzyme is suitable for this purpose. It is very important to digest the plasmid as completely as possible, by using an excess of enzyme and time. For best results, 1–2 μg of cleaved plasmid should be used in the electroporation experiment.
2. The recipient strain in the electroporation experiment must contain the plasmid pKD46 (4), which has the λ red system under the control of an arabinose-regulated promoter. This plasmid is also temperature-sensitive for replication and thus must be maintained in cells grown at 30°C or lower temperature. The plasmid can be prepared from strain BW25113 by standard miniprep methods and introduced into any desired strain by transformation, using chemically competant cells.
3. The recipient strain containing pKD46 is grown overnight in LB medium containing ampicillin at 30°C, then diluted 1/100 into 50 mL fresh medium containing ampicillin and 20 mM arabinose and grown at 30°C to mid-log phase. The cells are then washed in 30 mL ice-cold 10% glycerol three times and are resuspended in 100 μL of 10% glycerol. 50 μL of the highly concentrated cell suspension is then used in an electroporation experiment, with outgrowth of the cells in SOC medium at 37°C for at least 1 h. The sample is then plated onto solid LB medium (containing 1.5% w/v agar) containing the appropriate antibiotic associated with the synthetic module (chloramphenicol, gentamycin, or kanamycin). The isolated single colonies from these plates are repurified to single colony isolates on fresh plates that are incubated at 42°C. We have found that this step helps to ensure the complete loss of the pKD46 from the strains. The resulting single colony isolates are then checked on LB + ampicillin plates to be certain that they are ampicillin sensitive and thus do not contain the entire landing-pad plasmid that escaped linearization.
4. The presence of the synthetic module on the bacterial chromosome is confirmed by colony PCR using appropriate flanking sequences (13), by analysis of the phenotype, or both.
5. To move the chromosomally integrated module between bacterial strains (to build up complex devices with more than 1 module), bacteriophage P1vir-mediated generalized recombination is used. We use the procedures for P1vir lysate production and transduction experiments that have been contributed

by the Sauer group (10). Transduction experiments are performed on LB medium containing the appropriate antibiotics, and it is very important to repurify transductants to single colonies, so as to remove all traces of the bacteriophage.

### 3.3. Hysteresis Experiments with Genetic Toggle Switches

Genetic toggle switches have two stable steady states, and display hysteresis. In order to test the extent of bistability of a synthetic toggle switch, the expression level of the *lacZ* reporter is determined in cells that have experienced different induction histories. In the typical experiment, a single colony isolate of the strain to be tested in incubated overnight in two culture tubes in W-salts medium containing the desired carbon and nitrogen sources (such as, for example, glucose-glutamine-casein hydrolysate medium). One culture tube will contain no inducer; to the other the IPTG inducer is added to 0.4 mM, a saturating concentration. Thus, one culture is externally forced to remain in the induced state while the other is kept in the naive state. These overnight cultures are diluted the next day by factor of $10^5$–$10^6$ (depending on the growth rate of the cells) into several 125 mL culture flasks containing 15 mL of fresh medium with different amounts of IPTG, typically between 0 and 200 mM, and they are incubated until after they reach mid-log phase ($OD_{600}$ of 0.4). At this point, samples are removed from the flasks and the activity of the *lacZ* reporter is measured. The results of the experiment are a graph of reporter activity vs. inducer concentration, as shown in Fig. 2, from which the range of inducer concentrations, where the system exhibits bistability can be estimated.

### 3.4. Experiments with Genetic Oscillators

The genetic oscillators that we work with have a period of 9–12 h. This period of oscillation requires a culture to be growing for several days in order to see more than a few oscillations, i.e., a continuous culture. In order to run a continuous culture in which we could measure oscillations in reporter gene expression without continuous human monitoring of the culture, we have developed a computer-controlled system that is capable of maintaining an automated continuous culture (9). The automated system keeps a continuous culture at an approximately constant (user defined) optical density, takes a sample of the culture at desired intervals (usually every half an hour is adequate), and stores the samples at 4°C.

To start the experiment, a freezer culture of the strain to be examined is streaked out for single colonies on solid medium that is the same as to be used for the clock experiment. For example, a commonly used medium for our work is W-salts based succinate-glutamine-caseine hydrolysate medium as described above. After ~30 h incubation, a single colony is used to start an overnight liquid culture (250 mL) in the same medium to be used for the clock experiment, except containing 0.4 mM IPTG to entrain the cell population. This culture is incubated for 12 h at 30°C. After overnight induction, the cells are pelletted, washed with medium lacking IPTG, repelleted, resuspended in medium lacking IPTG and introduced into the growth chamber (flask) of the automated system. The automated system is able to adjust to optical density of the culture to the desired $OD_{600} = 0.4$, and maintain the culture at constant turbidity.

The automated system consists of three peristaltic pumps (one sampling pump, one media pump, and one cleaning pump), two diverter valves (one to flow alternatively ethanol or water through the cleaning line and another to alternatively flow culture or water through the sampling line) as well as an spectrometer to measure the optical density of samples as the culture flows through a Z-cell, and an automated sampler inside a refrigerator capable of holding 180 samples. The oscillator bacterial culture itself is maintained within a vented 1 L Erlenmeyer flask at a volume of 250 mL, with constant aeration via filtered air injection. About four to five 4-L bottles containing minimal media are prepared for each experiment. Medium is fed into the device using a sterile tubing system (9). The schematics of the system are shown in Fig. 3, and a more complete description of the system and its development are in ref. 9.

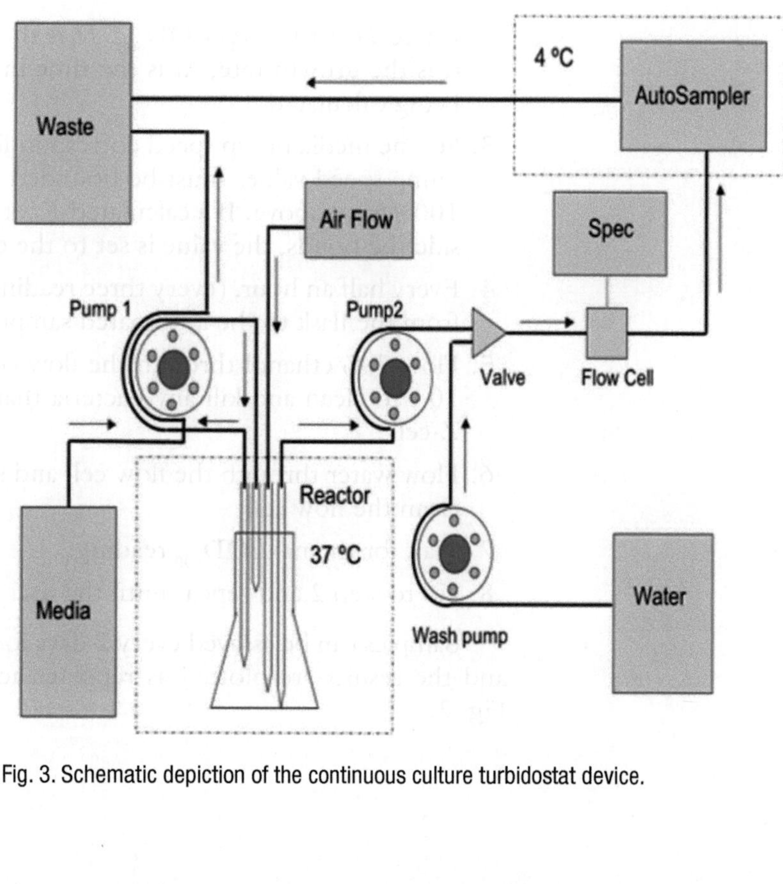


Fig. 3. Schematic depiction of the continuous culture turbidostat device.

The computer program that controls the automated system was written in Labview TM, and it implements the following algorithm:

1. Initialize all variables, set all the pumps off and set the diverter valves to their initial positions. Set the initial flow of media in at 50% (in terms of a percentage of the pump top speed). Set the target optical density as the user given value. Run the sampling pump to flow culture from the flask to the Z-cell, where an initial optical density reading is done
2. Every 10 min, run the sampling pump to flow culture from the flask to the Z-cell, and take an OD reading. Calculate the growth rate of the culture using the formula

$$D_i = D_{i-1}\text{Exp}\{(\alpha - f)\Delta t\}.$$

   where $D_i$ is the current $OD_{600}$, $D_{i-1}$ is the previous $OD_{600}$ reading, $f$ is the flow of media out (proportional to pump speed), $\Delta t$ is the time interval (set to 10 min), and $\alpha$ is the growth rate. Given the calculated value of the growth rate $\alpha$, calculate the flow necessary to make the culture reach the target $OD_{600}$ in the next time interval. That is

$$D_t = D_i\text{Exp}\{(\alpha - F)\Delta t\}.$$

   where $D_t$ is the target $OD_{600}$, $D_i$ is the current $OD_{600}$ reading, $\alpha$ is the growth rate, $\Delta t$ is the time interval, and $F$ is the flow being calculated.
3. Set the media pump speed corresponding to the flow $F$. Media pump speed values must be bounded by 10% from below and 100% from above. If a calculated $F$ corresponds to a value outside the bonds, the value is set to the closest limit.
4. Every half an hour, (every three readings) take 2 mL of culture from the flask to the automated sampler to be kept stored.
5. Flow 70% ethanol through the flow cell and sampling line for 10 s to clean and kill any bacteria that maybe attached to the Z-cell.
6. Flow water through the flow cell and sampling line for 60 s to clean the flow cell.
7. Wait for the next $OD_{600}$ reading.
8. Go to step 2 and repeat until the user stops the program.

Samples can be assayed every 2 days for β-galactosidase activity, and the results are plotted as reporter activity versus time, as in Fig. 2.

## 4. Notes

We have found that experiments with genetic oscillators as described here are quite reproducible, as long as the overnight induction period (entrainment procedure) is consistent. As the overnight induction period is lengthened, the amplitude of the initial *lacZ* values decreases, and the first peak of oscillating gene expression becomes delayed, such that a slightly different pattern of gene expression is obtained. Thus, the entrainment procedure must be consistently applied.

We have also found that one cannot take a sample of the culture at the end of a clock experiment and use it to start another clock experiment. When we have tried this, we have observed poor oscillations in the second run. This suggests that during the clock experiments, we are picking up mutations that eliminate the clock or one of its components, or the reporter gene, and these mutations allow the cells to grow faster and overtake the culture. Indeed, this may be part of the reason that we observe damping of oscillations over time. Thus, to determine how well any particular version of the clock oscillates, one must perform several experiments using single colonies derived from a freezer culture, and watch the oscillations of gene expression for the first few periods before a large number of mutants have accumulated. Remarkably, even though mutants accumulate during the clock runs, the damping of oscillations is quite reproducible from run to run for a given strain, as long as the entrainment procedure is consistent. Apparently, mutants are accumulated in a reproducible fashion in these runs. This is not unlikely, since the clock experiments are started with large cell populations.

The performance of the synthetic genetic oscillator depends upon the host cell in which it is integrated. We are using a derivative of *E. coli* strain 3.300 as the host cells for the device. This strain was obtained from the *E. coli* genetic stock center (CGSC). We have observed subtle, but real, differences in oscillatory performance when the clock was placed into different *E. coli* backgrounds, such as MG1655 or YMC10, but as yet we have no explanations for these differences.

The hysteresis experiments using toggle switch-containing strains are quite reproducible, as long as the initial dilution of the starting cultures is high as in our protocol. We have found that the approach to the steady state is very slow for induced cultures that have been shifted to lower concentrations of (or absence of) IPTG. Thus, if the overnight cultures of induced cells are not diluted sufficiently, they will not have reached the steady state by the time the culture $OD_{600}$ has reached 0.4, and the resulting levels of reporter gene expression that are measured will be considerably higher than the true steady state values. We have also found that the kinetic order of induction and de-induction can be very high for the devices shown in Fig. 1, and thus that accurate determination of these profiles requires examining as many different IPTG concentrations as possible. We routinely examine 12–15 different IPTG concentrations for each of the induced and uninduced cultures, with the IPTG values chosen based on preliminary experiments.

## Acknowledgments

The authors would like to thank our colleagues Mariette Atkinson, Michael Savageau, Avraham Mayo, Dong-Eun Chang, Daniel Forger, Shelly Leung, Steven Selinsky, Aaron Reifler, and David Schauder who participated in the development of the synthetic genetic devices used in our experiments and their characterization. This work was supported by grant R01-GM063642 from the NIH-NIGMS.

### References

1. Kaern M., Blake, W. J., and Collins, J. J. (2003) The Engineering of Gene Regulatory Networks. *Annu. Rev. Biomed. Eng.* **5**, 179–206.
2. Andrianantoandro, E., Basu, S., Karig, D. K., and Weiss, R. (2006) Synthetic biology: new engineering rules for an emerging discipline. *Mol. Syst. Biol.* **2**, 0028.
3. Sambrook, J., Fritsch, E. F., and Maniatis, T. (1989) Molecular Cloning. A laboratory Manual. 2nd Ed. Cold Spring Harbor Laboratory Press, Cold Spring Harbor, NY.
4. Datsenko, K. A., and Wanner, B. L. (2000) One-step inactivation of chromosomal genes in *Escherichia coli* K-12 using PCR products. *Proc. Natl. Acad. Sci. USA* **97**, 6640–6645.
5. Atkinson, M. R., Savageau, M. A., Meyers, J. T., and Ninfa, A. J. (2003) Development of genetic circuitry exhibiting toggle switch or oscillatory behavior in *Escherichia coli. Cell* **113**, 597–607.
6. Chang, D.-E., Leung, S., Atrkinson, M. R., Reifler, A., Forger, D., and Ninfa, A. J. (2010) Building biological memory by linking positive feedback loops. *Proc. Natl. Acad. Sci. USA* **107**, 175–180.
7. Gardner, T. S., Cantor, C. R., and Collins, J. J. (2000) Construction of a genetic toggle switch in *Escherichia coli. Nature* **403**, 339–342.
8. Elowitz, M. B., and Leibler, S. A synthetic oscillator network of transcriptional regulators. *Nature* **403**, 335–338.
9. Ninfa, A. J., Selinsky, S., Perry, N., Atkins, S., Xiu Song, Q., Mayo, A., Arps, D., Woolf, P., and Atkinson, M. R. (2007) Using two-component systems and other bacterial regulatory factors for the fabrication of synthetic genetic devices. *Methods Enzymol.* **422**, 488–512.
10. Sauer laboratory group. P1vir phage transduction. http://openwetware.org/wiki/Sauer:P1vir_phage_transduction.
11. Dagert, M., and Ehrlich, S. (1979) Prolonged incubation in calcium chloride improves the competance of Escherichia coli cells. *Gene* **6**, 23–28.
12. Oehler, S., Eismann, E. R., Kramer, H., and Muller-Hill, B. (1990) The three operators of the *lac* operon cooperate in repression. *EMBO J.* **9**, 973–979.
13. Ninfa, A. J., Ballou, D. P., and Benore, M. (2010). Fundamental Laboratory Approaches for Biochemistry and Biotechnology, 2nd Edition. John Wylie and Sons, New York.

# Chapter 18

# Studying Microbial Communities in Biofilms


## Katie Brenner



## Abstract

Most microorganisms in nature subsist as heterogeneous surface-associated communities called biofilms. In biofilms members of one or more microbial species live together for multiple generations, and this allows them to cooperate and co-adapt. The ability to reliably manipulate, characterize, and engineer microbial biofilms will enable controlled studies of ecosystem dynamics and unprecedented design opportunities for biological sensors and actuators. Biofilms can be grown in the laboratory, and spatial structure, gene expression, and productivity (total biomass accumulation) can be observed and quantified as a function of time using confocal laser scanning microscopy. This chapter details the materials and methods necessary to grow and study engineered microbial communities in biofilms.



**Key words:** Microbial biofilm, Confocal microscopy, Community, Consortium, Flow


## 1. Introduction

It is important to study engineered microbes and microbial communities growing as biofilms for several reasons. First, we can gain accurate insight into behavior of microbes only if we study them in a natural context, and the majority of microbes in nature spend most of their time in biofilms (1, 2). Second, microbes in biofilms interact over long periods of time; growing engineered organisms in biofilms allows us to determine how engineered behaviors such as communication impact cooperation, competition, and co-adaptation of a community through multiple generations (3–6). Finally, an important aspect of communal behavior is the formation of three-dimensional structures. We can learn about microbial communication and collaboration by observing these structures (1, 7–10). By studying a biofilm's three-dimensional structure using confocal microscopy, we can directly measure the impact of microbial interactions upon biofilm persistence and productivity (11).


Wilfried Weber and Martin Fussenegger (eds.), *Synthetic Gene Networks: Methods and Protocols*, Methods in Molecular Biology, vol. 813, DOI 10.1007/978-1-61779-412-4_18, © Springer Science+Business Media, LLC 2012

In the laboratory, we mimic the natural biofilm environment by exposing surface-attached biofilms to a constant flow of nutrient solution over periods ranging from days to weeks. It can be challenging to design a flow system that ensures repeatable biofilm growth (12). The goal of this chapter is to provide a well-tested set of materials and methods for growing biofilms in the laboratory. Materials and methods are specified for *setting up the initial flow system, routinely setting up the flow system, sample preparation, inoculation, initial maintenance, data collection, and terminating experiments.* Methods for sterilizing the flow system and for general maintenance throughout the biofilm experiment are also provided. Some experimental details such as timing of observations, temperature, and rate of flow must be determined by the user. In early biofilm experiments, while optimizing these variables, it is wise to perform at least six independent replicates for each control and each experimental case so that repeatable results are most likely to be obtained.

## 2. Materials

These materials are required for basic biofilm flow experiments in which biofilms grow to a thickness of no more than 100 μm, with laminar flow and no recycling of the effluent. Unless noted, suppliers and manufacturers are unimportant and a similar part may be substituted. Please see Fig. 1 for a diagram of the flow system.

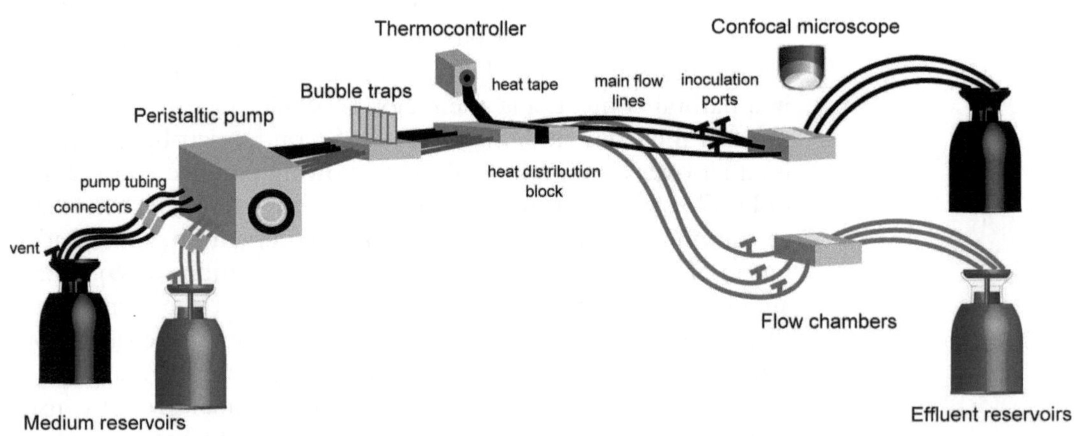


Fig. 1. The biofilm flow system. Drawn by the peristaltic pump, medium that feeds the biofilms flows out of Pyrex bottles serving as medium reservoirs, through Tygon tubing that connects to pump tubing as it flows past the pump, into Norprene tubing that connects to the bubble traps, and back into Tygon tubing after the bubble traps. After passing through a heat strip that guards against contamination, long stretches of tubing lead to inoculation ports and segments and into the flow chambers. Effluent flows out of these chambers into effluent reservoirs. Biofilm behavior is observed with a confocal microscope at the flow chamber.

### 2.1. Setting Up the Initial Flow System

General equipment

1. Peristaltic pump with one channel for each replicate (Watson Marlowe 205U/CA, 020.5716.00A, see Note 1).
2. As needed, a laboratory cart with shock-absorbing wheels is useful to transport the biofilm experiments from bench to autoclave and from laboratory to microscope facility.
3. A heat-distribution block can be made by a machine shop: an aluminum block 2 cm × 4 cm × 12 cm with holes evenly spaced and drilled through it in the middle of and perpendicular to the 2 cm × 12 cm plane. The diameter of the holes matches the outer diameter of the tubing and the number of holes is the maximum number of flow lanes that will be run at a given time (usually equal to the number of channels on the peristaltic pump). Tubing bringing medium to the biofilms will pass through this block and, when the block is heated, the medium will encounter a 4-cm stretch of intense heat.
4. Thermocontroller (Barnstead/Thermolyne Corp International, Heating Controller Series 45500, CN45515).
5. Heating tape (Barnstead/Thermolyne Corp International, BS0051-020).

For each set of triplicate experiments

6. Two Pyrex standard, orange-lid, autoclavable glass bottles (one each for the sterile medium reservoir and the effluent reservoir; choose bottle size based upon pump speed setting – bottles should hold adequate medium to last overnight).
7. Two syringe filters to create a vent on each medium and effluent reservoir – Nalgene PTFE Syringe Filters, 13 mm, 0.2 μm pores (VWR, 28196-262).
8. Two connectors, one for each of these vents – female luer hose barbs, 1/16 in. (Cole Parmer, A-06359-27).
9. Two 3 cm lengths of Tygon ABW00002 tubing (see details below) to create the vent.
10. One triplicate flow chamber (Stovall Life Sciences Inc., ACCFL0001 – this exact part is recommended, but see Note 2).

For each flow line

11. Tygon 3350 ABW00002 sanitary gas-permeable tubing (VWR, 63009-254; ID 1/16 in., OD 1/8 in. – this exact part, from any supplier, is recommended). Each flow line requires the following lengths of ABW00002:

    Two 0.75 m (lead and effluent lines).

    One 2 m (main flow line).

One 3 cm (removable inoculation segment).

12. One silicone peroxide-cured pump tube (Cole Parmer, A-07616-38; ID 1.65 mm – pack of 6).
13. One bubble trap (or use multichannel bubble traps to save space) (Biosurface Technologies, BSTFC34 – this exact part is recommended but see Note 3).
14. Norprene food tubing to connect to the bubble traps (Cole Parmer, EW-95710-10; 50 ft/pack, ID = 1/8 in., see Note 4). Two pieces 5 cm long for each flow line.
15. Connectors (note packet sizes and buy accordingly):

    Two ID = 1/8–1/16 in. reducing connectors (Cole Parmer, EW-06362-11).

    One ID = 1/16 in. three-way connectors (Cole Parmer, A-06365-88).

    One ID = 1/16 in. straight barbed connector (Cole Parmer, A-06365-11).
16. One inoculation port (Fisher Scientific, 03-215-5 – pack of 100).

#### 2.2. Routinely Setting Up the Flow System

1. One autoclaved graduated cylinder (same volume as Pyrex reservoir bottles).
2. Medium to feed biofilms at inoculation (at least 100 ml per flow line).
3. If biofilm adhesion requires a surface coating, have a stock solution on hand at appropriate temperature.
4. One large presterilized petri dish into which all flow chambers for use in the experiment can fit.

#### 2.3. Sample Preparation

1. Petri dishes prepared with agar medium (choose an experimentally appropriate growth substrate) and any antibiotics for use in confirming live cell counts.

For each flow line, you will need

2(a) First time experiments: one 50 ml centrifuge-safe sterile test tube (Corning, 430290).

2(b) Repeat experiments: one 15 ml centrifuge-safe sterile test tube (Corning, 430052).

3. One 2 ml sterile Eppendorf tube (USA Scientific, 1620-2700).
4. 500 ml sterile 0.9% NaCl stock solution (replenish as necessary).
5. Appropriate antibiotic stock solutions, if needed.
6. Cultures or suspensions of organisms to be inoculated (for example, 10 ml overnight cultures grown to saturation). Unless experiments dictate otherwise, grow cultures in the

same medium and at the same temperature that they will first encounter in the biofilm environment.

### 2.4. Inoculation

For each flow line

1. One 1 ml syringe (Cole Parmer, BD, YO-07940-00).
2. One 22 guage 1 in. needle (Cole Parmer, BD, EW-07940-58).
3. One small Kimwipe.
4. One sterile alcohol swab (BD, 326895).

Keep on hand

5. One roll of laboratory tape.
6. Tubing clamps (one for each replicate or triplicate, depending upon type – some types of clamps must be threaded onto the tubing during initial setup).
7. Stopwatch with multiple timers.
8. Permanent marker.

### 2.5. Initial Maintenance

For each flow line

1. One tubing clamp.
2. One sterile alcohol swab.
3. One sterile razor blade.

### 2.6. Data Collection

1. Confocal microscope with a microscope-slide stage.
2. COMSTAT or other image-processing software (see COMSTAT information at http://www.comstat.dk).

### 2.7. Terminating Experiments

1. Bleach dilution solution made with three parts sterile water to one part standard bleach (6% sodium hypochlorite). (Make enough to fill all medium reservoirs).
2. Sterile water (enough to fill all medium reservoirs).
3. Sterile alcohol swabs (BD, 326895).
4. Optional: 3% hydrogen peroxide solution.

## 3. Methods

### 3.1. Setting Up the Initial Flow System

The first time biofilm experiments are performed, follow these instructions to set up the flow system. This system should then be rebuilt by replacing all tubing, syringe filters, and inoculation ports after it is used 15–20 times. In the meantime, replace parts as necessary (for example, flow chambers and inoculation ports will need more frequent replacement). Please see Fig. 1 for a diagram of the flow system.

1. Drill four 1/8 in. holes in the lids of all Pyrex bottles that will be used as reservoirs in the initial system. Three of these holes are for tubes to feed (medium reservoirs) or exhaust (effluent reservoirs) the biofilms, and the fourth hole is for a vent. (If running 15 replicates at a time, with one medium and one effluent reservoir for every three replicates, there will be a total of ten bottles with holes in lids.)
2. To create each vent, insert a 3 cm length of Tygon ABW00002 through one of the holes in each lid (see Note 5). 1 cm of tubing should hang into the bottle, 2 cm should hang outside. On the outside end of the tubing, place a female luer hose barb with the barrel-end attached to the Nalgene PTFE syringe filter.
3. Into the remaining three holes of each drilled lid, insert one of the 0.75 m lengths of Tygon tubing. The inside end of tubes in the medium reservoir bottles should be approximately 2 cm longer than the height of the reservoir bottles so that the tubes hang down and rest on the bottoms of the bottles parallel to the bottom face. These inside ends should be cut diagonally to minimize the formation of air bubbles. In contrast, the inside ends of tubes inserted into the effluent bottles should hang down into the bottles only 2–3 cm so that, even when the effluent bottles are nearly full, the tubes will hang above the effluent; these effluent tubes should never become submerged (if they do, experiments must be terminated because of the risk of contamination).
4. To the outside end of each medium reservoir tube, attach a straight barbed connector and then a pump tube.
5. To each pump tube, attach a 1/16 to 1/8 in. conversion connector and then a 5 cm length of Norprene tubing.
6. Attach the open end of the Norprene tubing to an inlet on a bubble trap. At this stage it is useful to number the tubes and to keep them in numerical order.
7. Attach another 5 cm length of Norprene tubing to the outlet of the bubble trap.
8. Attach a 1/8 to 1/16 in. connector to the open end of this Norprene tubing, and to it attach the 2 m length of Tygon tubing – this is the main flow line.
9. Thread the open end of this 2 m length of tubing through a hole in the heat distribution block. Thread tubes through the holes in the block in numerical order.
10. Attach a three-way connector to the open end of this main flow line.
11. Attach the 3 cm inoculation segment to the open in-line end of the three-way connector.

12. Attach an inoculation port to the perpendicular end of the three-way connector. For easier access, fold the rim of the inoculation port down so that the septum is exposed.
13. Attach one of the barbed ports of a triplicate flow chamber to the open end of the 3 cm inoculation segment. Attach tubes in numerical order so that each medium reservoir will feed all three lanes of one flow chamber.
14. Attach one of the 0.75 m tubes to the barbed outlet of the flow lane. It is critical that replicates of the same experiment flow into the same effluent collection vessel to minimize the possibility of cross-contamination.
15. Repeat steps 4–14 for each of the flow lanes.
16. Wrap the heating tape tightly around the heat distribution block perpendicular to the direction in which the tubing passes through it, and plug tape into the thermocontroller.

### 3.2. Sterilizing the Flow System

Sterilize the flow system each time you finish an experiment to preserve your system and ensure accurate results for the next experiement.

1. Remove flow chambers from the system. Remove stopcocks, but not filters, from the bubble traps (see Note 6). Remove heating tape from the heat distribution block (aluminum heat distribution block remains in place and can be autoclaved).
2. Cover each of the resulting open ends of Tygon tubes, including the entire inoculation segment and the attached inoculation port, with aluminum foil.
3. Place the inflow parts (consisting of medium reservoirs through foil-covered inoculation segments) and the effluent parts (consisting of foil-covered ends of effluent tubing through the effluent reservoirs) into autoclavable tubs. Also, include any extra bottles necessary for media preparation and the graduated cylinder.
4. Autoclave 30 min, dry cycle.

### 3.3. Routinely Setting Up the Flow System

The user will most often complete procedures 3.3 and 3.4 concurrently, followed immediately by procedure 3.5.

1. Replace any necessary parts and sterilize the entire flow setup as described in procedure 3.2.
2. Sterilize flow chambers in a UV hood (it is useful to place them together in a large, presterilized petri dish).
3. If any surface coating is to be used, prepare the appropriate solution in a sterile environment. To coat a single lane of a flow chamber, inject 200 μl of the solution through the barbed connector into the lane. Place the chambers inside a large presterilized petri dish and incubate as necessary.

4. Attach pump tubing to pump with the pump-stay arms, keeping tubes in numerical order.
5. Return stop-cocks to the tops of each bubble trap. Stop-cocks should be in the open position – turned in-line with the trap.
6. Thread heating tape tightly around the heat distribution block, plug in, and set thermocontroller to medium (for example, controller setting “4”).
7. Working in a sterile environment, attach flow chambers to inoculation segments and effluent lines. Keep them in numerical order so that the correct medium will feed the correct flow lane. All inoculation ports should face upward perpendicular to the flow chambers for easy access. Now the flow system is completely assembled, sterile, and air-tight.
8. Prepare adequate medium or media to feed biofilms for the first 24 h or for the period until you intend to change media, and distribute into each medium reservoir using the sterile graduated cylinder. This should be done in a sterile environment – bottle lids with dangling tubes should not be set aside, but rather held approximately 5 cm from the bottle mouth with tubing dangling back into the bottles while measured medium is poured from a graduated cylinder into each reservoir.
9. Once medium reservoirs are full, begin flow to fill the bubble traps (see Note 7). Fill each bubble trap with at least 3 ml medium. Do not overfill traps and thereby saturate filters (replace any accidentally saturated filters with new, sterile filters). Using the Watson Marlowe pump specified, at speed 4.0, filling traps to proper volume takes 10–15 min.
10. When bubble traps are full and stopcocks are closed, adjust pump speed to experimental flow rate.
11. Run pump until tubing throughout the flow system has been flushed. Effluent reservoirs should contain at least 20 ml of medium.

### 3.4. Sample Preparation

1. Have ready the liquid cultures or suspensions of organisms to be inoculated. These are usually monocultures, unless experiments dictate otherwise.
2. Measure the optical density of each culture or suspension.
3. Place 14 ml of 0.9% NaCl solution into 50 ml (*first time experiments*) or 15 ml (*repeat experiments, assuming no more than 1 ml is found in* step 6a) centrifuge-safe tubes and label them. Make one tube for each culture. Add antibiotics or other additives as needed.

   *The first time a given experiment is run (for repeat experiments, skip to* step 6b*):*

4. Place 1 ml of each liquid culture/suspension from step 1 into the centrifuge tube prepared in step 3.

5. Measure the optical density of this dilution. If needed, add more of the culture to the NaCl solution to achieve the desired optical density for inoculation (e.g., $OD_{600} = 0.07$).

6(a) *In first-time experiments*, record the final amount of culture used to reach the desired optical density.

6(b) *In repeat experiments*, add the amount of culture found in step 6a to the vial prepared in step 3 and measure optical density to confirm. More or less of the dilution solution may be prepared, maintaining the ratio of 0.9% NaCl to culture, as needed.

7. Plate dilutions of the NaCl suspension to confirm live cell counts.
8. Number or otherwise label a 2 ml Eppendorf tube for each flow lane to be inoculated, and place them in a tube rack in order of inoculation.
9. Pipette 1.5 ml of culture dilution solution prepared in step 6b into the properly numbered Eppendorf tube prepared in step 8. If organisms or cultures are to be mixed in the biofilm, pipette the correct ratios of each into the Eppendorf tube at this step.

### 3.5. Inoculation

The inoculation procedure must be completed carefully but quickly so that all flow lanes incubate for the same amount of time. Incubation time determines how many bacteria initially adhere to the substrate to seed the biofilm, which impacts long-term biofilm development. As per these instructions, prepare all materials ahead to speed inoculation. This procedure assumes that the user has just completed procedures 3.3 and 3.4 for routine setup of the flow system and sample preparation.

1. Attach syringes to needles, retaining needle caps. Number or label each with a permanent marker, one for each lane to be inoculated.
2. Fill each appropriately labeled syringe with a little more than 1 ml of the dilution solution from the correspondingly labeled Eppendorf tube. Recap the needle and tap the barrel of the syringe while it is upright to remove air bubbles from the contents.
3. Using tube clamps, clamp all of the tubes just upstream of the heat distribution block.
4. Immediately stop the pump. Adjust pump speed to an adequate flushing speed (e.g., 4.0) but do not restart the pump.
5. Detach all pump-stays from the top of the pump so that no tubes are in contact with the pump, but leave the tubes engaged in the stays so that stays can be quickly reattached to resume flow.
6. Remove alcohol swab(s) from packaging and cleanse all inoculation ports.

7. Inoculate the three flow lanes of one flow chamber with solution from the correspondingly numbered syringes. To inoculate one lane, pierce the inoculation port and allow the needle to penetrate into the three-way connector. Then, quickly but with even speed, inject the contents of the syringe into the line, being careful to stop after injecting 1 ml in order to avoid injecting air bubbles into the line (do not completely empty the syringe).
8. As soon as all three lanes of one flow chamber are inoculated, set a timer on the stopwatch for 4 min or your chosen incubation time. Each chamber must incubate for this same period of time.
9. Turn the flow chamber glass-side-down so that the glass is facing the work surface.
10. Use laboratory tape to secure the chamber glass-side-down so that it remains flat and is not disturbed as you inoculate the remaining chambers.
11. Repeat steps 7–10 for each flow chamber. Watch the time so that if one of the timers is nearing zero, you are not in the middle of inoculating another chamber.
12. When the incubation timer for a flow chamber reaches zero, reattach the pump stays that are associated with flow lines for that chamber to the pump. Be sure to restart the pump after reattaching the first set of pump stays; simply leave it running while reattaching all other sets.
13. Immediately remove the tube clamp(s) from the three flow lines that are now attached to the pump and flowing.
14. Reset the same timer for 4 min. The flow chamber should remain glass-side-down for the next 4 min while any planktonic cells are flushed.
15. When the timer reaches zero for this 4 min flushing step, remove tape and place chamber glass-side-up. Replace tape to keep chamber stable.
16. When all chambers are inoculated, incubated, flushed, and glass-side-up, set the pump to proper speed for experiments.
17. Check the whole setup to ensure that there are no leaks. In particular, check that the inoculation ports are very tightly attached to the three-way connectors and that the bubble traps are closed.
18. Check the whole setup again 2–4 h later to be sure there are no leaks.

***3.6. Initial Maintenance (12–24 h After Inoculation)***

1. Clamp all tubes 10 cm upstream of the flow chambers.
2. Immediately stop the pump.

Working quickly, one flow lane at a time, in a sterile environment if possible

3. Cleanse tubing immediately upstream of the inoculation segment with a fresh alcohol swab (use a clean swab for each line).
4. Use a sterile razor blade to make a blunt cut through the main flow line tubing that is immediately upstream of the three-way connector and inoculation segment.
5. Quickly, detach the three-way connector and inoculation segment from the flow chamber and discard them.
6. Attach the blunt-cut end of the main flow line tubing to the inlet barb on the flow chamber.
7. Cleanse the newly reattached tubing and surroundings with the alcohol swab and discard it.
8. Repeat steps 3–7 for all flow lanes.
9. Start pump.
10. Immediately remove tubing clamps.

***3.7. General Maintenance Throughout the Biofilm Experiment***

1. As often as your pump speed and reservoir size dictates, replace medium in reservoirs with freshly prepared medium and discard contents of the effluent reservoirs (take samples if necessary for the experiment, and discard remaining effluent into a beach solution). It is best to work in a sterile environment. Medium reservoirs can either be refilled from a larger container, or replaced entirely (be very careful not to contaminate the dangling tubing when transferring the lid and tubing from an old reservoir bottle to a new one).
2. When replacing medium, also remove any air bubbles from the system by holding each flow chamber perpendicular to the usual direction of flow, with the outflow end of the chamber upward. Wait until air bubbles flow out of the chamber.
3. If experiments are long term, repeat the "initial maintenance" procedure as often as needed to keep the main flow line tubing clear of significant biofilm growth. Cut off 5–10 cm upstream of the flow chamber each time.
4. If experiments are long-term, the effluent tubing may become overgrown with biofilm. In this case, build entirely new effluent sections (including tubing, lids with holes in them, and reservoir bottles) and autoclave them. To attach them, use steps 1 and 2 of procedure 3.6 to stop the pump, then detach the effluent sections, re-attach new ones, and start the pump again using steps 9 and 10 of that procedure.

### 3.8. Data Collection

Data collection protocols will be different for each experiment but this is a general guideline. This data collection method will yield at least nine image stacks for every experimental condition.

1. Every 6 h, collect a stacked image in at least three locations within each flow lane; one each near the inlet, middle, and outlet of the flow lane. The exact imaging location should be chosen randomly to avoid bias.
2. Examine all images in each stack by hand. Any images that contain glare (suggesting that the stack began within the glass substrate) must be eliminated, and the stack renumbered accordingly. If blurring is evident (suggesting movement of the stage or camera during the image collection) eliminate the entire stack.
3. Import stacks into COMSTAT for processing. If an outlier becomes evident during processing with COMSTAT, that entire image stack may be eliminated from all calculations. At least six stacked images should always be included in every calculation for every condition.

### 3.9. Terminating Experiments

1. Replace all medium reservoirs with bottles containing the bleach dilution solution.
2. Flow the bleach solution through the setup for at least 4 h.
3. Replace the bleach reservoirs with bottles containing sterile water.
4. Flow the water through the setup for at least 4 h.
5. After the bleach and water rinse steps, flow all liquid out of the flow setup so that bubble traps are empty and no liquid remains in tubing.
6. Detach flow chambers from the setup.
7. If desired, place flow chambers in a 3% hydrogen peroxide bath for an hour (see Note 8). To achieve the cleansing affect, be sure that the flow lanes fill with solution when submerged. Jiggle and tap flow chambers until lanes fill completely.

## 4. Notes

1. To expand the number of replicates, it is possible to use a single peristaltic pump channel to provide flow to more than one replicate by using a connector to split the tubing after the pump. However, this is inadvisable because it is difficult to achieve equal flow rates and volumes in each half of the split tube.

2. There are circular indentations in the molded plastic bottoms of the Stovall ACFL00001 flow cells. These depressions are deep enough that they provide shelter from the flow and can retain planktonic cells. In some sensitive experiments, this retention can be a problem. It is important to monitor these regions of the base with the confocal microscope, and to turn the flow chambers upside-down one to two times daily for 5 min (or more/less depending upon flow rate) when conducting sensitive experiments to help wash planktonic cells out of those depressions.
3. There are several suppliers for bubble traps but we find that the BST traps are autoclavable and hold up well. After 10–15 week-long experiments, the old syringe parts should be removed, the holes through which medium flows in the traps can be carefully reamed with a drill to remove deposits, and new syringe parts restored to the trap. A 3/16 in. hollow metal punch can be used to cut the hole in the black rubber part of the syringe. This treatment can keep the trap in good working order for several years.
4. The size of Tygon tubing (ABW00007) that is equivalent to the Norprene Food Tubing does not autoclave well – it changes size enough that medium leaks from around the bubble traps after autoclaving. Thus, it is best to use the Norprene Food tubing listed to connect to the bubble traps.
5. Inserting tubing into the holes in bottle lids is much easier if the tubing is cut on an angle. Additionally, clipping the tubing on a slight angle provides a wider access point for fluid flowing into the lead tubes.
6. Never autoclave the stopcocks (tops of the bubble traps). It is helpful to keep extras on hand in case you accidentally do so. The stopcocks are male luer 1-way polycarb stopcocks (Cole Parmer, U-30600-01, pack of 10).
7. Air bubbles threaten biofilm experiments. To minimize them, be sure to keep the entire flow setup on one level (do not place reservoirs or effluent containers on the lower level of a double-shelf cart, for example).
8. If carefully maintained and well-cleaned, the Stovall flow chambers can be used more than once, but they cannot be autoclaved. At most, use them only three to four times before discarding. For prolonged (several week) or critical (leading to published data) experiments, begin with new flow chambers and discard them immediately after use.

## References


1. Costerton JW, et al. (1994) Biofilms, the customized microniche. *J Bacteriol* **176**, 2137–42.
2. Shapiro JA (1998) Thinking about bacterial populations as multicellular organisms. *Annu Rev Microb* **52**, 81–104.
3. West SA, Diggle SP, Buckling A, Gardner A, Griffins AS (2007) The social lives of microbes. *Annu Rev Ecol Evol S* **38**, 53–77.
4. Wingreen NS, Levin SA (2006) Cooperation among microorganisms. *PLoS Biol* **4**, e299.
5. Wuertz S, Okabe S, Hausner M (2004) Microbial communities and their interactions in biofilm systems: An overview. *Water Sci Technol* **49**, 327–36.
6. Xavier JB, Foster KR (2007) Cooperation and conflict in microbial biofilms. *Proc Natl Acad Sci USA* **104**, 876–81.
7. Johnson CR, Boerlijst MC (2002) Selection at the level of the community: The importance of spatial structure. *Trends Ecol Evol* **17**, 83–90.
8. Johnson CR, Seinen I (2002) Selection for restraint in competitive ability in spatial competition systems. *Proc Biol Sci* **269**, 655–63.
9. Hassell MP, Comins HN, May RM (1994) Species coexistence and self-organizing spatial dynamics. *Nature* **370**, 290–292.
10. Sternberg C, et al. (1999) Distribution of bacterial growth activity in flow-chamber biofilms. *Appl Environ Microbiol* **65**, 4108–17.
11. Heydorn A, et al. (2000) Quantification of biofilm structures by the novel computer program comstat. *Microbiology* **146 (Pt 10)**, 2395–407.
12. Goeres DM, et al. (2005) Statistical assessment of a laboratory method for growing biofilms. *Microbiology* **151**, 757–62.

# Chapter 19

## Quantitative Analysis of the Spatiotemporal Dynamics of a Synthetic Predator–Prey Ecosystem

***Stephen Payne, *Robert Phillip Smith, and Lingchong You**

### Abstract

A major focus in synthetic biology is the rational design and implementation of gene circuits to control dynamics of individual cells and, increasingly, cellular populations. Population-level control is highlighted in recent studies which attempt to design and implement synthetic ecosystems (or engineered microbial consortia). On the one hand, these engineered systems may serve as a critical technological foundation for practical applications. On the other hand, they may serve as well-defined model systems to examine biological questions of broad relevance. Here, using a synthetic predator–prey ecosystem as an example, we illustrate the basic experimental techniques involved in system implementation and characterization. By extension, these techniques are applicable to the analysis of other microbial-based synthetic or natural ecosystems.

**Key words:** Synthetic ecosystem, Quorum sensing, Predator–prey ecosystem, Microbial consortia, Biodiversity, Synthetic biology, Bidirectional communication, Gene circuit

## 1. Introduction

A major challenge of synthetic biology is to create systems with increasing complexity in a predictable manner. Many examples have been constructed and implemented in the last decade. However, these systems generally involve circuits operating at the single-cell level and include devices, such as switches (1–7), oscillatory networks (2, 8–13), transcriptional cascades (14), logic gates (15–17), cell counters (18), light detectors (19), and time-delay circuits (20, 21).

Increasing efforts have been made to construct gene circuits that program population dynamics. Several population-level systems have been constructed that result in predictable behaviors, such as coordinated oscillations (22, 23), pattern formation (24, 25), edge

* These authors contributed equally to this work.

Wilfried Weber and Martin Fussenegger (eds.), *Synthetic Gene Networks: Methods and Protocols*,
Methods in Molecular Biology, vol. 813, DOI 10.1007/978-1-61779-412-4_19, © Springer Science+Business Media, LLC 2012

detectors (26), targeted consensus consortia (27), programmed population control (28), logic gates (29), and predator–prey dynamics (30, 31). A common strategy of these circuits is the use of direct or indirect communication within one population or between two or more populations. This communication is often achieved by rewiring bacterial quorum-sensing genes (32–34). A canonical example of quorum sensing is the well-characterized *luxR/luxI* system from *Vibrio fischeri*. The LuxI protein synthesizes diffusible acyl homoserine lactone (AHL) molecules which accumulate both inside and outside of the cell (32). The AHL concentration in the medium increases with cell density. At sufficiently high concentrations, AHL can bind to and activate the LuxR receptor protein, which in turn activates transcription downstream of the *lux* promoter.

Our laboratory has recently co-opted two quorum sensing systems in the creation of a synthetic predator–prey ecosystem in the bacterium *Escherichia coli* (30). In this system, quorum-sensing modules from *V. fischeri* and *Pseudomonas aeruginosa* (*lasR/lasI* system) allow two separate populations to communicate bidirectionally. The predator cells kill the prey cells by activating transcription of a toxic protein, whereas the prey cells rescue the predator cells by activating transcription of an antitoxin protein. Implementation of this circuit in a microchemostat resulted in oscillatory behavior of both populations that resembled canonical predator–prey dynamics (30).

Synthetic ecosystems, such as the predator–prey system, present unique opportunities for exploring questions regarding natural biological systems. For instance, synthetic systems have been used to study biodiversity (35, 36), cooperation (37, 38), and fitness optimization in fluctuating environments (39). Using synthetic systems to uncover ecological principles has several advantages. Since a synthetic system is composed of well-characterized components that have defined functions and since the system can be studied in a controllable, defined environment, the system is unaffected by numerous biotic and abiotic factors that can obscure the core dynamics in a natural setting. Furthermore, the ability to modulate both the environment and the synthetic circuit allows investigators to draw conclusions which are difficult, if not impossible, to make by studying natural ecosystems. Finally, given the rapid growth rate of bacteria (the cell chassis of choice for most synthetic biologists), synthetic ecosystems can be studied over shorter, more manageable time scales.

In the following chapter, we describe methods for characterizing the predator–prey ecosystem first described in Balagadde et al. (30) and later studied in Song et al. (35). We first focus on the characterization of circuit dynamics in liquid medium. We then describe how to measure and characterize spatiotemporal dynamics in solid medium. Furthermore, we describe how to quantify cell motility and signal diffusivity, both of which are critical parameters required for modeling of the system in the solid phase. We extend these experiments and modeling methods to specifically examine biodiversity of the predator–prey system. Although the chapter focuses on methods implemented for characterizing this model predator–prey ecosystem, the principles and experiments can be extended to examine a wide range of synthetic ecosystems.

## 2. Materials

1. LB liquid medium.
2. Agar.
3. M9 medium: 1× M9 salts (48 mM $Na_2HPO_4$, 22 mM $KH_2PO_4$, 862 mM NaCl, 19 mM $NH_4Cl$) 0.4% glucose, 0.1% casamino acids, 0.1% thiamine, 1 mM $MgSO_4$, 0.1 mM $CaCl_2$. Supplement with 100 mM 3-(*N*-morpholino)propanesulfonic acid (MOPS) and adjust the pH to 7.0 using 5 M KOH.
4. TBK medium: 1% tryptone, 0.5% yeast extract, and 0.7% KCl. Supplement with 100 mM MOPS and adjust pH to 7.0 using 5 M KOH.
5. 100 μg/mL stock of chloramphenicol (prepared in 100% ethyl alcohol and stored at −20°C).
6. 50 μg/mL stock of kanamycin (stored at −20°C).
7. 100 mM stock of isopropyl β-D-1-thiogalactopyranoside (IPTG) (stored at −20°C).
8. 1 mM stock of 3-oxohexanoyl-homoserine lactone (3OC6HSL, prepared in 100% ethyl acetate and stored at −20°C).
9. 1 mM stock of *N*-3-oxododecanoyl-HSL (3OC12HSL, prepared in 100% ethyl acetate and stored at −20°C).
10. 12 mL culture tubes.
11. 25 mm×71 mm×1 mm glass slides.
12. 24 mm×24 mm coverslips.
13. 8-cm diameter Petri dish.
14. 250 mL conical flask.
15. A microplate reader.
16. A fluorescence microscope with filters that examine both GFP (excitation: 470/40 nm, emission: 525/50 nm) and RFP (excitation: 546/12 nm, emission: 605/75 nm).
17. Gel imager.

## 3. Methods

### 3.1. The Synthetic Predator–Prey Ecosystem

While the protocols below are applied in the context of the synthetic predator–prey ecosystem first described in Balagadde et al. (30) and further characterized in Song et al. (35), they can be readily generalized to examine other ecosystems consisting of programmed microbes (38) or assemblies of natural microbial isolates (40). The predator–prey ecosystem consists of two different cell strains, a predator and a prey, which communicate bidirectionally via AHL molecules to modulate each other's growth. To distinguish between the two populations, the predator constitutively expresses GFPuv(lva), and the prey constitutively expresses mCherry.

The logic of the circuit is presented in Fig. 1. Each strain contains two plasmids which separately confer resistance to chloramphenicol and kanamycin. Activation of the circuit using IPTG causes the predator cells to express a toxic protein, CcdB, which results in cell death. In addition, transcription from the $P_{Ltet\text{-}01}$ promoter results in expression of the LuxR receptor protein, which binds to 3OC6HSL, and the LasI protein, which synthesizes 3OC12HSL. The presence of IPTG causes the prey cells to express the LasR receptor protein, which binds to 3OC12HSL, and the LuxI protein, which synthesizes 3OC6HSL.

Upon activation of the circuit, the predator population dies as a result of CcdB expression. As the prey population increases in

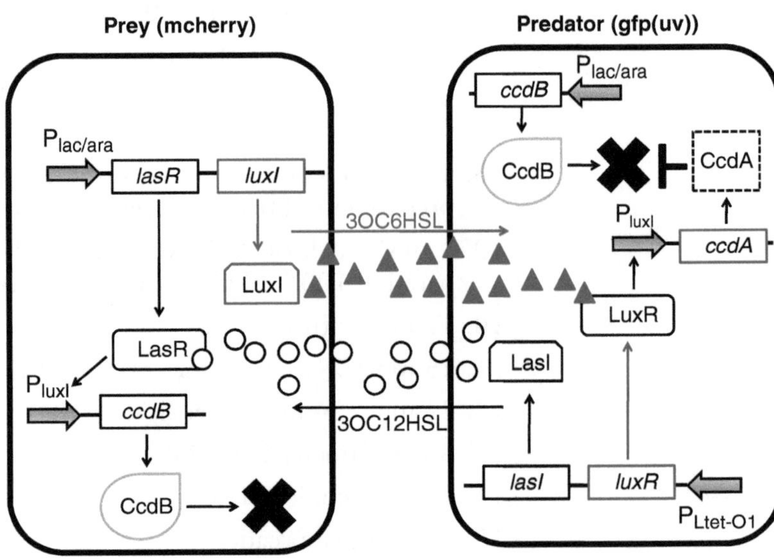


Fig. 1. The synthetic predator–prey ecosystem. A $P_{LtetO\text{-}1}$ promoter controls expression of the luxR receptor and the lasI protein, while a $P_{lac/ara}$ promoter controls expression of CcdB toxin protein, the luxI protein, and the lasR receptor. A $P_{luxI}$ promoter controls expression of the CcdA antitoxin protein and the CcdB toxin protein in the predator and prey cells, respectively. The AHL molecules 3OC6HSL and 3OC12HSL are represented by *dark triangles* and *hollow circles*, respectively. To identify individual populations, the predator cells express GFPuv(lva), while the prey cells express mCherry.

density, the concentration of 3OC6HSL increases in the medium. At a sufficiently high concentration, 3OC6HSL binds to and activates the LuxR receptor in the predator cells. Activated LuxR drives expression of an antitoxin protein, CcdA, via a $P_{luxI}$ promoter. CcdA binds to and neutralizes CcdB, thus rescuing the predator population. As the predator population increases in density, the concentration of 3OC12HSL increases in the medium. At a sufficiently high concentration, 3OC12HSL binds to the lasR receptor protein and, via a $P_{luxI}$ promoter, drives expression of CcdB in the prey cells. As a result, the density of the prey population, and subsequently the concentration of 3OC6HSL, decreases. As the concentration of 3OC6HSL decreases, CcdB is reactivated in the predator cells. This circuit logic results in oscillatory dynamics between both populations which resemble the canonical predator–prey relationship. This dynamic was observed when the ecosystem was studied in a microchemostat (30). We note that the methods and materials required to assemble, maintain, and study a synthetic ecosystem in a microchemostat are beyond the scope of this chapter.

### 3.2. Initial Characterization of Circuit Dynamics in Liquid Medium

In the protocol below, we use optical density ($OD_{600}$) to measure the growth of the individual predator and prey strains. This protocol serves as a test to confirm that the killing and rescuing components of the circuit are functional.

1. Inoculate single colonies of both the predator and prey strains into separate 12 mL culture tubes containing 6 mL of TBK medium (see Note 1). Supplement the medium with 50 μg/mL chloramphenicol and 50 μg/mL kanamycin (henceforth referred to as appropriate antibiotics).
2. Vortex the medium and the cells briefly to ensure adequate mixing.
3. Separate each culture into three, 2 mL aliquots each placed in a separate 12 mL culture tube.
4. For each strain, three different conditions are examined: OFF, (i.e., no IPTG) with IPTG, and with IPTG and a saturating concentration of AHL. For the OFF condition, do not add any additional reagents to the 2 mL aliquot. For the IPTG condition, add IPTG to a final concentration of 1 mM. Finally, for the IPTG and AHL condition, add 1 mM IPTG and 100 nM of the appropriate AHL. Specifically, 3OC6HSL and 3OC12HSL are added to the predator and prey strain, respectively.
5. Shake the culture tubes at 250 revolutions per minute (RPM) at 37°C for 20 h in an air shaker (see Note 2).
6. Remove 200 μL of culture from each tube and place the cultures in a 96-well plate. Use a microplate reader to measure the cell density ($OD_{600}$) of each culture (see Note 3). The expectation is

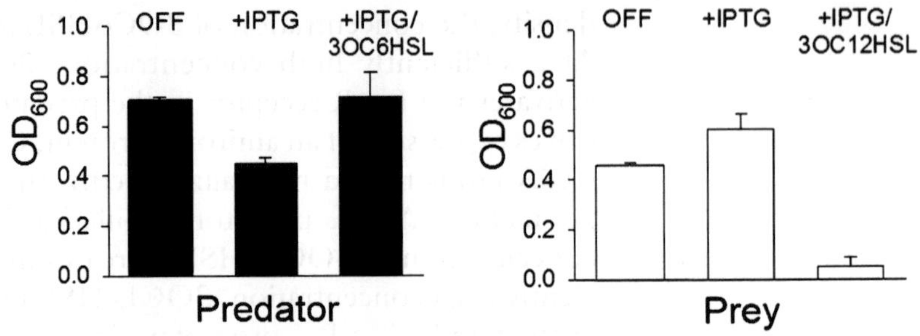


Fig. 2. Circuit dynamics in liquid medium. Each strain is cultured separately in order to examine the killing and rescuing functions of the circuit. Additional IPTG to the predator strain results in killing; however, killing can be stopped by the inclusion of a saturating amount of AHL (3OC6HSL). In contrast, IPTG allows prey cells to grow; however, addition of a saturating amount of AHL (3OC12HSL) results in killing.

that the OFF cultures should both grow to high densities ($OD_{600} > 0.5$). The prey, but not the predator, will grow to a high density in the presence of IPTG alone. In contrast, the predator, but not the prey, will grow to a high density in the presence of both IPTG and AHL. See Fig. 2 for results typically generated from this analysis.

### 3.3. Examining Biodiversity in Liquid Medium

In the following protocol, we describe quantification of the temporal dynamics of the predator–prey ecosystem in liquid medium. We focus on measuring biodiversity by quantifying the relative abundance of the two populations. We use population-specific fluorescent markers to distinguish each population and to calculate biodiversity.

1. Inoculate single colonies of the predator and prey populations into two separate 12 mL culture tubes containing 2 mL of M9 medium supplemented with the appropriate antibiotics (see Note 4).
2. Shake the tubes overnight at 250 RPM at 37°C in a shaker.
3. Remove 200 μL of each overnight culture. Place the cultures into separate wells of a 96-well plate and measure optical density ($OD_{600}$).
4. Using fresh M9 medium (supplemented with the appropriate antibiotics), adjust the $OD_{600}$ of both cultures to ~0.8.
5. Dilute 30 μL of each culture into 3 mL of M9 medium supplemented with the appropriate antibiotics and 1 mM IPTG in a 12 mL culture tube.
6. Shake the culture at 250 RPM at 37°C in an air shaker.
7. At 5, 12, and 24 h after inoculation, remove 1 μL of the culture (see Note 5).

8. Place the 1 μL culture on a 25 mm × 71 mm × 1 mm slide and add a 22 mm × 22 mm coverslip over the culture.
9. Examine the slide with a fluorescence microscope. A magnification lens of 100× (oil immersion) should be used to view the cells.
10. Using the phase contrast, GFP, and RFP filters in succession, capture images of five randomly selected fields of view which contain between 50 and 100 cells.
11. The biodiversity index (BI) is determined using a program that calculates the average mass ratio between the predator and populations. The program must extract the total number of colored green (predator) or red (prey) pixels within a single bacterium. The biodiversity index can then be calculated using the formula:

$$BI = 1 - \sum_{i=1}^{2} x_i^2 = 2x_1x_2$$

where $x_i$ is the fraction of the $i$th population in co-culture ($i = 1$ for predator and 2 for prey). In this case, $x_1$ is proportional to the number of green pixels, and $x_2$ is proportional to the number of red pixels.

### 3.4. The Predator–Prey Ecosystem in Solid Medium: Preparation of Soft Agar

In the protocol below, we describe the preparation of soft agar, which is used as a solid-phase environment in the predator–prey ecosystem.

1. Prepare 100 mL of M9 liquid medium (see Note 6).
2. Add agar to the M9 liquid medium such that the final concentration is between 0.1% and 0.5% (see Note 7).
3. Boil the solution in a microwave until the mixture contains no agar aggregates and is completely liquefied. This step requires one to microwave the culture in small ~10 s rounds to ensure that the mixture does not overflow. In between heating rounds, allow the culture to cool for 10 s. Then, using heat-resistant gloves, gently shake the flask in a circular motion by hand. Once the majority of bubbles dissipate, place the flask back in the microwave and restart the heating process (see Note 8).
4. Remove the flask with heat-resistant gloves and allow the medium to cool to ~50°C. This cooling process generally takes ~10–20 min. If the investigator needs more time for the cooling process, place the flask on a preset hotplate to prevent solidification.
5. The molten agar prepared in this section is used directly in the following sections.

### 3.5. Examining Biodiversity of Well-Mixed Populations in Solid Medium

In the following protocol, we describe how to measure biodiversity in solid medium when the predator and prey populations are mixed. In addition to biodiversity, we note that these measurements are also useful for characterizing other aspects of the spatiotemporal dynamics of the system, such as spatial heterogeneity of the two populations.

1. One day prior to the experiment, prepare M9 solid medium, containing 1.5% agar, the appropriate antibiotics, and 1 mM IPTG. Using a pipette, transfer 10 mL of the agar to an 8-cm diameter Petri dish and allow this to set overnight. This solid medium will serve as the bottom agar in the following steps.
2. Inoculate single colonies of the predator and prey strains in separate 12 mL culture tubes containing 2 mL of M9 medium (supplemented the appropriate antibiotics). This step is henceforth referred to as an overnight culture.
3. Prepare molten agar as described above.
4. Adjust the overnight culture to an $OD_{600}$ of ~0.8 using fresh M9 medium supplemented with the appropriate antibiotics.
5. Add 30 μL of the predator and prey cultures to 3 mL of molten agar.
6. Supplement the molten agar with 1 mM IPTG and the appropriate antibiotics.
7. Gently mix the molten agar to ensure that both the predator and prey cells are well mixed. Do not vortex as this creates bubbles that can obscure the results.
8. Gently spread the molten agar over plates containing the bottom agar made in step 1.
9. Incubate the Petri dishes in a 37°C incubator. The plates should be placed in a flat-bottom box with a small amount of water on the bottom. The box should be sealed with cellophane to keep the moisture (see Note 9).
10. At intermittent time points, remove a small piece of the top agar. Place the piece of agar into a 12 mL culture tube containing 3 mL of M9 medium.
11. Vortex the tube for 5 min to free the bacteria from the agar.
12. Measure biodiversity using a fluorescent microscope as described in Section 3.3, "Biodiversity in Liquid Medium."

### 3.6. Quantifying Effective Cell Diffusivity

In the protocol below, we describe an experimental setup which allows one to quantify cell motility in solid culture. This is an important parameter required for modeling biodiversity in the solid phase. Note that motility for each strain is measured separately.

1. Prepare an overnight culture as described above.
2. Prepare molten agar as described above.

3. Using a pipette, transfer 8 mL of the molten agar into an 8-cm Petri dish. Add appropriate antibiotics (see Note 10).
4. Allow the plates to solidify at room temperature for approximately 2 h.
5. Add 5 μL of the undiluted overnight liquid culture to the center of the soft agar surface (see Note 11).
6. Incubate the Petri dishes in a 37°C incubator. The plates should be placed in a flat-bottom box with a small amount of water on the bottom. The box should be sealed with cellophane to keep moisture (see Note 9).
7. For at least four time points during incubation (see Note 12), remove the plates. Image the plates using a gel imager with settings optimized to observe a contrast between the area of cell growth and background.
8. Measure the diameter of the cell growth at each time point either manually using a ruler or computationally using gel imaging software.
9. To estimate dispersion of the cell growth, use a partial differential equation (PDE) of the form:

$$\frac{\partial C}{\partial t} = \mu_{\text{max}} C + D_{\text{cell}} \nabla^2 C.$$

where $C$ is cell density; $t$ is time; $\mu_{\text{max}}$ is the maximum growth rate of the population (determined by liquid-phase examination of the culture's growth dynamics); and $D_{\text{cell}}$ is effective cell diffusivity. One can fit $D_{\text{cell}}$ to best match the expansion of growth observed experimentally (see Note 13).

10. The estimated value of $D_{\text{cell}}$ can be used to model the solid-phase system dynamics.

### 3.7. Quantifying Effective Diffusivity of Extracellular Signals

In the protocol below, we describe a method to estimate the diffusivity of AHL (3OC6HSL and 3OC12HSL). Since these parameters can alter the dynamics of the ecosystem, it is important for one to estimate their values by combining experimental and computational methods.

1. Prepare an overnight culture of a strain harboring a gene circuit that contains a fluorescent protein coding region downstream of a promoter sensitive to the extracellular signal of interest. These strains are often referred to as reporter strains.
2. Prepare molten agar as described above and supplement with the appropriate antibiotics.
3. While the agar is still molten, add 200 μL of the undiluted overnight culture (see Note 14). Gently mix but do not vortex.
4. Using a pipette, transfer 8 mL of the mixture into an 8-cm diameter Petri dish.

5. Allow 2 h for the medium to solidify at room temperature.
6. Once the plates have solidified, add 5 μL of the appropriate AHL to the center of the agar surface (see Note 15).
7. Incubate the Petri dishes in a 37°C incubator. The plates should be placed in a flat-bottom box with a small amount of water on the bottom. The box should be sealed with cellophane to keep the moisture (see Note 9).
8. For at least four time points during incubation (see Note 12), remove the plates. Image the plates using a low magnifying fluorescent imager, such as the Kodak Image Station 2000MM used in Song et al. (35), with settings optimized to observe a contrast between the fluorescent growth and background.
9. Estimate the diffusion of the extracellular signal using a system of PDEs of the following forms:

$$\frac{\partial S}{\partial t} = D_s \nabla^2 S - d_s S$$

$$\frac{\partial F}{\partial t} = k_F C \frac{S}{K_s + S} - d_F F \quad .$$

$$\frac{\partial C}{\partial t} = \mu_{\max} C(1 - \frac{C}{C_m}) - d_c C$$

where $S$ is the extracellular signal concentration; $F$ is the fluorescent signal concentration; $C$ is cell density; and $t$ is time. The kinetic constants are as follows: $D_S$ is the effective diffusivity of the extracellular signal; $d_S$ is the degradation rate of the extracellular signal; $k_F$ is the synthesis rate of fluorescent signal; $K_S$ is the half-activation rate of the inducible promoter by the extracellular signal; $d_F$ is the degradation rate of fluorescent signal; $\mu_{\max}$ is the cellular growth rate; $C_m$ is the cellular growth capacity; and $d_C$ is the cellular death rate. One can fit $D_S$ to best match the spatial expansion of the fluorescent signal (see Notes 16 and 17).

10. The value of $D_S$ can then be utilized in the modeling of the solid-phase dynamics of the synthetic ecosystem (see Note 18).

### 3.8. Examining Biodiversity of Partitioned Populations in Solid Medium

In the following protocol, we describe how to measure biodiversity in solid medium when the predator and prey populations are partitioned. By relating the biodiversity index to the experimental conditions, one can make general conclusions regarding the effect of motility and habitat partitioning on biodiversity in the ecosystem.

1. Prepare overnight cultures of the predator and prey cells separately.
2. Prepare molten M9 medium containing 0.2% agar as described above (see Notes 6 and 7).

3. Upon melting and boiling agar by microwave and cooling to ~50°C, add 1 mM IPTG and appropriate antibiotics.
4. Spread 10 mL of molten agar onto an 8-cm Petri dish.
5. Using fresh M9 medium (supplemented with the appropriate antibiotics) dilute the overnight cultures to $OD_{600} = 0.55$.
6. After solidification of the agar, add 10 μL of the diluted predator and prey cells on the agar surfaces at varying distances: 0, 1, or 2 cm (see Note 11).
7. Incubate the Petri dishes in a 37°C incubator for 20 h. The plates should be placed in a flat-bottom box with a small amount of water on the bottom. The box should be sealed with cellophane to keep the moisture (see Note 9).
8. Use a plate imager, such as the Kodak Image Station 2000MM described in Song et al. (35), to measure green and red fluorescence. See Fig. 3 for results typically obtained using this experimental method.
9. For each experimental setup, process the images to obtain the relative predator and prey densities from the green and red fluorescence measurements, respectively, using imaging software of the investigator's choice (see Note 16).

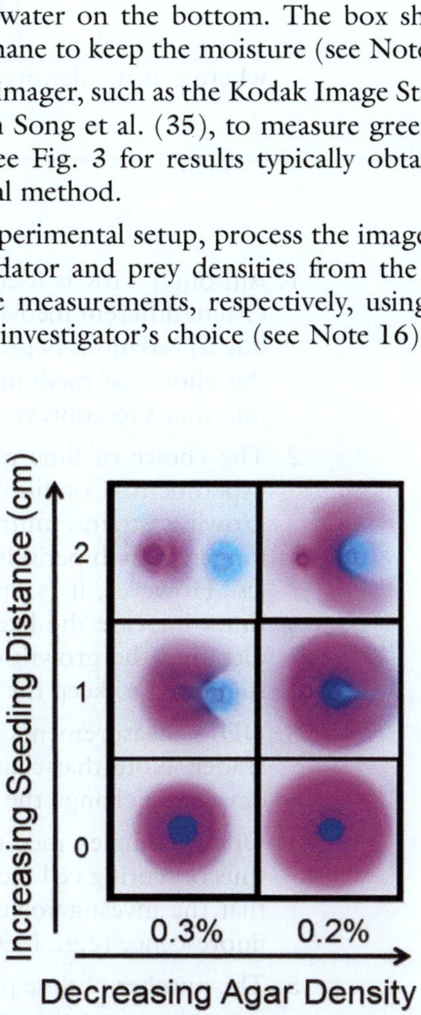


Fig. 3. Experimentally obtained predator–prey patterns after 20 h of cellular growth in 0.2% and 0.3% agar plates with varying seeding segregation distances (0, 1, and 2 cm). The predator area of growth is on the *left* (*purple*), while the prey area of growth is on the *right* (*blue*). These patterns illustrate that increasing habitat segregation and decreasing cellular motility increases biodiversity in the synthetic predator–prey ecosystem.

10. Calculate the biodiversity index (BI) using the formula:

$$BI = 1 - \sum_{i=1}^{2} x_i^2 = 2x_1x_2$$

where $x_i$ is the fraction of the $i$th population in co-culture ($i=1$ for predator and 2 for prey). In this case, $x_i$ can be calculated as average densities over a spatial domain ($\Omega$) using the following:

$$x_i = \frac{\int_\Omega p_i \, dxdy}{\int_\Omega p_1 \, dxdy + \int_\Omega p_2 \, dxdy}.$$

where $p_i$ is the density distribution in $\Omega$ for population $i$.

## 4. Notes

1. Although TBK is used here, the choice of medium can vary. Often, different media will generate different cellular dynamics due to variations in growth rate, carrying capacity, etc. As such, the choice of medium will depend on the specific questions one wants to address.
2. The choice of time point can vary based on the goal of the experiment. Conditions that favor slow growth (e.g., slow-growing strains, nutrient-poor medium) will often require longer growth periods in order to observe the circuit dynamics. However, it is important to note that longer induction times increase the likelihood of circuit-unresponsive mutants dictating the growth dynamics of the culture. It is therefore suggested to keep the growth period as short as possible.
3. $OD_{600}$ measurements are made using 200 μL in a microplate reader. Note that changing the volume (i.e., path length) or device can change the $OD_{600}$ value.
4. Often, complex medium, such as LB, will emit fluorescence, thus obscuring cell measurements. As such, it is recommended that the investigator choose a medium with low background fluorescence (e.g., M9 medium).
5. The number of time points for which samples are taken can be varied based on the specific goal of the experiment.
6. The choice of medium can vary depending on the goal and experimental setup. The investigator must be mindful of both the growth rate and diffusivity of the cell strain utilized. If both of these parameters are relatively high, it is useful to choose a medium that is less rich in nutrients (e.g., M9

medium). If the investigator is using a plate imager, such as the Kodak Image Station 2000MM used in Song et al. (35), he must be mindful of background fluorescence from the medium. Darker medium, such as LB, emits significant background levels, especially when excited at wavelengths on the order of 400–500 nm. Lastly, the buffering of the medium to a particular pH may be necessary for ecosystems utilizing an extracellular signal (e.g., AHL) to ensure a relatively constant rate of degradation over the course of cellular growth. This can be done by buffering with MOPS or other weak acids, depending on the specific pH range of interest (28).

7. The appropriate agar concentration generally varies from ~0.1% to 0.5% agar and is dependent on the motility of the cell strain utilized. Below 0.1% agar, the resulting soft agar solution is easily breakable, giving rise to a heterogeneous mixture with liquid cracks. Above 0.5% agar, even motile strains cannot move effectively, giving rise to tightly packed microcolonies. To find the optimal agar density, a preliminary experiment involving the spread of the culture on a Petri dish containing many different agar densities may be necessary.
8. The exact duration of heating depends on the volume and agar concentration but typically lasts 1.5–3 min for agar concentrations of ~0.1–0.5% in a volume of 50–100 ml.
9. Growing cultures in the presence of moisture is essential to ensure that the structural integrity of the environment is not altered during the course of cellular growth. Without sufficient moisture, evaporation of the solution will occur, thereby increasing the effective agar concentration.
10. The investigator must be mindful of the specific reagents he is using to ensure that it is not prone to deactivation at temperatures of ~50°C.
11. The volume of liquid culture added to the soft agar surface depends on two factors. First, when working at agar concentrations <0.2%, one should use no more than 5 μL culture to avoid collapsing of the soft agar surface. For >0.2% agar, one can use larger volumes of overnight culture. In addition, for slow-growth conditions, it may be helpful to start with a higher culture volume (or a lower dilution factor).
12. By increasing the number of time points taken, the investigator increases the confidence he can have in his estimated parameter value. At least four time points should be taken to ensure a baseline of confidence in this method.
13. When modeling this system, one must be mindful of plausible confounding factors. For example, if cellular growth is inhibited

by the production of a certain protein *P*, one approach to account for the effect is by assuming:

$$\mu = \frac{\mu_{max}}{1+\theta P}.$$

where $\mu_{max}$ is the cellular growth rate without inhibition and θ is the metabolic burden of *P* on growth (7). However, we note that the specific form of equation used to describe the metabolic burden may depend on the specific protein under consideration. If the synthetic gene circuit has a negligible effect on cell motility and growth, one can avoid such complications by using the corresponding wild-type strain without the synthetic gene circuit. The general form of this equation assumes that the cells are still in exponential phase.

14. The appropriate amount of cells added depends on the concentration of cells in the overnight culture, the sensitivity of the inducible promoter to the extracellular signal, the toxicity of the extracellular signal to the reporter strain, and the amount of extracellular signal initially added. Typically, this process requires some tuning by the investigator.
15. The concentration of this extracellular signal will depend on the other parameters described in Note 16.
16. To relate cell density to fluorescence, one should construct a calibration curve. First, one should pipette varying dilutions of predator or prey cells into molten agar (or liquid culture) and spread the mixture onto a Petri dish. Then, one can measure the fluorescence emitted from each of the dishes. By relating average fluorescence to the dilution factor of the cells, one can obtain the calibration curve for the predator or prey cells.
17. Only the signal diffusivity is fitted in this equation. Additional parameters, such as $\mu_{max}$ and the signal degradation rate, were determined separately (e.g., $\mu_{max}$ and cellular diffusivity) or drawn from literature (e.g., degradation rate of AHL molecules).
18. Indirect measurements of $D_s$ are often quite crude. The development of more direct and precise methodologies to measure the diffusion of AHL is required.

## Acknowledgments

This work was partially supported by the National Institutes of Health (5R01CA118486), a National Science Foundation CAREER award (LY), a DuPont Young Professorship (LY), a David and Lucile Packard Fellowship (LY), and a Department of Homeland Security Graduate Fellowship (SP).

## References


1. Gardner, T. S., Cantor, C. R., and Collins, J. J. (2000) Construction of a genetic toggle switch in Escherichia coli, *Nature 403*, 339–342.
2. Atkinson, M. R., Savageau, M. A., Myers, J. T., and Ninfa, A. J. (2003) Development of Genetic Circuitry Exhibiting Toggle Switch or Oscillatory Behavior in Escherichia coli, *Cell 113*, 597–607.
3. Ham, T. S., Lee, S. K., Keasling, J. D., and Arkin, A. P. (2008) Design and Construction of a Double Inversion Recombination Switch for Heritable Sequential Genetic Memory, *PLoS ONE 3*, e2815.
4. Lou, C., Liu, X., Ni, M., Huang, Y., Huang, Q., Huang, L., Jiang, L., Lu, D., Wang, M., Liu, C., Chen, D., Chen, C., Chen, X., Yang, L., Ma, H., Chen, J., and Ouyang, Q. (2010) Synthesizing a novel genetic sequential logic circuit: a push-on push-off switch, *Mol Syst Biol 6*.
5. Kobayashi, H., Kærn, M., Araki, M., Chung, K., Gardner, T. S., Cantor, C. R., and Collins, J. J. (2004) Programmable cells: Interfacing natural and engineered gene networks, *Proceedings of the National Academy of Sciences of the United States of America 101*, 8414–8419.
6. Kramer, B. P., Viretta, A. U., Baba, M. D.-E., Aubel, D., Weber, W., and Fussenegger, M. (2004) An engineered epigenetic transgene switch in mammalian cells, *Nat Biotech 22*, 867–870.
7. Tan, C., Marguet, P., and You, L. (2009) Emergent bistability by a growth-modulating positive feedback circuit, *Nat Chem Biol 5*, 842–848.
8. Stricker, J., Cookson, S., Bennett, M. R., Mather, W. H., Tsimring, L. S., and Hasty, J. (2008) A fast, robust and tunable synthetic gene oscillator, *Nature 456*, 516–519.
9. Elowitz, M. B., and Leibler, S. (2000) A synthetic oscillatory network of transcriptional regulators, *Nature 403*, 335–338.
10. Fung, E., Wong, W. W., Suen, J. K., Bulter, T., Lee, S.-G., and Liao, J. C. (2005) A synthetic gene-metabolic oscillator, *Nature 435*, 118–122.
11. Tigges, M., Denervaud, N., Greber, D., Stelling, J., and Fussenegger, M. (2010) A synthetic low-frequency mammalian oscillator, *Nucl. Acids Res. 38*, 2702–2711.
12. Swinburne, I. A., Miguez, D. G., Landgraf, D., and Silver, P. A. (2008) Intron length increases oscillatory periods of gene expression in animal cells, *Genes & Development 22*, 2342–2346.
13. Tigges, M., Marquez-Lago, T. T., Stelling, J., and Fussenegger, M. (2009) A tunable synthetic mammalian oscillator, *Nature 457*, 309–312.
14. Hooshangi, S., Thiberge, S., and Weiss, R. (2005) Ultrasensitivity and noise propagation in a synthetic transcriptional cascade, *Proceedings of the National Academy of Sciences of the United States of America 102*, 3581–3586.
15. Anderson, J. C., Voigt, C. A., and Arkin, A. P. (2007) Environmental signal integration by a modular AND gate, *Mol Syst Biol 3*.
16. Win, M. N., and Smolke, C. D. (2008) Higher-Order Cellular Information Processing with Synthetic RNA Devices, *Science 322*, 456–460.
17. Zhan, J., Ding, B., Ma, R., Ma, X., Su, X., Zhao, Y., Liu, Z., Wu, J., and Liu, H. (2010) Develop reusable and combinable designs for transcriptional logic gates, *Mol Syst Biol 6*.
18. Friedland, A. E., Lu, T. K., Wang, X., Shi, D., Church, G., and Collins, J. J. (2009) Synthetic Gene Networks That Count, *Science 324*, 1199–1202.
19. Levskaya, A., Chevalier, A. A., Tabor, J. J., Simpson, Z. B., Lavery, L. A., Levy, M., Davidson, E. A., Scouras, A., Ellington, A. D., Marcotte, E. M., and Voigt, C. A. (2005) Synthetic biology: Engineering Escherichia coli to see light, *Nature 438*, 441–442.
20. Weber, W., Kramer, B. P., and Fussenegger, M. (2007) A genetic time-delay circuitry in mammalian cells, *Biotechnology and Bioengineering 98*, 894–902.
21. Weber, W., Stelling, J., Rimann, M., Keller, B., Daoud-El Baba, M., Weber, C. C., Aubel, D., and Fussenegger, M. (2007) A synthetic time-delay circuit in mammalian cells and mice, *Proceedings of the National Academy of Sciences of the United States of America 104*, 2643–2648.
22. Danino, T., Mondragon-Palomino, O., Tsimring, L., and Hasty, J. (2010) A synchronized quorum of genetic clocks, *Nature 463*, 326–330.
23. Marguet, P., Tanouchi, Y., Spitz, E., Smith, C., and You, L. (2010) Oscillations by Minimal Bacterial Suicide Circuits Reveal Hidden Facets of Host-Circuit Physiology, *PLoS ONE 5*, e11909.
24. Basu, S., Gerchman, Y., Collins, C. H., Arnold, F. H., and Weiss, R. (2005) A synthetic multicellular system for programmed pattern formation, *Nature 434*, 1130–1134.
25. Sohka, T., Heins, R., and Ostermeier, M. (2009) Morphogen-defined patterning of Escherichia coli enabled by an externally

tunable band-pass filter, *Journal of Biological Engineering 3*, 10.

26. Tabor, J. J., Salis, H. M., Simpson, Z. B., Chevalier, A. A., Levskaya, A., Marcotte, E. M., Voigt, C. A., and Ellington, A. D. (2009) A Synthetic Genetic Edge Detection Program, *137*, 1272–1281.
27. Brenner, K., Karig, D. K., Weiss, R., and Arnold, F. H. (2007) Engineered bidirectional communication mediates a consensus in a microbial biofilm consortium, *Proceedings of the National Academy of Sciences of the United States of America 104*, 17300–17304.
28. You, L., Cox, R. S., Weiss, R., and Arnold, F. H. (2004) Programmed population control by cell-cell communication and regulated killing, *Nature 428*, 868–871.
29. Tamsir, A., Tabor, J. J., and Voigt, C. A. (2011) Robust multicellular computing using genetically encoded NOR gates and chemical 'wires', *Nature 469*, 212–215.
30. Balagadde, F. K., Song, H., Ozaki, J., Collins, C. H., Barnet, M., Arnold, F. H., Quake, S. R., and You, L. (2008) A synthetic Escherichia coli predator-prey ecosystem, *Mol Syst Biol 4*.
31. Weber, W., Daoud-El Baba, M., and Fussenegger, M. (2007) Synthetic ecosystems based on airborne inter- and intrakingdom communication, *Proceedings of the National Academy of Sciences of the United States of America 104*, 10435–10440.
32. Miller, M. B., and Bassler, B. L. (2001) Quorum Sensing in Bacteria, *Annual Review of Microbiology 55*, 165–199.
33. Pai, A., Tanouchi, Y., Collins, C. H., and You, L. (2009) Engineering multicellular systems by cell-cell communication, *Current Opinion in Biotechnology 20*, 461–470.
34. Waters, C. M., and Bassler, B. L. (2005) Quorum Sensing: Cell-to-Cell Com-munication in Bacteria, *Annual Review of Cell and Develop mental Biology 21*, 319–346.
35. Song, H., Payne, S., Gray, M., and You, L. (2009) Spatiotemporal modulation of biodiversity in a synthetic chemical-mediated ecosystem, *Nat Chem Biol 5*, 929–935.
36. Kerr, B., Riley, M. A., Feldman, M. W., and Bohannan, B. J. M. (2002) Local dispersal promotes biodiversity in a real-life game of rock-paper-scissors, *Nature 418*, 171–174.
37. Chuang, J. S., Rivoire, O., and Leibler, S. (2009) Simpson's Paradox in a Synthetic Microbial System, *Science 323*, 272–275.
38. Shou, W., Ram, S., and Vilar, J. M. G. (2007) Synthetic cooperation in engineered yeast populations, *Proceedings of the National Academy of Sciences of the United States of America 104*, 1877–1882.
39. Acar, M., Mettetal, J. T., and van Oudenaarden, A. (2008) Stochastic switching as a survival strategy in fluctuating environments, *Nat Genet 40*, 471–475.
40. Kim, H. J., Boedicker, J. Q., Choi, J. W., and Ismagilov, R. F. (2008) Defined spatial structure stabilizes a synthetic multispecies bacterial community, *Proceedings of the National Academy of Sciences of the United States of America 105*, 18188–18193.

# Chapter 20

# *Drosophila* S2 Schneider Cells: A Useful Tool for Rebuilding and Redesigning Approaches in Synthetic Biology

Jianying Yang and Michael Reth

## Abstract

Synthetic biology is an engineering approach to biology. A synthetic biologist wants to describe biological molecules and their subdomains as well-defined parts of a molecular machine. To achieve this goal, synthetic biologists rebuild minimal functional biological systems from well-defined parts or they design new molecules that do not exist in nature but have new and useful functions. In short, these engineering approaches can be summarized as "rebuild, alter, and understand." The *Drosophila* S2 Schneider cell is a useful tool for both rebuilding and redesigning approaches. S2 cells are phagocytic cells that easily take up large amounts of DNA from the cell culture. They, thus, have a high cotransfection rate, allowing the coexpression of up to 12 different proteins. We have developed a transient transfection protocol allowing the rapid and parallel analysis of wild-type and altered forms of a biological system. This chapter describes our methods to rebuild and better understand mammalian signaling systems in the evolutionary distant environment of Drosophila S2 cells.

**Key words:** S2 cells, Rebuilding, Protein complex

## 1. Introduction

*Drosophila* Schneider cells were isolated from late-stage *Drosophila melanogaster* embryos over 30 years ago (1). They are easily maintained at room temperature and do not require an incubator with a $CO_2$ supply. Among drosophilists, S2 continue to provide a good system to study many functional aspects of Drosophila cell biology. For example, the fact that these cells can be grown in large amounts in culture and that they also have an efficient silencing mechanism (RNAi) has permitted knockdown experiments for functional analysis of *Drosophila* proteins (2, 3). In the meantime, vectors for constitutive and inducible expression of proteins have been developed in the S2 cell system (4–8). Especially, the inducible vectors

Wilfried Weber and Martin Fussenegger (eds.), *Synthetic Gene Networks: Methods and Protocols*,
Methods in Molecular Biology, vol. 813, DOI 10.1007/978-1-61779-412-4_20, 

have turned out to be very useful since they allow regulated gene expression. These vectors are now used for the production of large amounts of proteins required for protein crystallographic studies (9, 10). In these applications, both transiently transfected S2 cells and selected stable lines are commonly used (11, 12). The same standard transient transfection protocol using large amounts of plasmids (10–20 μg per $10^6$ cells) resulting in a high transfection efficiency (>30%) can also be used to establish stable lines. When our lab became interested in conducting rebuilding approaches, we were particularly intrigued by the high transient transfection efficiency of S2 cells. We asked whether this would also hold true for the transfection of several different vectors at the same time. Therefore, we modified the protocol not using microgram amounts of one plasmid, but rather nanogram amounts of multiple plasmids for the transient transfection. We noticed that, indeed, S2 cells not only take up large amounts of a single vector but they could also take up many different vectors without compromising the transfection efficiency. Most notably, expression of the proteins from different cotransfected vectors appears to be closely correlated (13, 14). Thus, in a single transient transfection experiment, one can rebuild a complex system composing up to 12 different components in S2 cells with a reasonably high efficiency. This unique feature made the S2 cells an ideal system for a detailed study of protein function in signaling subsystems. Using this approach, we have discovered the oligomeric nature of the B cell antigen receptor (BCR) on resting cells and have better characterized the interaction of the protein tyrosine kinase Syk with the signaling subunits of this receptor (13, 15). Herein, we describe our method for using S2 cell in synthetic rebuilding approaches.

## 2. Materials

1. Schneider's S2 cells: The cells can be obtained from the American Type Culture Collection (http://www.atcc.org/) or purchased from Invitrogen as *Drosophila* S2 Cells (Invitrogen, cat. no. R690-07).
2. Schneider's *Drosophila* Powder Medium, revised (SERVA, cat. no. 47521). Store powder dry at +2°C to +8°C.
3. Liquid S2 cell medium: All the steps should be done while stirring slowly. To prepare 2 L of medium, add ca. 1,600 mL millipore water (pure water from Millipore lab water purification system) in a 4-L flask and dissolve 48.8 g powder medium. Add 0.8 g sodium bicarbonate ($NaHCO_3$), the pH of the medium should be around 4.7. Adjust the pH to 9.2 ± 0.2 with NaOH pellets. Then, adjust the pH to 6.7 ± 0.2 with 37%

HCL. Dissolve 1.59 g calcium chloride dihydrate ($CaCl_2 \cdot 2H_2O$) in 100 mL millipore water. Add the calcium chloride solution dropwise to the medium slowly to avoid precipitation. Add millipore water to 2 L. Filter the medium through a 0.22-μm filter, aliquot in glass bottles (400 mL per bottle).

The prepared liquid S2 cell medium can be stored at +2°C to +8°C in the dark for ca. 8 weeks.

4. Complete S2 cell medium: Add FCS (heat inactivated) to the liquid S2 cell medium to a final concentration of 10% (v/v) and L-Glutamine to a final concentration of 2 mM.
5. Conditioned medium: Collected from well-growing, nice-looking (viability higher than 90%) S2 cells.

   Prior to preparation, the cells should be tested for their transfection efficiency using, for example, a GFP expression vector. Ideally, S2 cells should have a transfection efficiency above 30%. Grow the cells to around 70% confluency. Collect the medium and centrifuge for 20 min at 500 × *g* to remove the cells. Filter the supernatant with a 0.22-μm filter, make aliquots, and freeze the prepared conditioned medium at −20°C for no longer than 1 year. The conditioned medium should be thawed at room temperature before use. If precipitates are observed, centrifuge again to remove the precipitates.
6. Schneider's *Drosophila* Medium, liquid (Invitrogen, cat. no. 11720067).
7. Serum-free medium (for transfection): Add L-Glutamine to Invitrogen's Schneider's *Drosophila* Medium to a final concentration of 2 mM.
8. Culture dishes and plates (Greiner Bio-One) (see Note 1).
9. Laminar flow hood for tissue culture.
10. Pipets, hemocytometer, and other standard equipment for tissue culture.
11. Materials for molecular cloning and plasmid preparation.
12. FuGENE HD transfection reagent (Roche cat. no. 04709705001).
13. Cell culture-grade dimethyl sulfoxide (DMSO).
14. 100 mM $CuSO_4$, prepared using millipore water and filtered through a 0.22-μm filter. Store at +2°C to +8°C for ca. 1 year.
15. 27°C incubator.
16. 1× PBS buffer.
17. 2% paraformaldehyde (PFA), for 100 mL, add 98 mL 1 × PBS to a 100 mL bottle, add 50 μL of 1N NaOH and 2g paraformaldehyde into the PBS. Fill a 2000 mL beaker with about 400 mL $H_2O$ and put a thermometer into the beaker. Heat the

water on a hot stir plate to 64°C. Put the PBS bottle into the beaker with the warm water and stir the paraformaldehyde until it is clear. Remove the 2% PFA from the beaker on the hot plate and cool it down to room temperature. Immediately make 1 mL aliquots and freeze at −20°C.

18. Fluromount-G or Dapi-fluromount-G (SouthernBiotech, cat. no. 0100-01 or 0100-20).

## 3. Methods

### 3.1. S2 Cell Culture

We started our work with S2 cells by obtaining a cell line as a generous gift from Dr. K. Karjalainen (then at the Basel Institute of Immunology, Switzerland). S2 cells can be cultured with Schneider's *Drosophila* medium supplemented with 10% FCS at 27°C in a water-saturated atmosphere with atmospheric $CO_2$ levels in the dark (see Note 2). To maintain S2 cells in culture, cells should not be split below $5 \times 10^5$ cells/mL, as low-density cultures grow very slowly or arrest completely. For routine culture, we split the cells every 2 days to give approximately $0.5–1 \times 10^6$ cells/mL. If the cells need to be kept in culture for 3 days without splitting, we suggest making several different dilutions starting from about $0.5 \times 10^6$ cells/mL. We normally grow S2 cells in 15-cm petri dishes. Since S2 cells do not attach very well to the dish, they can be resuspended by gently pipetting up and down. Trypsinization is not required. Alternatively, using a cell lifter at the beginning might save some work. As with any cell line, it is highly recommended to keep track of the number of passages that have taken place since the cells were thawed (see Note 3).

#### 3.1.1. Freezing S2 Cells

It is important to maintain adequate frozen culture stocks. Keeping S2 cells too long in culture can reduce their transfection efficiency (see Notes 3 and 4).

To freeze cells:

1. Grow cells to a density of $3–5 \times 10^6$ cells/mL (log phase) in 20–30 mL of medium in a 15-cm petri dish. Collect the cells and centrifuge them for 5 min at $300 \times g$.
2. Prepare freezing medium: 90% FCS and 10% DMSO. Filter the freezing medium through a 0.22-μm filter.
3. Resuspend the cell pellet in freezing medium at a final cell concentration of $0.5–1 \times 10^7$ cells/mL.
4. Aliquot the cells into vials (0.5 mL per vial) for freezing.
5. Place the vials into a container with dry ice, and allow freezing for 24–48 h at −80°C before transferring to liquid nitrogen.

*3.1.2. Thawing S2 Cells*

1. Thaw a vial of frozen S2 cells rapidly by exposing it to running hot water.
2. Immediately after the cells are thawed, move the vial to tissue culture hood and sanitize the vial with 70% ethanol.
3. Pipet the cells into a fresh 15-mL falcon tube containing 5 mL complete S2 cell medium (prewarmed to room temperature).
4. Pellet the cells by centrifuging at 300 × *g* for 5 min.
5. Resuspend the cells in 6 mL of fresh complete S2 cell medium and distribute the cells to three wells in a 6-well plate as follows:

| | Well no. | | |
|---|---|---|---|
| | 1 | 2 | 3 |
| Resuspended cells | 1 mL | 2 mL | 3 mL |
| Complete S2 cell medium | 2 mL | 1 mL | – |
| Conditioned medium | 1 mL | 1 mL | 1 mL (final concentration: 25%, see Note 5) |

6. Place the cells in an incubator and replace the medium every 2 days.

Normally, after 4–6 days, the cells start to expand. Reduce the amount of conditioned medium after several passages.

### 3.2. Using S2 Cell for Synthetic Rebuilding Approaches

To rebuild a part of the cell machinery or a signaling pathway in S2 cells, one must first clone the related genes of interest into S2 cell expression vectors and introduce the expression plasmids into the S2 cell by transfection.

*3.2.1. Choice of Expression Vector*

Expression vectors with a variety of different promoters and polyadenylation signals have been tested in *Drosophila* S2 cells. While both the *Drosophila* hsp70 promoter and the *Drosophila* metallothionein (Mtn) promoters are strongly inducible, the hsp70 promoter has considerable basal activity in S2 cells (6, 16). Although the mRNA stability is somewhat affected by various polyadenylation signals, the polyadenylation signals derived from late SV40, *Drosophila* Mtn, early SV40, or *Drosophila* alcohol dehydrogenase (Adh) all result in significant levels of target gene protein expression (4, 5). We are currently using the expression vector pRmHa3, which is based on pUC18 and contains a *Drosophila* Mtn promoter and *Drosophila* Adh polyadenylation signal (5).

*3.2.2. Construct the Expression Plasmids*

1. Kozak sequence: *Drosophila* uses a different Kozak sequence than vertebrates (17). However, for synthetic rebuilding approach, overexpression of the target protein normally should

be avoided. For this reason, an optimization of the Kozak sequence is not always necessary. We, therefore, often insert the cDNA or genes of interest directly into the multicloning site of pRmHa3 without modifying the sequence flanking the natural ATG. In many cases, carrying over a short (<100 base) native 5′ UTR with the cDNA still allows reasonable expression of the target protein (see Note 6).

2. Signal peptide: We suggest using the original signal peptide from the genes of interest unless a problem with protein expression is encountered (see Note 7).
3. Tags: Most of commonly used tags, such as flag-tag, HA-tag, work fine in S2 cells (see Note 8). The position of the tag has to be tested with the special genes of interest.

   All the constructs should be verified by sequencing prior to use.

*3.2.3. Preparing the Expression Plasmids for Transfection*

It is critical to prepare the plasmids using a commercial kit to obtain high-quality plasmids necessary for achieving a high transfection efficiency and cotransfection rate. The plasmid DNA concentration should be determined using the 260-nm absorption (estimates of DNA content based on the intensity of gel bands are not sufficiently accurate). Determine the DNA purity using a 260/280-nm ratio; the optimal ratio is 1.8.

Prepare the plasmid DNA solution in sterile Tris/EDTA (TE) buffer or sterile water, and store at −20°C for no longer than 2 years. A working solution of the plasmid DNA should be prepared at a concentration of 0.1 μg/μL and also stored at −20°C (see Note 9).

*3.2.4. Transfect the S2 Cells with Fugene6 HD*

One day prior to the transfection, plate out S2 cells in 3 ml of complete S2 cell medium at $1 \times 10^6$ cells/well in 6-well plates. Transfect the S2 cells 18–24 h later.

1. For each transfection, add 100 μL serum-free medium (see Note 10) in an Eppendorf tube.
2. Add corresponding plasmids (300 ng each) to each Eppendorf tube and mix well by pipetting (see Notes 11 and 12).
3. Add 4 μL of Fugene HD to each tube and vortex briefly. Incubate for 25 min at room temperature (see Notes 13 and 14).
4. During the incubation time, remove the culture medium of the S2 cells and add 2 mL fresh serum-free medium to wash the cells.
5. After the incubation, remove the washing medium from the cells, add 900 μl serum-free medium to the DNA–Fugene HD mixture, and transfer to the cells (see Note 15).

6. Incubate the cells for 18 h and replace the Fugene HD-containing medium with 3 mL of complete S2 cell medium.
7. 8–24 h later, start to induce the expression of the target proteins.

*3.2.5. Induce the Expression*

One of the greatest features of the S2 cell system is the inducible expression of the target proteins. Since we are using the pRmHa3 expression vector, which contains the *Drosophila* Mtn promoter, the expression of the genes of interest can be regulated by adding or removing $CuSO_4$ to the culture medium. To achieve the desired expression level for the target proteins, a pilot experiment should be done with various amounts of $CuSO_4$ (final concentration ranges from 0.1 to 2 mM) and different induction times (from 6 to 24 h) (see Note 16). In certain case, one might even want to stop the expression after a defined time by removing the $CuSO_4$ from the culture and track the cells for additional times. A general consideration is to express the target gene close to its physiological expression level.

*3.2.6. Analyze the Resulted S2 Cells*

Depending to the biological question being addressed, different methods or method combinations could be used to analyze the S2 cells after the rebuilding. A western blot assay can be used to study the steady-state expression level of the target protein(s). FACScan analysis of extra- or intracellularly stained S2 cells can give information about the distribution of the target protein(s). An imaging experiment can provide information about the subcellular localization and the relative abundance of the target protein(s) in different organelles. As an example, we describe below a detailed protocol for examining S2 cells using a microscope.

Prepare the S2 Cells for Imaging Experiment

The subcellular localization of target proteins can be studied by expressing fluorescence protein (FP)-tagged genes in S2 cells and examining the resulting cells by confocal or epifluorescence microscopy. Both living or fixed S2 cells can be used in this approach.

1. After inducing expression of the target genes, recover the S2 cells from a 6-well plate using a cell lifter.
2. Transfer the cells to a FACS tube. Centrifuge at 300 × g for 5 min to collect the cells.
3. Wash the cells twice with 1× PBS, and resuspend the cells in 500 μL PBS.
4. Transfer 100 μL of the cells to another FACS tube. Perform FACSscan analysis to verify the transfection efficiency (see Note 17).
5. To image living S2 cells, count the cells and dilute to $1 \times 10^6$ cells/mL in 400 μL of PBS. Mix well, transfer to 35-mm

μ-Dish, and visualize the cells directly using a microscope equipped with an inverted stage.

6. To fix the cells prior to imaging, count the cells, transfer $5 \times 10^5$ cells to an Eppendorf tube, and centrifuge at $3,000 \times g$ for 5 min.
7. Remove the supernatant, and resuspend the cells in 500 μl of 2% PFA (see Note 18). Keep the cells in the dark at room temperature for 8–10 min.
8. Centrifuge at $3,000 \times g$ for 10 min to collect the cells.
9. Resuspend the cells in 40 μL of Fluromount-G or Dapi-fluromount-G. Mix well and transfer to a microscope slide. Place a coverslip over the cells and slightly press the center of the coverslip with the blunt end of a pencil to allow it to settle down evenly (see Note 19).
10. Keep the slides at a clean and dark place overnight to dry the slides (see Note 20).
11. Visualize the slides using either an inverted or conventional microscope.

## 4. Notes

1. We use culture dishes and plates from Greiner Bio-One. S2 cells partially attach to culture dishes and attach reasonably well to the 6-well plates. This makes it easy to change the medium during the transfection and is not too difficult to recover the cells, especially with the help of a cell lifter. We do not recommend using culture dishes and plates from Corning since the cells attach tightly to the surface. Extensive use of the cell scraper or trypsinization can resuspend the cells in this situation, but subsequent culture and transfection are suboptimal.
2. In the past, we successfully used Schneider's *Drosophila* Medium from Invitrogen to grow our S2 cells. However, after their recent changes in the composition of the medium, our S2 cells grow more slowly in this medium and have a low transfection efficiency. This is probably because we do not grow our S2 cells in suspension to high densities but rather grow them partially attached at a relatively low concentration. Currently, we use complete S2 cell medium prepared from the SERVA powder medium to grow our S2 cells.
3. During the passage of the S2 cells, we noticed changes in the endogenous protein phosphorylation level by western blot probably due to changes in the expression level of endogenous S2 cell kinases. Newly thawed cells normally have less

background. Thus, depending on the question being asked, it may be important to avoid comparing data from newly thawed cells with cells that have been in culture for long periods. Generally, cells should be discarded 8–10 weeks after thawing or whenever an unexpected high-background phosphorylation level is encountered.

4. Antibiotics, such as penicillin, are often used in S2 cell culture for protein production on a preparative scale. However, we have found that it slows down the growth of S2 cell in our culture conditions and reduces the transfection efficiency. However, without the protection of the antibiotics, the cells are more easily contaminated. Thus, all solutions and equipment coming into contact with living cells must be sterile and verified frequently. Aseptic techniques should be rigorously applied.
5. Since S2 cells may release insect growth factors into the medium, the use of conditioned medium when thawing cells could help the cells to recover.
6. When comparing the protein expression level from wild-type and mutant constructs, identical upstream and downstream sequences should be used when constructing the expression vectors.
7. The influenza virus hemagglutinin signal sequence (18) and the N terminus of human tissue plasminogen activator (tPA) (19) have been shown to be efficient signal sequences in S2 cell system.
8. We noticed that the anti-flag M2 antibody recognizes an endogenous S2 protein. It runs at around 47 kD on a reducing SDS-PAGE.
9. It is suggested to thaw the working solution of the plasmids on ice before the experiment and keep it on ice during the transfection. To control the quality of the plasmids, we normally run an agarose gel after thawing the plasmids. If a plasmid shows faint additional bands or strong degradation, a new working solution has to be prepared.
10. Serum-free medium is made from the Invitrogen's liquid Schneider's *Drosophila* Medium. Using medium prepared from SERVA power medium in this step very often leads to the death of most cells.
11. Depending on the goals of the experiment, it is normally useful to monitor the transfection efficiency. We do this by cotransfecting an EGFP expression plasmid.
12. The total amount of the DNA used for each transfection should be equalized with the empty vector pRmHa-3.
13. To achieve high transfection efficiency, do not allow undiluted FuGENE HD to come into contact with any plastic surfaces

other than pipette tips. Directly add the FuGENE HD reagent into the serum-free medium. Do not use siliconized pipette tips or tubes. A minimum 25-min incubation time is required to achieve a good transfection efficiency. We observed a slight increase in the transfection efficiency when the incubation time is extended to 45 min.

14. We previously used Cellfectin from Invitrogen for our transfection experiments. However, Invitrogen discontinued Cellfectin and introduced the new Cellfectin 2. We experienced low transfection efficiencies with the new Cellfectin 2. This is also probably related to the way we culture our cells.
15. S2 cells are quite sensitive to tonicity changes. Thus, it is recommended not to remove the medium from all wells of all plates if one has many samples in one experiment. It is best to process one plate (six samples) at a time.
16. Beginning around 0.5 mM, $CuSO_4$ shows considerable toxicity in S2 cells. Thus, lower $CuSO_4$ concentrations and longer induction time is a better choice compared with higher $CuSO_4$ concentrations and shorter induction times.
17. An FACScan analysis provides not only information regarding the transfection efficiency, but also the distribution of the expression level in the cells. Thus, one could anticipate the signal intensity for the microscopy experiment.
18. Use PFA that has never been previously thawed.
19. Fluromount-G is very viscous, so cut the head of the pipette tips before pipetting. To avoid the formation of air bubbles on the slide, resuspend the cells in 5 μl PBS before adding the Fluromount-G. Precooling the slide might also help to reduce the amount of air bubbles.
20. The prepared slides can be stored in the dark for several weeks. Keeping the slides at a lower temperature may elongate their lifetime.

## Acknowledgments

We thank Peter Nielsen for critical reading of this manuscript. This work is supported by the DFG through SFB 746 and EXC 294. This work was funded by the EU through grant FP7/2007-2013 (#201106: SYBILLA).

## References


1. Schneider, I. 1972. Cell lines derived from late embryonic stages of Drosophila melanogaster. *Journal of embryology and experimental morphology* 27:353–365.
2. Cheng, L. W., J. P. Viala, N. Stuurman, U. Wiedemann, R. D. Vale, and D. A. Portnoy. 2005. Use of RNA interference in Drosophila S2 cells to identify host pathways controlling compartmentalization of an intracellular pathogen. *Proc Natl Acad Sci USA* 102:13646–13651.
3. Caplen, N. J., J. Fleenor, A. Fire, and R. A. Morgan. 2000. dsRNA-mediated gene silencing in cultured Drosophila cells: a tissue culture model for the analysis of RNA interference. *Gene* 252:95–105.
4. Angelichio, M. L., J. A. Beck, H. Johansen, and M. Ivey-Hoyle. 1991. Comparison of several promoters and polyadenylation signals for use in heterologous gene expression in cultured Drosophila cells. *Nucleic Acids Res* 19:5037–5043.
5. Bunch, T. A., Y. Grinblat, and L. S. Goldstein. 1988. Characterization and use of the Drosophila metallothionein promoter in cultured Drosophila melanogaster cells. *Nucleic Acids Res* 16:1043–1061.
6. Di Nocera, P. P., and I. B. Dawid. 1983. Transient expression of genes introduced into cultured cells of Drosophila. *Proc Natl Acad Sci USA* 80:7095–7098.
7. Kirkpatrick, R. B., R. E. Matico, D. E. McNulty, J. E. Strickler, and M. Rosenberg. 1995. An abundantly secreted glycoprotein from Drosophila melanogaster is related to mammalian secretory proteins produced in rheumatoid tissues and by activated macrophages. *Gene* 153:147–154.
8. Steller, H., and V. Pirrotta. 1985. A transposable P vector that confers selectable G418 resistance to Drosophila larvae. *EMBO J* 4:167–171.
9. Gauthier, L., K. J. Smith, J. Pyrdol, A. Kalandadze, J. L. Strominger, D. C. Wiley, and K. W. Wucherpfennig. 1998. Expression and crystallization of the complex of HLA-DR2 (DRA, DRB1*1501) and an immunodominant peptide of human myelin basic protein. *Proceedings of the National Academy of Sciences of the United States of America* 95:11828–11833.
10. Hennecke, J., A. Carfi, and D. C. Wiley. 2000. Structure of a covalently stabilized complex of a human alphabeta T-cell receptor, influenza HA peptide and MHC class II molecule, HLA-DR1. *EMBO J* 19:5611–5624.
11. Bernard, A. R., T. A. Kost, L. Overton, C. Cavegn, J. Young, M. Bertrand, Z. Yahia-Cherif, C. Chabert, and A. Mills. 1994. Recombinant protein expression in a Drosophila cell line: comparison with the baculovirus system. *Cytotechnology* 15:139–144.
12. Deml, L., H. Wolf, and R. Wagner. 1999. High level expression of hepatitis B virus surface antigen in stably transfected Drosophila Schneider-2 cells. *J Virol Methods* 79:191–203.
13. Rolli, V., M. Gallwitz, T. Wossning, A. Flemming, W. W. A. Schamel, C. Zürn, and M. Reth. 2002. Amplification of B cell antigen receptor signaling by a Syk/ITAM positive feedback loop. *Mol Cell* 10:1057–1069.
14. Wossning, T., and M. Reth. 2004. B cell antigen receptor assembly and Syk activation in the S2 cell reconstitution system. *Immunol Lett* 92:67–73.
15. Yang, J., and M. Reth. 2010. Oligomeric organization of the B-cell antigen receptor on resting cells. *Nature* 467:465–469.
16. Rubin, D. M., A. D. Mehta, J. Zhu, S. Shoham, X. Chen, Q. R. Wells, and K. B. Palter. 1993. Genomic structure and sequence analysis of Drosophila melanogaster HSC70 genes. *Gene* 128:155–163.
17. Cavener, D. R. 1987. Comparison of the consensus sequence flanking translational start sites in Drosophila and vertebrates. *Nucleic Acids Res* 15:1353–1361.
18. Schetz, J. A., O. J. Kim, and D. R. Sibley. 2003. Pharmacological characterization of mammalian D1 and D2 dopamine receptors expressed in Drosophila Schneider-2 cells. *J Recept Signal Transduct Res* 23:99–109.
19. Culp, J. S., H. Johansen, B. Hellmig, J. Beck, T. J. Matthews, A. Delers, and M. Rosenberg. 1991. Regulated expression allows high level production and secretion of HIV-1 gp120 envelope glycoprotein in Drosophila Schneider cells. *Biotechnology (N Y)* 9:173–177.

# Chapter 21

# Synthetic Gene Networks in Plant Systems

## Astrid Junker and Björn H. Junker

## Abstract

Synthetic biology methods are routinely applied in the plant field as in other eukaryotic model systems. Several synthetic components have been developed in plants and an increasing number of studies report on the assembly into functional synthetic genetic circuits. This chapter gives an overview of the existing plant genetic networks and describes in detail the application of two systems for inducible gene expression. The ethanol-inducible system relies on the ethanol-responsive interaction of the AlcA transcriptional activator and the AlcR receptor resulting in the transcription of the gene of interest (GOI). In comparison, the translational fusion of GOI and the glucocorticoid receptor (GR) domain leads to the dexamethasone-dependent nuclear translocation of the GOI::GR protein. This chapter contains detailed protocols for the application of both systems in the model plants potato and *Arabidopsis*, respectively.

**Key words:** Plant, *Arabidopsis thaliana*, Potato, Phytobrick, Inducible gene expression, Dexamethasone, Ethanol

## 1. Introduction

In comparison with well-established model systems in synthetic biology, such as yeast or bacteria, plants are distinctive in different physiological and structural aspects. Plants are sessile organisms which have to cope with differing environmental conditions. They are characterized by a high structural complexity with numerous different, specialized cell types that build a plant body. Cell fate determination as the basis of the establishment of this body plan follows strict rules, and more and more of the involved molecular components (e.g., genes) are and will be identified. Although plant cells gain high degrees of specificity, many types of cells maintain the capacity to regenerate whole plants (pluri- or totipotency). Therefore, the well-established transformation methods for a wide range of plant species make the plant system feasible for synthetic biology methods.

Wilfried Weber and Martin Fussenegger (eds.), *Synthetic Gene Networks: Methods and Protocols*,
Methods in Molecular Biology, vol. 813, DOI 10.1007/978-1-61779-412-4_21, 

Plant-specific cells or tissues of interest for synthetic biology are, for instance, cells of a storage compartment or cells with special biosynthetic activities. Synthetic tools are or have been applied in order to make use of these plant features in synthetic gene circuits with different functionalities. Plant seeds, for instance, have been used already for production and storage of pharmaceutical proteins (antibodies, ref. 1). In terms of metabolic engineering, plant-specific metabolic pathways, such as secondary metabolism or cell wall metabolism, are tuned in order to achieve an increased synthesis of drugs, polymers, and other metabolites required for, e.g., biofuel production. Furthermore, the versatile interactions between plants and environment make the plant system suitable for the development of synthetic biosensors (2).

In the future, the plant system might be feasible for creation of different cell types with specific functions, corresponding to the highest level of synthetic biology, the cellular level. First steps in this direction were made with the construction of a recombinant chloroplast genome (3).

Table 1 gives a series of examples for components of synthetic gene circuits which have been established in plant systems referred to as "phytobricks." This term is in analogy with the bacterial "biobricks" in the "registry of standard biological parts" (http://www.partsregistry.org/Main_Page). In order to follow one of the engineering principles, hierarchization, synthetic components are grouped according to their different points of action as defined by Young and Alper (30): process units (transcription and translation) and process streams (DNA, RNA, protein).

Modern DNA engineering can be put on the level of de novo synthesis of DNA which recently came up to the genome level for unicellular organisms (31). In plants, recombinant DNA technologies have been applied for the construction of a chloroplast genome (3). Otherwise, plant DNA engineering mainly involves promoter engineering (reviewed in ref. 4). For example, the code for target specificity during the interaction of transcription activator-like (TAL) effector proteins of pathogenic *Xanthomonas* bacteria with specific DNA target sequences in the plant cell nucleus bears a great potential for engineering of synthetic regulatory circuits (6).

The construction of transcription initiation control gates is well-established in plant model systems. Several chemically inducible expression systems on the basis of non-plant regulatory systems were applied for accurate temporal control of transgene expression (Table 1), especially for transgenes which compromise plant development when expressed under the control of constitutive promoters. The ethanol- and dexamethasone-inducible systems are in the focus of the methodical part of this chapter.

Transactivation systems ensure exact spatial control of transgene expression by enhancer elements adjacent to the tDNA integration site. These drive the expression of the mGAL4-VP16 transcriptional

**Table 1**
**Examples for synthetic components used in plants**

| | System | Components | References |
|---|---|---|---|
| *DNA engineering* | | | |
| Recombinant DNA | Recombinant chloroplast genome | BGM vector for integration into *Bacillus subtilis* genome | (3) |
| | Synthetic promoters | Transcription factor-binding sites<br>Minimal promoter sequences | (4, 5) |
| | Synthetic promoters | Tandem repeats of TAL effector proteins<br>Recognition sequences in target DNA | (6) |
| *Transcription engineering* | | | |
| Chemically inducible systems (for reviews, see ref. 7, 8) | Ethanol/acetaldehyde | AlcR transcription factor<br>AlcA promoter | (9, 10) |
| | Estradiol | XVE chimeric transcription factor<br>*E. coli* LexA repressor<br>VP16 transactivating domain<br>*Homo sapiens* estradiol receptor domain<br>LexA operator | (11–13) |
| | Dexamethasone | pOp/LhG4<br>pOp chimeric promoter (two lac operators and minimal 35 S promoter)<br>LhG4 transcription activator (high-affinity DNA-binding mutant of *lac* repressor and *Saccharomyces cerevisiae* GAL4 transcription activation domain) | (14–16) |
| | Quorum sensing | *Chimeric transcription activator*<br>TraR receptor (*A. tumefaciens*) which binds 3-oxooctanyl-L-homoserine lactone<br>VP16 transactivating domain | (17) |
| | Ecdysone | *Chimeric transcription activator*<br>Ecdysone receptor ligand-binding domain<br>GAL4 and LexA DNA-binding domain<br>VP16 transactivating domain<br>*Responsive promoter*<br>GAL4 or LexA responsive elements<br>35 S minimal promoter | (18) |

(continued)

**Table 1 (continued)**

| | System | Components | References |
|---|---|---|---|
| Transactivation | Enhancer trap lines | mGAL4-VP16 transcriptional activator<br>Promoter containing GAL4 upstream activator sequences (UASs) | (19–21) |
| *RNA engineering (regulation of RNA activity, stability, or localization)* | | | |
| mRNA abundance | Artificial microRNAs | Natural miRNA precursor sequence<br>Target-specific region replaced by artificial 21mer | (22) |
| *Translation engineering (regulation of translation efficiency)* | | | |
| Translation efficiency | Synthetic riboswitch | Natural and synthetic riboswitches from *E. coli* and *B. subtilis*, optimized for function in plants/plastids | (23) |
| *Protein engineering (regulation of protein stability, connectivity, and localization)* | | | |
| Nuclear translocation | Dexamethasone inducible | Rat glucocorticoid receptor domain | (24, 25) |
| | Cytokinin inducible | Bacterial PhoB-VP64 with histidin kinase domain (phosphorylation)<br>Synthetic *Plant Pho* promoter | (26) |
| Protein stability | Elastin-like polypeptides | Highly repetitive polypeptides | (27–29) |

activator, which in turn activates a promoter containing GAL4 upstream activator sequences (UASs) controlling the transgene expression. Systems for combined spatial and temporal control have been developed (32, 33).

In terms of RNA engineering, one example is the construction of artificial microRNAs (mRNAs) which is a valuable tool in order to target the mRNA of single genes or gene families for degradation. Therefore, the target-specific site within the microRNA backbone is replaced by the artificial sequence and downregulation of plant genes has been proven (22).

Translational control of gene expression is mainly based on specific RNA secondary structures (stems and loops) that influence translation efficiency. RNA riboswitches represent an inducible system of translation control which perform their regulatory action upon ligand binding. A synthetic riboswitch for plants on the basis of bacterial forms has been engineered for inducible expression of chloroplast genes (23).

Regarding protein engineering, different protein fusions have an effect on protein location, protein stability, or protein state. For example, fusing the glucocorticoid receptor domain to the protein of interest is sufficient to achieve the inducible nuclear translocation of the fusion protein. This system is well-established in plants and was used for the identification of transcription factor target genes (24). Another system for inducible protein translocation is based on bacterial proteins that translocate to the nucleus in a cytokinin-dependent manner, where they activate transcription after binding to a synthetic plant promoter (26). The fusion of highly repetitive sequences derived from mammalian elastin proteins has been proven to enhance protein stability and therefore protein yield in plant-based expression systems (34).

The following paragraphs describe in detail the experimental workflows for the application of the ethanol- and dexamethasone-inducible systems in potato and Arabidopsis, respectively. Both systems differ mechanistically, whereas the ethanol-inducible system simply represents a transcription-initiation control gate. The constitutively expressed AlcA transcription factor only interacts with the AlcR promoter in the presence of acetaldehyde or ethanol resulting in the transcription of the gene of interest (GOI). The dexamethasone-inducible system additionally affects protein location as a protein design component. The fusion protein of the GOI (e.g., a transcription factor) and the glucocorticoid receptor domain is constitutively expressed in the cytosol, where it is stabilized by heat-shock proteins (HSPs). Upon dexamethasone treatment, a conformational change leads to the split off of the HSP and the subsequent translocation of the fusion protein to the nucleus, where it binds its target promoters.

Both mechanisms are depicted schematically in Fig. 1. Figure 1a, c makes use of conventional arrow diagrams. In order to follow another engineering principle, standardization, the Systems Biology Graphical Notation (SBGN, ref. 37) was employed comparatively for the representation of both induction mechanisms (Fig. 1b, d). SBGN comprises three languages for unambiguous visualization of biological networks, analogous to wiring diagrams in the engineering field. The SBGN Entity Relationship language has been applied for visualization of plant regulatory networks (38).

## 2. Materials

### 2.1. Arabidopsis

#### 2.1.1. Cloning

1. Chemocompetent *Escherichia coli* cells DH5α (39).
2. Plasmid DNA of pDONR vectors (pDONR201, Kanamycin$^r$; pDONR207, Gentamycin$^r$; pDONR223, Spectinomycin$^r$) (Invitrogen, Groningen, the Netherlands).
3. Plasmid DNA of p35S::R1R2::ΔGR (Kanamycin$^r$, Gateway binary vector for translational fusion to the glucocorticoid receptor (GR) domain) (24).

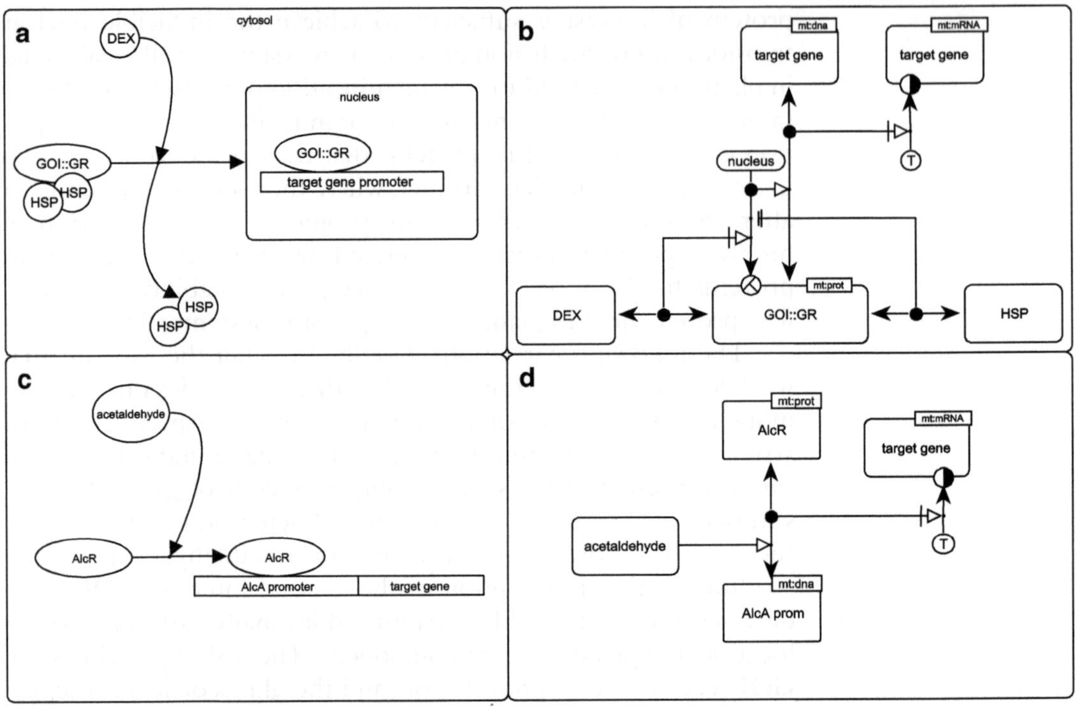


Fig. 1. (**a**) Conventional arrow diagram depicting the mechanism of the dexamethasone-inducible system. The GOI::GR protein is constitutively expressed in the cytosol. Heat-shock proteins (HSPs) act as chaperones and stabilize the GOI::GR protein. Dexamethasone treatment induces conformational changes after the HSP split-off and thereby allows the nuclear translocation of the GOI::GR protein. (**b**) SBGN Entity Relationship (ER) diagram depicting the mechanism of the dexamethasone-inducible system. The interaction (*filled circle*) of GOI::GR and HSP inhibits (‖–), whereas the interaction of Dex and GOI::GR necessarily stimulates ((◁|–)) the location 'nucleus' for the GOI::GR protein. The location of the GOI::GR protein in the nucleus stimulates (◁–) the interaction of target gene DNA and the GOI::GR protein which in turn necessarily stimulates the existence of the target gene mRNA. (**c**) Conventional arrow diagram depicting the mechanism of the ethanol-inducible system. Acetaldehyde (or ethanol) provokes the interaction of the AlcR transcriptional activator with the AlcA promoter leading to transcription of the target gene. (**d**) SBGN Entity Relationship diagram depicting the mechanism of the ethanol-inducible system. Acetaldehyde or ethanol stimulates the interaction between AlcR and AlcA. This interaction in turn necessarily stimulates the existence of the target gene mRNA. Diagrams were drawn using the SBGN-ED Add-on (35) for the VANTED software (36). The full specification of the ER language can be accessed through doi:10.1038/npre.2010.3719.2.

4. Gateway BP and LR clonase mix (Invitrogen, Groningen, the Netherlands).
5. TE buffer; 10 mM Tris/HCl, 1 mM EDTA, pH 8.0, sterilize by autoclaving.
6. Primers for amplification of the cDNA of your gene of interest (see Note 2).

*2.1.2. Production of Competent Agrobacterium Cells*

1. YEB medium; 5 g/l Bacto beaf extract, 1 g/l yeast extract, 5 g/l peptone, 14.6 mM sucrose, 15 g/l agar for solid medium, 0.2 mM magnesium sulfate solution, pH 7.2, sterilize by autoclaving.
2. TE buffer.

*2.1.3. Transformation of Agrobacterium*

1. Competent cells of the *Agrobacterium tumefaciens* strain pGV2260 (40).
2. YEB medium.
3. Rifampicin solution dissolved in DMSO (stock solution 10 mg/ml) (Duchefa, the Netherlands).

*2.1.4. Transformation of Arabidopsis*

1. Arabidopsis seeds (wild-type Columbia 0).
2. Infection medium; 2.652 g/l Murashige & Skoog micro- and macrosalt mixture (Duchefa, the Netherlands), 14.6 mM sucrose, 1% (v/v) vitamin solution, 0.04% (v/v) Silwet L-77, pH 5.7.
3. Silwet L-77 (Union Carbide Chemicals and Plastics).
4. Vitamin solution; 0.3 mM thiamin × HCl, 0.24 mM pyridoxine×HCl, 0.41 mM nicotinic acid, 234.4 mM 2-(*N*-Morpholino) ethanesulfonic Acid*$H_2O$ (MES), 55.5 mM inositol, dissolve in dd$H_2O$.
5. Kanamycin solution dissolved in water. Store sterile-filtered stocks of 50 mg/ml at –20°C.

*2.1.5. Induction Procedures*

1. Dexamethasone (Sigma–Aldrich) dissolved in ethanol or DMSO according to the way of application (see Notes 9 and 11). Store ethanolic stocks as 10–50 mM and DMSO stocks at 100 mM solution at –20°C.

***2.2. Potato***

*2.2.1. Cloning*

1. Plasmid DNA of the p35S:alcR,alcA:GUS-in-pBin19 binary vector for ethanol-inducible expression of the β-glucuronidase (GUS) gene (41).

*2.2.2. Transformation of Potato*

1. Murashige & Skoog micro-and macrosalt mixture (Duchefa, the Netherlands).
2. α-Naphthylphthalamic acid (NAA) solution dissolved in DMSO (stock solution 2 mg/ml) (Duchefa, the Netherlands).
3. Giberellic acid 3 (GA3) solution dissolved in water (dissolve powder first in a few droplets of ethanol) (stock solution 1 mg/ml) (Duchefa, the Netherlands).
4. Zeatinribose.
5. Claforan solution dissolved in water. Claforan inhibits Agrobacterium growth after cocultivation.

*2.2.3. Induction Procedure*

1. Soil (Einheitserde P, Tantau, Uetersen, Germany).
2. Fertilizer (Hakaphos Rot, Compo, Münster, Germany).
3. Acetaldehyde (Fluka Chemie GmbH, Buchs, Switzerland).

## 3. Methods

### 3.1. Application of the Dexamethasone-Inducible System in Arabidopsis

#### 3.1.1. Cloning

1. Amplify by PCR the coding sequence of your GOI from cDNA of the corresponding tissue, where the gene is expressed (see Notes 1 and 2).
2. Check the size of the PCR product by gel electrophoresis and purify the PCR product.
3. Perform BP recombination reaction (mix 15–150 ng purified PCR product with 150 ng DONR vector and 1 μl BP clonase mix, add TE buffer to 5 μl; incubate for 1 h at room temperature or 24 h at 14°C) (see Note 3).
4. Transform the whole volume into *E. coli*.
5. Extract plasmid DNA from positive *E. coli* cells and check for integration of the GOI cDNA by PCR (see Note 4).

#### 3.1.2. Production of Competent Agrobacterium Cells

1. Plate Agrobacterium cells from the glycerol stock on YEB plates.
2. Incubate the plates for 1–2 days at 28°C.
3. Pick cells from a colony.
4. Incubate cells in 6 ml liquid YEB medium for 1–2 days at 28°C under constant shaking on an orbital shaker (200 rpm).
5. Add 100 ml YEB medium to 2 ml of liquid culture.
6. Incubate at 28°C with 200 rpm until $OD_{600}$ 0.5–1.0.
7. Incubate culture for 5 min on ice in the centrifuge can.
8. Centrifuge culture for 20 min at 4°C with 6,000 × *g*.
9. Discard the supernatant.
10. Resuspend the cells in 5 ml TE buffer (see Note 5).
11. Centrifuge for 10 min at 4°C with 6,000 × *g*.
12. Discard the supernatant.
13. Resuspend the cell pellet in 10 ml YEB medium.
14. Make aliquots of 200 μl per tube.
15. Shock freeze cells in liquid nitrogen.
16. Store competent cells at –80°C.

#### 3.1.3. Transformation of Agrobacterium tumefaciens

The freeze-and-thaw method (42) is commonly used for the transformation of *Agrobacterium tumefaciens*.

1. Add around 1 μg plasmid DNA to the competent *Agrobacterium* cells.
2. Incubate for 5 min on ice, 5 min in liquid nitrogen, and finally 5 min at 37°C.
3. Add 1 ml YEB medium.

4. Incubate cells under shaking for 4 h at 28°C.
5. Centrifuge shortly and discard the supernatant.
6. Add 100 μl fresh YEB medium.
7. Plate cells on plates containing selective YEB medium (with rifampicin f.c. 50 μg/ml) and incubate for 1–2 days at 28°C.

#### 3.1.4. Transformation of Arabidopsis

The floral dip method (43) is widely used for transformation of *Arabidopsis thaliana*.

1. Grow *Arabidopsis thaliana* ecotype Columbia 0 plants in the greenhouse in small planting pots filled with soil mixed with substrate. Place pots in a tray with an inlay made of felt.
2. Cover seeds (1–2 per pot) loosely with a 0.5-cm layer of soil.
3. Grow plants for 3 weeks under short-day conditions (8 h light, 16 h dark).
4. Grow the plants for further 3 weeks under long-day conditions (16 h light, 8 h dark).
5. Cut the emerging bolts to induce growth of secondary bolts.
6. Grow transgenic *Agrobacterium tumefaciens* culture in liquid YEB medium at 28°C until $OD_{600} > 2.0$.
7. Harvest cells by centrifugation and resuspend the cell pellet in three volumes of infiltration medium supplemented with 0.01% (v/v) Silwet L-77.
8. Dip Arabidopsis inflorescences into the *Agrobacterium tumefaciens* suspension for about 1 min.
9. After short shaking, incubate the treated plants horizontally in covered bowls for 24 h to keep a high humidity.
10. Uncover plants after 24 h and set them upright.
11. Harvested seeds were from dry siliques.
12. Surface sterilize seeds by treatment with 70% (v/v) ethanol for 2 min and 5% (w/v) sodium hypochloride solution for 10 min.
13. Wash sterile seeds with $ddH_2O$ seeds repeatedly and plate them onto selection plates containing MS medium supplemented with the appropriate antibiotics (see Note 6).
14. Seal plates with microporous tape and incubate overnight at 4°C to break the dormancy.
15. Transfer plates to a growth chamber at 23°C with long-day conditions and leave them for 2 weeks.
16. Transfer viable plants to soil, grown them up, and collect their seeds. Analyze transgenic plants by PCR or Northern hybridization (see Notes 7 and 8).

*3.1.5. Induction Procedures*

Induction in agar media

1. Surface sterilize seeds of transgenic and wild-type Arabidopsis lines.
2. For permanent induction, plate seeds on plates with MS medium supplemented with dexamethasone and/or the appropriate antibiotics (see Note 9). For control treatments, add equivalent volumes of DMSO.
3. For short-term induction, transfer imbibed seeds or seedlings at the appropriate time point from MS medium without Dex to Dex-containing MS plates.

Induction in liquid media

1. Surface sterilize seeds of transgenic and wild-type Arabidopsis lines.
2. Plate seeds on solid MS medium supplemented with the appropriate antibiotics.
3. Transfer seedlings from solidified medium to conical flasks (250 ml) containing 100 ml MS medium supplemented with dexamethasone (usually, at a concentration of 10–30 μM) (see Note 9). For control treatments, add equivalent volumes of DMSO.
4. Place the flasks on an orbital shaker (50–110 rpm) in a growth chamber with a 16-h photoperiod at 21°C (see Note 10).
5. After the appropriate period of incubation, harvest seedlings and shock freeze them in liquid nitrogen. Samples might be kept at −80°C until preparation.

Soil drenching

1. Grow Arabidopsis plants in the greenhouse.
2. For watering, use tap water supplemented with dexamethasone of the appropriate concentration (usually, at a concentration of 20–30 μM) (see Note 11). For control treatments, add equivalent volumes of ethanol.
3. Pour supplemented water into the tray. Perform induction either as single application of a bigger volume (1 L) of supplemented water or continuously apply smaller volumes over several days.

Painting

1. Grow Arabidopsis plants in the greenhouse in small planting pots filled with soil mixed with substrate.
2. At the appropriate time point, perform induction with a fine paintbrush which was dipped into the dexamethasone solution (usually, at a concentration of 20–30 μM) supplemented with 0.02% Silwet L-77.
3. Brush over one side of each leaf.

Spraying

1. Grow Arabidopsis plants in the greenhouse in small planting pots filled with soil mixed with substrate.
2. Prepare spraying solution (20–30 μM Dex and 0.01% (w/v) Tween-20 as a wetting agent).
3. Spray leaves or other parts of the plant and cover other parts with plastics.

### 3.2. Application of the Ethanol-Inducible System in Potato

#### 3.2.1. Cloning

1. The cloning procedure starts with the vector p35S:alcR,alcA:GUS-in-pBin19 (see Note 12).
2. For exchanging the GUS reporter against your GOI, first cut out the alcA:GUS cassette with *Hin*dIII (keep the open vector for step 4) and ligate this cassette into a standard cloning vector (e.g., pUC19).
3. From the resulting plasmid, cut out the GUS gene with *Bam*HI and integrate your GOI.
4. Cut out the alcA:GOI cassette with *Hin*dIII and ligate it into the open vector from step 2.
5. Protocols for the production of competent *Agrobacterium* cells and the *Agrobacterium* transformation are the same as in the Arabidopsis part.

#### 3.2.2. Transformation of Potato

1. If possible, retrieve an in vitro culture of *Solanum tuberosum* plants cv. *Desiree*. Propagate plants by cutting the shoot in pieces, each containing one leaf, using a sterile scalpel. Transfer cuttings to MS medium containing 2% (w/v) sucrose, 0.8% (w/v) select agar, and 125 μg/ml Claforan.
2. Grow plants for 4 weeks under the following conditions: 22°C, 50% relative humidity, 3000 Lux, and a 16-h light and 8-h dark regime.
3. Prepare leaf discs.
4. Incubate leaf discs in liquid MS medium containing 2% (w/v) sucrose.
5. Add 50 μl of the recombinant Agrobacterium strain from an exponentially growing culture.
6. Mix for 5 min and then incubate in the dark for 2 days.
7. For callus induction, transfer leaf discs to MS medium supplemented with 1.6% (w/v) glucose, 2 mg/l zeantinribose, 20 μg/ml NAA, 20 g/l GA3, 0.5 g/l Claforan, 50 mg/I Kanamycin, and 0.8% (w/v) agar.
8. Transfer leaf discs after 1 week to the same medium with reduced Claforan concentration (0.25 g/l).
9. Put discs or developing calli on fresh medium plates every 10 days.

10. Cut off shoots after shoot formation and transfer them to MS medium containing 2% (w/v) sucrose and 250 μg/ml Claforan (after 3–4 transfers).

*3.2.3. Induction Procedure*

1. Transfer 4-week-old plants from in vitro culture to soil in 2.5-L pots with a minimum daytime illumination of 250-μmol photons $m^{-2}$/s at 22°C and a relative humidity of 60%.
2. Depending on plant age, supply the plants with 50–250 ml of water and 0.7 g/L fertilizer.
3. Start the induction when the plant has produced several tubers (usually, around 10 weeks after planting).
4. Detach any automatic watering system, if applicable.
5. Around 3 h after the start of the next light phase, supply 200 ml 0.2% (w/v) acetaldehyde in water to the soil of each pot (see Note 13).
6. Repeat step 3 at the end of the light phase.
7. Harvest exactly 24 h after the first application of inducer: remove a tuber from the soil, punch out a longitudinal cylinder, and by using a vegetable slicer, start slicing the cylinder from the region of stolon attachment (usually, the top). The first slice contains the skin, and the second slice is the only sample that is taken from this tuber (see Note 14).
8. Put the slice into a preweighted plastic tube, and snap freeze the tube in liquid nitrogen.
9. Run the respective extraction and measurement method to prove the expression of your GOI.

## 4. Notes

1. Because the induction mechanism is based on nuclear translocation, this DEX-inducible system is only applicable for proteins that act in the nucleus, mainly transcription factors.
2. For translational fusion, amplify the GOI without stop codon. The use of the Gateway recombination technology requires special attachment site sequences. The forward primer should contain the attB1 recombination site ggggacaagtttgtacaaaaaa gcaggct and the reverse primers the attB2 recombination site ggggaccactttgtacaagaaagctgggt at the 5′ end, respectively.
3. Choose the appropriate DONR vector with a different antibiotics resistance than your expression vector, in this case pDONR 207 or pDONR223.

4. Recheck selection-positive *E. coli* and Agrobacterium colonies by PCR amplifying a fragment, including the GOI::GR fusion.
5. All materials and buffers should have a temperature of ~4°C.
6. For plating seeds on solid MS medium, collect sterile seeds in a petri dish within a small amount of sterile water and use a pipette with a yellow pipette tip (–20-μl volume) to place seeds on the medium.
7. After about 1 week of selection, viable plants (transformants) can be recognized as green seedlings with long roots. Negative plants are small, yellowish, and have short roots.
8. Perform segregation analysis to determine single-insertion and homozygous plants in the T1 and further generations. Alternatively, perform Southern blot analysis to determine the transgene copy number. Plate 50–100 seeds of the T0 transgenic line on selective medium and count green versus yellowish seedlings after 1–2 weeks. A 3:1 (25% negative plants) segregation in the first generation marks single-insertion lines (or sometimes multiple insertions at the same locus). Homozygous lines can be obtained by repeating segregation analysis in the next generations (100% transformants).
9. Use DMSO as solvent for dexamethasone for in vitro treatments to prevent possible negative effects of EtOH on plant growth.
10. In liquid induction experiments, seedlings may be incubated for 2–3 days in liquid medium after transfer from solid medium and prior to addition of dexamethasone or DMSO.
11. In soil-drenching experiments, use ethanol as solvent for dexamethasone to prevent accumulation of DMSO in the soil.
12. If tuber-specific inducible expression of the GOI is desired, the 35 S promoter can be exchanged with the B33 patatin promoter (44).
13. Acetaldehyde boils at about 20°C, so store the stock at 4°C. When pipetting acetaldehyde, pipet the liquid up and down several times to allow the pipet tip to cool down. Other inducers, such as ethanol or acetone, might be used as well, but acetaldehyde was shown to induce most rapidly and to have the least side effects on metabolism (10).
14. Induction in potato tubers is spatially inhomogeneous. Induction usually occurs in a patchy pattern right under the skin (presumably under the lenticels) and directly under the point of stolon attachment (10).

## References

1. Fiedler, U., and Conrad, U. (1995) High-level production and long-term storage of engineered antibodies in transgenic tobacco seeds. *Bio-Technology* **13,** 1090–93.
2. Antunes, M. S., Ha, S. B., Tewari-Singh, N., Morey, K. J., Trofka, A. M., Kugrens, P., Deyholos, M., and Medford, J. I. (2006) A synthetic de-greening gene circuit provides a reporting system that is remotely detectable and has a re-set capacity. *Plant Biotechnology Journal* **4,** 605–22.
3. Itaya, M., Fujita, K., Kuroki, A., and Tsuge, K. (2008) Bottom-up genome assembly using the *Bacillus subtilis* genome vector. *Nature Methods* **5,** 41–43.
4. Venter, M. (2007) Synthetic promoters: Genetic control through cis engineering. *Trends in Plant Science* **12,** 118–24.
5. Rushton, P. J., Reinstadler, A., Lipka, V., Lippok, B., and Somssich, I. E. (2002) Synthetic plant promoters containing defined regulatory elements provide novel insights into pathogen- and wound-induced signaling. *Plant Cell* **14,** 749–62.
6. Boch, J., Scholze, H., Schornack, S., Landgraf, A., Hahn, S., Kay, S., Lahaye, T., Nickstadt, A., and Bonas, U. (2009) Breaking the code of DNA binding specificity of TAL-type III effectors. *Science* **326,** 1509–12.
7. Moore, I., Samalova, M., and Kurup, S. (2006) Transactivated and chemically inducible gene expression in plants. *Plant Journal* **45,** 651–83.
8. Padidam, M. (2003) Chemically regulated gene expression in plants. *Current Opinion in Plant Biology* **6,** 169–77.
9. Roslan, H. A., Salter, M. G., Wood, C. D., White, M. R. H., Croft, K. P., Robson, F., Coupland, G., Doonan, J., Laufs, P., Tomsett, A. B., and Caddick, M. X. (2001) Characterization of the ethanol-inducible *alc* gene-expression system in *Arabidopsis thaliana*. *Plant Journal* **28,** 225–35.
10. Junker, B. H., Chu, C. C., Sonnewald, U., Willmitzer, L., and Fernie, A. R. (2003) In plants the *alc* gene expression system responds more rapidly following induction with acetaldehyde than with ethanol. *Febs Letters* **535,** 136–40.
11. Curtis, M. D., and Grossniklaus, U. (2003) A Gateway cloning vector set for high-throughput functional analysis of genes in planta. *Plant Physiology* **133,** 462–69.
12. Zuo, J. R., Niu, Q. W., and Chua, N. H. (2000) An estrogen receptor-based transactivator XVE mediates highly inducible gene expression in transgenic plants. *Plant Journal* **24,** 265–73.
13. Böhmdorfer, G., Tramontan, A., Luxa, K., and Bachmair, A. (2010) A synthetic biology approach allows inducible retrotransposition in whole plants. *Systems and Synthetic Biology* **4,** 133–38.
14. Aoyama, T., and Chua, N. H. (1997) A glucocorticoid-mediated transcriptional induction system in transgenic plants. *Plant Journal* **11,** 605–12.
15. Craft, J., Samalova, M., Baroux, C., Townley, H., Martinez, A., Jepson, I., Tsiantis, M., and Moore, I. (2005) New pOp/Lhg4 vectors for stringent glucocorticoid-dependent transgene expression in Arabidopsis. *Plant Journal* **41,** 899–918.
16. Moore, I., Galweiler, L., Grosskopf, D., Schell, J., and Palme, K. (1998) A transcription activation system for regulated gene expression in transgenic plants. *Proceedings of the National Academy of Sciences of the United States of America* **95,** 376–81.
17. You, Y. S., Marella, H., Zentella, R., Zhou, Y. Y., Ulmasov, T., Ho, T. H. D., and Quatrano, R. S. (2006) Use of bacterial quorum-sensing components to regulate gene expression in plants. *Plant Physiology* **140,** 1205–12.
18. Padidam, M., Gore, M., Lu, D. L., and Smirnova, O. (2003) Chemical-inducible, ecdysone receptor-based gene expression system for plants. *Transgenic Research* **12,** 101–09.
19. Gardner, M. J., Baker, A. J., Assie, J. M., Poethig, R. S., Haseloff, J. P., and Webb, A. A. R. (2009) Gal4 GFP enhancer trap lines for analysis of stomatal guard cell development and gene expression. *Journal of Experimental Botany* **60,** 213–26.
20. Johnson, A. A. T., Hibberd, J. M., Gay, C., Essah, P. A., Haseloff, J., Tester, M., and Guiderdoni, E. (2005) Spatial control of transgene expression in rice (*Oryza sativa* L.) using the Gal4 enhancer trapping system. *Plant Journal* **41,** 779–89.
21. Laplaze, L., Parizot, B., Baker, A., Ricaud, L., Martiniere, A., Auguy, F., Franche, C., Nussaume, L., Bogusz, D., and Haseloff, J. (2005) Gal4-GFP enhancer trap lines for genetic manipulation of lateral root development in arabidopsis thaliana. *Journal of Experimental Botany* **56,** 2433–42.

22. Schwab, R., Ossowski, S., Riester, M., Warthmann, N., and Weigel, D. (2006) Highly specific gene silencing by artificial microRNAs in Arabidopsis. *Plant Cell* **18,** 1121–33.
23. Verhounig, A., Karcher, D., and Bock, R. (2010) Inducible gene expression from the plastid genome by a synthetic riboswitch. *Proceedings of the National Academy of Sciences of the United States of America* **107,** 6204–09.
24. Baudry, A., Heim, M. A., Dubreucq, B., Caboche, M., Weisshaar, B., and Lepiniec, L. (2004) TT2, TT8, and TTG1 synergistically specify the expression of banyuls and proanthocyanidin biosynthesis in *Arabidopsis thaliana*. *Plant Journal* **39,** 366–80.
25. Gunl, M., Liew, E. F., David, K., and Putterill, J. (2009) Analysis of a post-translational steroid induction system for gigantea in Arabidopsis. *BMC Plant Biology* **9**:141.
26. Antunes, M. S., Morey, K. J., Tewari-Singh, N., Bowen, T. A., Smith, J. J., Webb, C. T., Hellinga, H. W., and Medford, J. I. (2009) Engineering key components in a synthetic eukaryotic signal transduction pathway. *Molecular Systems Biology* **5**:270.
27. Meyer, D. E., and Chilkoti, A. (1999) Purification of recombinant proteins by fusion with thermally-responsive polypeptides. *Nature Biotechnology* **17,** 1112–15.
28. Floss, D. M., Sack, M., Arcalis, E., Stadlmann, J., Quendler, H., Rademacher, T., Stoger, E., Scheller, J., Fischer, R., and Conrad, U. (2009) Influence of elastin-like peptide fusions on the quantity and quality of a tobacco-derived human immunodeficiency virus-neutralizing antibody. *Plant Biotechnology Journal* **7,** 899–913.
29. Floss, D. M., Schallau, K., Rose-John, S., Conrad, U., and Scheller, J. (2010) Elastin-like polypeptides revolutionize recombinant protein expression and their biomedical application. *Trends in Biotechnology* **28,** 37–45.
30. Young, E., and Alper, H. (2010) Synthetic biology: Tools to design, build, and optimize cellular processes. *Journal of Biomedicine and Biotechnology.* 130781.
31. Gibson, D. G., Glass, J. I., Lartigue, C., Noskov, V. N., Chuang, R., Algire, M. A., Benders, G. A., Montague, M. G., Ma, L., Moodie, M. M., Merryman, C., Vashee, S., Krishnakumar, R., Assad-Garcia, N., Andrews-Pfannkoch, C., Denisova, E. A., Young, L., Qi, Z., Segall-Shapiro, T. H., Calvey, C. H., Parmar, P. P., Hutchison III, C. A., Smith, H. O., and Venter, J. C. (2010) Creation of a bacterial cell controlled by a chemically synthesized genome. *Science* **329,** 52–56.
32. Jia, H., Van Loock, B., Liao, M., Verbelen, J. P., and Vissenberg, K. (2007) Combination of the alcr/alca ethanol switch and Gal4/VP16-UAS enhancer trap system enables spatial and temporal control of transgene expression in arabidopsis. *Plant Biotechnology Journal* **5,** 477–82.
33. Sakvarelidze, L., Tao, Z., Bush, M., Roberts, G. R., Leader, D. J., Doonan, J. H., and Rawsthorne, S. (2007) Coupling the Gal4 UAS system with alcR for versatile cell type-specific chemically inducible gene expression in Arabidopsis. *Plant Biotechnology Journal* **5,** 465–76.
34. Scheller, J., Leps, M., and Conrad, U. (2006) Forcing single-chain variable fragment production in tobacco seeds by fusion to elastin-like polypeptides. *Plant Biotechnology Journal* **4,** 243–49.
35. Czauderna, T., Klukas, C., and Schreiber, F. (2010) Editing, validating and translating of SBGN maps. *Bioinformatics* **26,** 2340–41.
36. Junker, B. H., Klukas, C., and Schreiber, F. (2006) Vanted: A system for advanced data analysis and visualization in the context of biological networks. *BMC Bioinformatics* **7,** 109.
37. Le Novère, N., Hucka, M., Mi, H. Y., Moodie, S., Schreiber, F., Sorokin, A., Demir, E., Wegner, K., Aladjem, M. I., Wimalaratne, S. M., Bergman, F. T., Gauges, R., Ghazal, P., Kawaji, H., Li, L., Matsuoka, Y., Villeger, A., Boyd, S. E., Calzone, L., Courtot, M., Dogrusoz, U., Freeman, T. C., Funahashi, A., Ghosh, S., Jouraku, A., Kim, S., Kolpakov, F., Luna, A., Sahle, S., Schmidt, E., Watterson, S., Wu, G. M., Goryanin, I., Kell, D. B., Sander, C., Sauro, H., Snoep, J. L., Kohn, K., and Kitano, H. (2009) The systems biology graphical notation. *Nature Biotechnology* **27,** 735–41.
38. Junker, A., Hartmann, A., Schreiber, F., and Bäumlein, H. (2010) An engineer's view on regulation of seed development. *Trends in Plant Science* **15,** 303–07.
39. Sambrook, J., Fritsch, E. F., and Maniatis, T. (2001) Molecular cloning: A laboratory manual, Cold Spring Harbour Laboratory Press, New York.
40. Deblaere, R., Bytebier, B., Degreve, H., Deboeck, F., Schell, J., Vanmontagu, M., and Leemans, J. (1985) Efficient octopine Ti plasmid-derived vectors for Agrobacterium-mediated gene-transfer to plants. *Nucleic Acids Research* **13,** 4777–88.
41. Sweetman, J. P., Chu, C. C., Qu, N., Greenland, A. J., Sonnewald, U., and Jepson, I. (2002) Ethanol vapor is an efficient inducer of the alc

gene expression system in model and crop plant species. *Plant Physiology* **129,** 943–48.

42. Höfgen, R., and Willmitzer, L. (1988) Storage of competent cells for Agrobacterium transformation. *Nucleic Acids Research* **16,** 9877.
43. Clough, S. J., and Bent, A. F. (1998) Floral dip: A simplified method for agrobacterium-mediated transformation of *Arabidopsis thaliana. Plant Journal* **16,** 735–43.
44. Rocha-Sosa, M., Sonnewald, U., Frommer, W., Stratmann, M., Schell, J., and Willmitzer, L. (1989) Both developmental and metabolic signals activate the promoter of a class-I patatin gene. *Embo Journal* **8,** 23–29.

# Chapter 22

# Design and Construction of Synthetic Gene Networks in Mammalian Cells

**Maria Karlsson, Wilfried Weber, and Martin Fussenegger**

## Abstract

Advances in the development of molecular tools for the inducible control of transcription, translation, and protein degradation are the basis for the rapidly emerging design and construction of synthetic gene networks in mammalian cells.

In this chapter, we describe such tools and how they can be integrated into a synthetic gene network with desired functionality. The network design and construction process is illustrated in the form of a detailed protocol for the implementation of a logic NOR gate based on an inducible promoter combined with an inducible protein degradation system.

**Key words:** Auxin, Boolean logic, Degron, Inducible expression, NOR gate

## 1. Introduction

Recent advances in mammalian synthetic biology (1) have resulted in the design and construction of genetic networks that perform complex computational functions, like Boolean algebra (2, 3), that act as tight (4), hysteretic (5), bistable (6), or time-delayed (7) switches or that show tunable oscillating (8, 9) gene expression. At the core of these systems are modular molecular tools that enable the control of gene expression via inducible promoters or via the inducible degradation of messenger RNA (mRNA) and proteins.

### *1.1. Controlling the Activity of Promoters*

Promoters with trigger-inducible activity are the backbone of genetic networks. They can be interconnected with each other in mutually repressing or activating configurations in order to design the required network topologies (10). In principle, three promoter

Wilfried Weber and Martin Fussenegger (eds.), *Synthetic Gene Networks: Methods and Protocols*,
Methods in Molecular Biology, vol. 813, DOI 10.1007/978-1-61779-412-4_22, © Springer Science+Business Media, LLC 2012

types have prevailed in mammalian synthetic biology: promoters that are responsive to a transcriptional activator A ($P_{Act}$, Fig. 1a), promoters that can be silenced by a transcriptional repressor R ($P_{Rep}$, Fig. 1b), and hybrid promoters ($P_{Hyb}$) that are responsive to combinations of multiple transactivators and repressors (3) (Fig. 1c). For each promoter type, several mutually compatible versions are available that can be combined with each other to form synthetic gene networks (Table 1). When using hybrid promoters responsive to an activator and a repressor, the silencing activity is commonly dominant.

### 1.2. Controlling mRNA Levels

Besides modulating gene expression at the level of transcription, the inducible degradation of mRNA provides a second level of control. For this purpose, (inducible) promoters can be applied to express short hairpin RNAs (shRNAs) complementary to the mRNA sequence that is to be degraded (11). shRNAs can be produced under the control of constitutive or inducible promoters (11) or by placing the shRNA sequence into an intron that is released during mRNA splicing (12) (Fig. 1d). The use of intronically encoded shRNAs is an elegant approach as in this configuration one promoter can be used to simultaneously produce one protein and to prevent the production of another (12). Given the high specificity of shRNAs towards their target sequence, they provide a flexible tool for constructing synthetic gene networks with multiple mutual interactions, as for example in networks performing complex Boolean logic operations (2).

### 1.3. Controlling Protein Levels

The trigger-inducible degradation of proteins represents a direct means to control protein levels while avoiding the time delay resulting from detours via promoter modulation (13). The two most prominent inducible protein degradation systems are based on a destabilized variant of human FKBP12 and on the *Arabidopsis*-derived, auxin-inducible degron (AID) system. For FKBP12, two mutations (F36V, L106P) were shown to destabilize FKBP12 and its protein fusion partners, whereas the addition of the FKBP ligand Shld1 protected FKBP12 from degradation (14). The auxin-responsive protein degradation system is based on the AUX/IAA17-derived, AID and the F-box protein TIR1 in combination with ubiquitin ligases. Upon addition of auxin, AID and TIR1 dimerize, thereby resulting in AID ubiquitination and its subsequent degradation together with the fused protein of interest (POI) (13) (Fig. 1e).

In this chapter, we provide a detailed protocol for the design and construction of a synthetic gene network functioning as a Boolean NOR gate by combining a tetracycline-activatable promoter with an auxin-inducible protein degradation system (Fig. 2). We describe how the individual parts can be functionally validated in mammalian cells and how the performance of the overall network can be assessed.

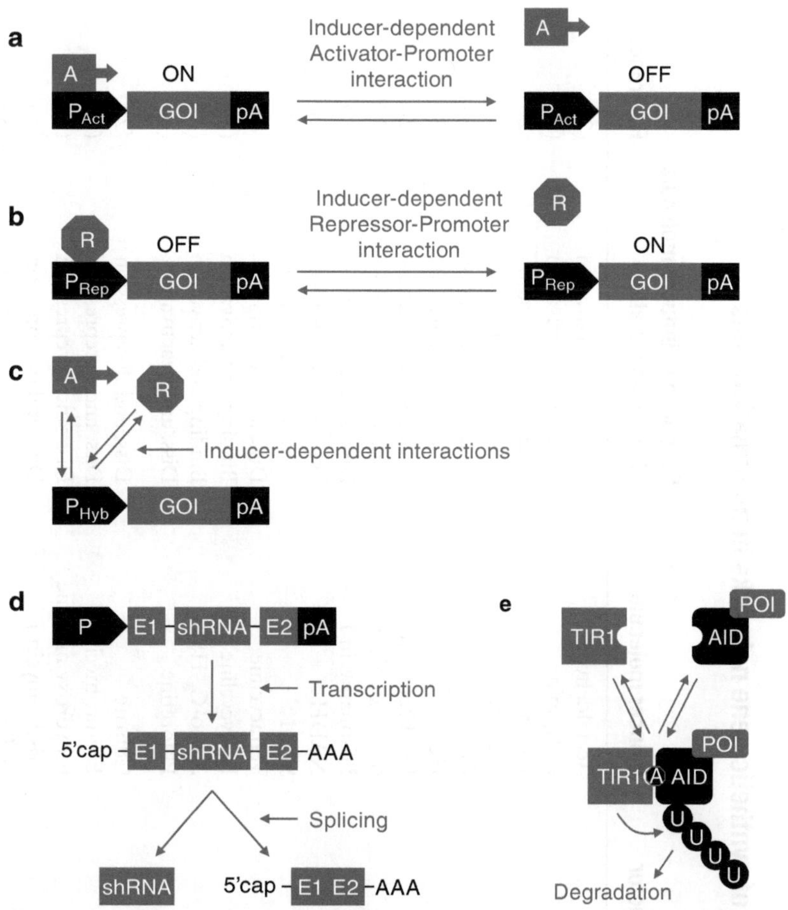


Fig. 1. Molecular control tools. (**a**) Activation-based regulatable promoters. An activatable promoter $P_{Act}$ is induced via the binding of a transactivator A to express the gene of interest (GOI). By modulating the binding of A to $P_{Act}$ by changing the concentration of a small-molecule inducer, the activity of the promoter can be adjusted to the required levels. (**b**) Repression-based regulatable promoters. A repressible promoter $P_{Rep}$ is silenced via the binding of a repressor protein R. Without repressor protein, the promoter is constitutively active, expressing the GOI. By modulating the binding of B to $P_{Rep}$ by changing the concentration of a small-molecule inducer, the activity of the promoter can be adjusted to the required levels. (**c**) Multi-input hybrid promoters. Hybrid promoters are constructed by introducing binding sites for small-molecule-responsive activators and repressors into the promoter architecture. The resulting hybrid promoter $P_{Hyb}$ can be induced via binding of an activator A and can be silenced via binding of a repressor R. Binding of the repressor R commonly results in transcriptional silence irrespective of whether an activator A is bound or not. (**d**) Intronically encoded short hairpin RNA (shRNA) for mRNA degradation. An (inducible) promoter P is used to transcribe a messenger RNA comprising two exons (E1, E2) as well as an intron encoding an shRNA. Following splicing, the joined exons are translated resulting in the desired protein while the released shRNA triggers the degradation of complementary mRNA sequences. (**e**) Auxin-inducible protein degradation. The protein of interest (POI) is fused to the auxin-inducible degron (AID). Upon addition of auxin (A), AID–POI heterodimerizes with the F-box protein TIR1, resulting in AID ubiquitination (U) and the subsequent proteasomal destruction of AID–POI.

**Table 1**
**Toolbox of regulatable promoters for the construction of synthetic gene networks in mammalian cells**

| Promoter type | P | Transcription factor | Inducer molecule | Response to inducer molecule (Diss. = dissociation) | Reference |
|---|---|---|---|---|---|
| $P_{Act}$ | $P_{AIR}$ | AlcR | Acetaldehyde | Binding and activation | (21) |
| | $P_{ART}$ | ARG | L-Arginine | Binding and activation | (22) |
| | $P_{BIT}$ | BIT | Biotin | Binding and activation | (23) |
| | $P_{CR5}$ | cTA | Cumate | Diss. and deactivation | (24) |
| | $P_{CR5}$ | rcTA | Cumate | Binding and activation | (24) |
| | $P_{CTA}$ | CTA | Heat | Diss. and deactivation | (25) |
| | $P_{ETR}$ | ET | Erythromycin | Diss. and deactivation | (26) |
| | $P_{NIC}$ | NT | 6-hyroxy nicotine | Diss. and deactivation | (27) |
| | $P_{PIP}$ | PIT | Pristinamycin I | Diss. and deactivation | (16) |
| | $P_{ROP}$ | REDOX | NADH | Diss. and deactivation | (28) |
| | $P_{SPA}/P_{SCA}$ | SPA/SCA | SCB1 | Diss. and deactivation | (29) |
| | $P_{TET}$ | tTA | Tetracycline | Diss. and deactivation | (30) |
| | $P_{TET}$ | rtTA | Doxycycline | Binding and activation | (31) |
| | (*tra* box)$_{1-7}$-$P_{CMVmin}$ | TraR-VP16 | 3-oxo-$C_8$-HSL | Binding and activation | (32) |
| | $P_{TtgR1}$ | $TtgA_1$ | Phloretine | Diss. and deactivation | (33) |
| $P_{Rep}$ | $P_{CuO}$ | CymR | Cumate | Diss. and derepression | (24) |
| | $P_{ETR}ON8$ | E-KRAB | Erythromycin | Diss. and derepression | (26) |
| | $P_{NIC}$ | NS | 6-hydroxy nicotine | Diss. and derepression | (34) |
| | $P_{PIR}ON$ | PIP-KRAB | Pristinamycin I | Diss. and derepression | (16) |
| | $P_{SCA}ON8$ | SCS | SCB1 | Diss. and derepression | (35) |
| | $P_{TetO}$ | TetR | Tetracycline | Diss. and derepression | (36) |
| | $P_{UREX8}$ | mUTS | Uric acid | Diss. and derepression | (37) |
| $P_{Hyb}$ | $tetO_7$-$ETR_8$-$P_{hCMVmin}$ | tTA | Tetracycline | Diss. and deactivation | (3) |
| | $tetO_7$-$PIR_3$- | E-KRAB | Erythromycin | Diss. and derepression | |
| | $ETR_8$-$P_{hCMVmin}$ | tTA | Tetracycline | Diss. and deactivation | (3) |
| | $scbR_8$-$PIR_3$- $P_{hCMVmin}$ | PIP-KRAB | Pristinamycin I | Diss. and derepression | |
| | | E-KRAB | Erythromycin | Diss. and derepression | |
| | | SCA | SCB1 | Diss. and derepression | (3) |
| | | PIP-KRAB | Pristinamycin I | Diss. and derepression | |

BioLogic NOR Gate

| Input 1 (A) | 0 | 0 | 1 | 1 |
|---|---|---|---|---|
| Input 2 (✱) | 0 | 1 | 0 | 1 |
| Output (GOI) | 1 | 0 | 0 | 0 |

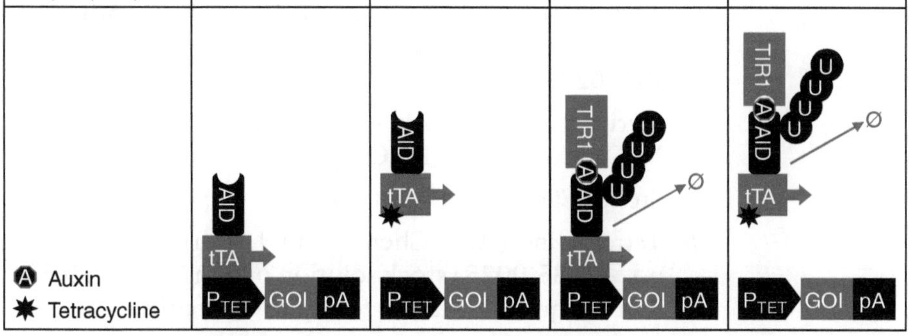


Fig. 2. Design of the synthetic NOR gate. In an NOR gate, the output signal is 0 if at least one input signal is 1. NOR logic was implemented by fusing the auxin-inducible degron (AID) to the tetracycline transactivator (tTA). In the absence of both auxin (A) and tetracycline (✱), tTA–AID binds and activates its cognate promoter $P_{TET}$ resulting in the transcription of the gene of interest (GOI). Upon addition of tetracycline, binding of tTA–AID to $P_{TET}$ is prevented, thereby abolishing GOI expression. GOI expression is also silent upon addition of auxin: auxin triggers dimerization of TIR1 with tTA–AID, resulting in AID ubiquitination (U) and the subsequent depletion of tTA–AID.

## 2. Materials

### *2.1. Cell Culture*

#### *2.1.1. Materials/Hardware*

1. HEK293-T cells (15).
2. Incubator (HERAcell 240i $CO_2$ incubator, Thermo Fisher Scientific, Langenselbold, Germany) for cultivation of mammalian cells at 37°C in a humidified atmosphere containing 5% $CO_2$.
3. Plasmids: The tetracycline-responsive plasmid pMF111 (16) and the tTA-encoding plasmid pSAM200 (16).
4. Anion-exchange-based DNA purification kit (Genomed Jetstar 2.0 Midi, Genomed AG, Bad Oeynhausen, Germany, cat. no. 210 050).
5. Cell counting device (CASY® Model DT Innovatis AG, Bielefeld, Germany). Alternatively, a haemocytometer can be used.
6. Cell culture plasticware: Petri dish (Greiner Bio-One GmbH, Frickenhausen, Germany, cat. No. 664 160) and 24-well plate (Greiner Bio-One GmbH, Frickenhausen, Germany, cat. no. 662 160).

#### *2.1.2. Solutions/Reagents*

1. Dulbecco's modified Eagle's complete medium (DMEM): DMEM (PAN Biotech GmbH, Aidenbach, Germany, cat. no. P03-0710) supplemented with 10% foetal calf serum (FCS) (PAN Biotech GmbH, Aidenbach, Germany, cat. no. P281803)

and 1% penicillin/streptomycin (PAN Biotech GmbH, Aidenbach, Germany, cat. no. P06-07100).

2. Sterile phosphate-buffered saline (PBS) solution: 50 mM $NaH_2PO_4$, 300 mM NaCl, 10 mM imidazole, pH 8.0.
3. Trypsin solution: Trypsin–EDTA in PBS without $Ca^{2+}$ and $Mg^{2+}$ (PAN Biotech GmbH, Aidenbach, Germany, cat. no. P10-023500).
4. Sterile 2× HBS stock solution: 100 mM HEPES, 280 mM NaCl, 1.5 mM $Na_2HPO_4$, pH 7.1.
5. Sterile $CaCl_2$ stock solution, 1 M.
6. Tetracycline (AppliChem GmbH, Darmstadt, Germany, cat. no. A1685,0025) stock solution of 2 mg/ml in 99.5% ethanol. Store aliquots at −20°C for up to 6 months.
7. Auxin: 3-Indolacetic acid (IAA) (Sigma–Aldrich, Steinheim, Germany, cat. no. I2886) stock solution of 500 mM in 99.5% ethanol. Store aliquots at −20°C for up to 6 months.

### 2.2. SEAP Reporter Gene Assay

#### 2.2.1. Materials/Hardware

1. 96-well, flat-bottom transparent plate (Carl Roth GmbH+Co. KG, Karlsruhe, Germany, cat. no. 9293.1).
2. Microplate reader (e.g. Synergy™ H4 Multi-Mode Microplate Reader, BioTec, Winooski, VT).

#### 2.2.2. Solutions/Reagents

1. 2× secreted alkaline phosphatase (SEAP) buffer: 20 mM homoarginine, 1 mM $MgCl_2$, 21% (v/v) diethanolamine, pH 9.8. Store buffer at 4°C in the dark for up to 6 months.
2. pNPP solution: 120 mM para-nitrophenyl phosphate in 2× SEAP buffer. Store in aliquots at −20°C. Thaw only once.

### 2.3. Sodium Dodecylsulphate-PAGE

#### 2.3.1. Materials/Hardware

1. Sodium dodecylsulphate (SDS)-gel electrophoresis chamber, gel casting chambers, and accessories (e.g. Bio-Rad, Hercules, CA).

#### 2.3.2. Solutions/Reagents

1. 5× SDS loading buffer: 10% (v/v) glycerol, 50 mM Tris/HCl, 20 g/l SDS, 2 g/L bromphenol blue, 10% (v/v) β-mercapto-ethanol, pH 6.8. Stock solution should be stored at −20°C.
2. TEMED: *N,N,N′,N′*-Tetramethylethylenediamine (AppliChem GmbH, Darmstadt, Germany, cat. no. A1148,0250). Store in aliquots at 4°C.
3. Ammonium persulphate (APS) stock solution of 10% (v/v) in water. The stock solution should be stored in aliquots at −20°C. Avoid repeated freezing and thawing cycles. Thawed APS can be stored at 4°C for up to 1 week.
4. 4× resolving gel buffer (1.5 M Tris/HCl (pH 8.8), 0.4% SDS). Store at room temperature.

5. 4× stacking gel buffer (0.5 M Tris/HCl (pH 6.8), 0.4% SDS). Store at room temperature.
6. 30% Acrylamide/Bis solution: 29:1 (3.3% C) (Bio-Rad Laboratories, Hercules, CA, cat. no. 161-0156). Due to the neurotoxicity of acrylamide, all experiments involving the substance should be performed with gloves under a ventilated hood.
7. Isopropanol.
8. Prestained protein molecular weight marker (e.g. Fermentas GmbH, St. Leon-Rot, Germany, cat. no. SM0671).
9. Running buffer: 25 mM Tris/HCl, 0.192 mM glycine, 0.1% SDS.

### *2.4. Western Blotting*

#### *2.4.1. Materials/Hardware*

1. Western blot transfer apparatus (PerfectBlue Semi-Dry Electro Blotter, PEQLAB Biotechnologie GmbH, Erlangen, Germany, cat. no. 52-1010).
2. PVDF membrane (Millipore Corporation, Bedford, MA, cat. no. IPVH20200).
3. Gel-blotting filter paper: Whatman 3 MM (Carl Roth GmbH+Co. KG, Karlsruhe, Germany, cat. no. A126.1).

#### *2.4.2. Solutions/Reagents*

1. PBS–Tween: PBS solution supplemented with 0.005% (v/v) Tween.
2. Methanol. Due to methanol's toxic properties, experiments involving the substance should be carried out under a ventilated hood. The methanol can be reused several times.
3. Transfer buffer: 193 mM glycine, 25 mM Tris base, 10% (v/v) methanol. Do not adjust pH and store the buffer at room temperature.
4. Blocking buffer: 2% (v/v) non-fat dry milk in PBS.
5. Monoclonal mouse anti-VP16 antibody (Santa Cruz Biotechnology Inc., Santa Cruz, CA, cat. no. sc-7545).
6. Monoclonal mouse anti-beta-actin antibody (Santa Cruz Biotechnology Inc., Santa Cruz, CA, cat. no. sc-47778).
7. Peroxidase-coupled secondary antibody (anti mouse IgG x HRP, Amersham Life Science, NJ, cat. no. NA931V).
8. Amersham ECL Plus Western Blotting Detection Reagents (GE Healthcare Europe GmbH, Freiburg, Germany, cat. no. RPN2132).

## 3. Methods

### *3.1. Design of the Gene Network*

A Boolean NOR gate was implemented in mammalian cells by fusing the AID to the tetracycline-responsive transactivator (tTA) and combining the resulting tTA–AID fusion protein with a tetracycline-responsive promoter. A schematic representation of the NOR gate gene network is shown in Fig. 2. In the absence of both tetracycline and auxin, tTA binds to its cognate promoter ($P_{TET}$) and the gene of interest (GOI) is transcribed, i.e. the output signal is 1. Addition of tetracycline impedes binding between tTA and its cognate operator, whereas addition of auxin dimerizes tTA–AID with TIR1 and triggers AID ubiquitination, resulting in depletion of tTA–AID. Hence, if either of the two substances, tetracycline or auxin, is present, GOI expression is abolished and the output signal is 0.

For the establishment of such a genetic network in mammalian cells, the following general design principles should be taken into consideration.

1. Design of fusion proteins: The addition of a tag to a protein always involves the risk of interfering with the natural properties of the protein. This risk can be reduced by avoiding the fusion of a tag to a terminus, which is essential for the function or structure of the protein. For DNA-binding proteins, previous experience has shown that the fusion domain should be introduced at the terminus located furthest away from the DNA-binding domain in order to avoid interference with the protein–operator sequence binding. The DNA-binding domain can be identified from the crystal structure of the protein, if available, or obtained from the ExPASy proteomics server (http://www.expasy.org/prosite/). In the case of tTA, the DNA-binding domain is located at the N terminus, and hence AID was fused to the C terminus of the protein.
2. Introduction of epitope tags for protein analysis: In general, detection of proteins relies on epitope tags, which can be recognized by common antibodies. The fusion of tags to the protein, thus, enables protein detection by Western analysis (see Note 1).
3. The use of standardized restriction sites to facilitate exchange of DNA cassettes: The use of standardized restriction sites, which are unique within the vector, simplifies the cloning procedure for the construction of other synthetic gene networks by allowing the exchange of DNA fragments, such as the transcription factor, tags, and promoters, in a single step without the need for PCRs.
4. Enhancing protein expression in mammalian cells: In order to enable efficient protein expression in mammalian cells, the mammalian cell-compatible ribosome binding sequence (Kozak consensus sequence, CCACC) should be fused upstream of the GOI and subsequently cloned into a mammalian expression vector, such as pcDNA3.1 (Invitrogen, Carlsbad, CA, cat. no. V790-20).
5. Monitoring expression using reporter gene: Placing a reporter gene under the control of the target promoter enables analysis and verification of the network (see Note 2).

### 3.2. Construction and Validation of Building Blocks

To construct the BioLogic NOR gate in mammalian cells, the three following transcription units were constructed as described below. A more detailed protocol describing the basic DNA techniques can be found elsewhere (17).

#### 3.2.1. Construction of the Tetracycline- and Auxin-Responsive Transactivator

To construct the plasmid pLMK105 encoding for an auxin- and tetracycline-responsive transactivator, tTA–AID, the AID was C-terminally fused to tTA via PCR. The Kozak sequence was fused upstream (5′) of the tTA–AID fusion gene and subsequently cloned into a mammalian expression vector and placed under the control of a human elongation factor 1 alpha promoter ($P_{EF1alfa}$).

#### 3.2.2. Construction of the c-myc-Tagged TIR1 Protein

In order to construct the plasmid pMK52 encoding a c-myc-tagged TIR1 protein, the c-myc tag and the Kozak sequence were fused to the C and N terminus of the TIR1 protein, respectively, and subsequently placed in a mammalian expression vector under the control of $P_{EF1alfa}$.

#### 3.2.3. Construction of the Tetracycline-Responsive Promoter

To construct the plasmid pMF111 encoding the reporter gene SEAP under the control of the tetracycline-responsive promoter, the *tetO* operator sites were fused upstream of the minimal human cytomegalovirus promoter ($P_{hCMVmin}$) (see Note 3). Subsequently, the SEAP gene was fused downstream of the promoter (as previously described in ref. 16) (see Note 4).

In order to verify that the DNA-binding and tetracycline-responsive properties of tTA are maintained after its fusion with AID, HEK293-T cells were co-transfected with the tTA–AID fusion protein encoding plasmid pLMK105 and the tetracycline-responsive plasmid pMF111, as described in the following protocols. For comparison, the plasmid pSAM200 encoding tTA was transfected together with pMF111.

#### 3.2.4. Cell Cultivation and Cell Seeding

The following protocols involving cell culture experiments are adapted to human embryonic kidney cells (HEK293-T). The following protocols can be adapted to other cell types with minor changes.

1. Cultivate HEK293-T cells in DMEM complete medium at 37°C in a humidified atmosphere containing 5% $CO_2$.
2. When the cells reach 80–90% confluence, aspirate the medium from the Petri dish, wash the cells with 3 ml PBS, and subsequently incubate the plate in the presence of 3 ml trypsin solution at 37°C for approximately 5–10 min until the cells are detached.
3. Gently tap the plate and mix the cell suspension with 7 ml DMEM complete medium.
4. Centrifuge for 3 min at 450 × *g* and discard the supernatant.

5. Resuspend the cells in 7 ml DMEM complete medium and determine the cell density using a CASY® cell counting device.
6. Dilute the cell suspension to a density of 100,000 cells/ml and seed 0.5 ml into each well of a 24-well plate.
7. Incubate the plate at 37°C in a humidified atmosphere containing 5% $CO_2$ for 12–24 h prior to transfection.

*3.2.5. Cell Transfection*

The quantities given in the following transfection protocol are adapted to one well of a 24-well plate (see Note 5). The plasmids pLMK105, pSAM200, and pMF111 were isolated using an anion-exchange DNA purification kit according to the manufacturer's protocol.

1. Prepare the transfection mixture with a total volume of 50 μl, containing 0.25 mM $CaCl_2$ and a total plasmid DNA amount of 0.75 μg (see Note 6).
2. Prepare a falcon tube containing 50 μl 2× HBS.
3. Slowly add the transfection mix prepared during step 1 drop-wise to the 2× HBS solution under vortexing.
4. Incubate for 20 min at room temperature so that the calcium phosphate–DNA complex forms.
5. Mix the precipitate well by vortexing and add the suspension slowly and dropwise to the well while gently rocking the media in the plate.
6. Centrifuge the cell culture plate at 2,500 × *g* for 5 min in order to let the precipitate sediment onto the cells.
7. Incubate the plate for 1.5 h. (If the cells are not centrifuged, the incubation time should be increased to 5 h to allow the DNA complexes to settle).
8. Remove the media containing the precipitate and add fresh media. Cultivate the cells in the presence or absence of 2 μg/ml tetracycline.
9. Incubate the cell culture plate for at least 24 h prior to analysis.

*3.2.6. SEAP Reporter Gene Analysis*

SEAP expression levels were measured according to the following protocol (see Note 7).

1. Transfer 200 μl of the cell culture supernatant to an Eppendorf tube (see Note 8).
2. Inactivate endogenous phosphatases by heating the sample for 15 min at 65°C in a thermoshaker and centrifuge at 14,000 × *g* for 1 min in order to remove cell debris.
3. Pipet 100 μl 2× SEAP buffer into each well of a transparent 96-well flat-bottom plate and add 80 μl of the sample. Start the reaction by adding 20 μl pNPP solution and measure the increase in light absorbance at 405 nm every 30 s for approximately 60 min.

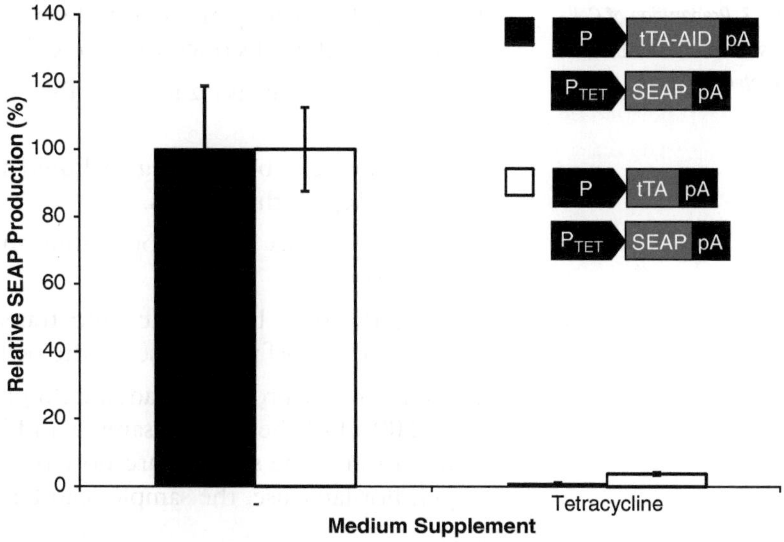


Fig. 3. Validation of the DNA-binding and tetracycline-responsive properties of tTA–AID. SEAP expression levels for HEK293-T cells co-transfected with the tTA-responsive reporter-gene-containing plasmid pMF111 and with either the tTA-containing plasmid pSAM200 or the plasmid pLMK105 encoding the fusion protein tTA–AID. Cells were incubated in the presence or absence of 2 μg/ml tetracycline for 48 h prior to quantification of the levels of SEAP activity in the cell culture medium.

4. Calculate the slope in optical density per minute (OD/min) and calculate the SEAP activity according to Lambert–Beers law; $E = \varepsilon \cdot c \cdot d$, where ε is the molar extinction coefficient of the reaction product *p*-nitrophenolate ($\varepsilon = 18{,}600\ M^{-1}/cm$), $c$ is the increase in *p*-nitrophenolate in $M\ min^{-1}$, and $d$ is the length in cm of the light path in the sample (usually, 0.5 cm in this configuration). Determine the activity in units per litre (U/L), taking into account the dilution of the sample in the buffer, as: SEAP activity(U/L) $= c \cdot 10^6 \cdot 200/80$ (Fig. 3).

In order to verify the expression of tTA–AID as well as the triggering of tTA–AID degradation by auxin in the presence of TIR1 (caused by dimerization of TIR1 with AID and subsequent ubiquitination), HEK293-T cells were co-transfected with the plasmid pMK52 encoding the TIR1 protein and the plasmid pLMK105 encoding tTA–AID. In addition, the tTA-encoding plasmid pSAM200 was co-transfected with pMK52 in order to confirm that the degradation of tTA–AID is AID specific and does not occur for unfused tTA. The transfections were carried out according to the protocol described in Subheading 3.2.4, step 5. The cells were cultivated in DMEM complete medium in the presence or absence of 500 μM IAA. Forty-eight hours after transfection, the cells were harvested and the protein levels were compared via western blot analysis according to the following protocol.

*3.2.7. Preparation of Cell Lysate for SDS-PAGE and Western Blotting*

1. Place cell culture plate on ice, discard the cell culture medium, and wash the cells once with ice-cold PBS.
2. Detach the cells using ice-cold PBS supplemented with 5 mM EDTA.
3. Lyse the cells by freezing in liquid nitrogen and thawing at 37°C. Repeat three times.
4. Leave the cells on ice for 2 min and centrifuge at 4°C for 13,200 × *g*.
5. Place the tube back on ice and transfer the supernatant to a new Eppendorf tube. Discard the pellet.
6. Denature the proteins by adding 25 μl of 5× SDS loading buffer to 100 μl of the protein sample and boil the mixture at 95°C for 5 min. The samples are now ready to be loaded onto the gel. For later use, the samples can be stored at −20°C.

*3.2.8. SDS-PAGE*

This protocol is adapted to the Bio-Rad system. It has been written as generically as possible and can with minor changes be adapted to other SDS-gel apparatus.

1. Ensure that the glass cassettes, combs, casting stand, and casting frames are clean and dry.
2. Set up the casting stand assembly. Make sure that the caster is set up properly in order to prevent the gel solution from leaking out when casting the gel.
3. Prepare the resolving gel by mixing 1.8 ml $dH_2O$, 1.3 ml 4× resolving gel buffer, and 2.2 ml 30% acrylamide/Bis solution in a falcon tube (see Note 9). In order to start the polymerization reaction, add 50 μl APS and 5 μl TEMED. Mix well and pour the gel into the cassettes so that approximately two-thirds are filled. Immediately cover the gel with isopropanol and allow the gel to solidify for 30 min.
4. Once the gel is polymerized, pour out the isopropanol and rinse the gel surface twice with water. Dry the area above the resolving gel with a tissue, taking care not to touch the surface of the gel.
5. Prepare the stacking gel by mixing 2.33 ml $dH_2O$, 1 ml 4× stacking gel buffer, and 0.66 ml 30% acrylamide/Bis solution in a falcon tube. Add 50 μl APS and 2.5 μl TEMED, mix well, and pour the gel onto the surface of the resolving gel. Insert the comb and allow the gel to polymerize for 1 h.
6. Once polymerized, gently remove the comb and rinse the wells twice with running buffer.
7. Place the gel cassettes in the electrode assembly and fill the inner and outer chambers with running buffer.

8. Load the gel using a Hamilton syringe. Include one well for a prestained molecular weight standard and, if available, one well with a positive control.
9. Assemble the chamber and connect it to a power supply. Separate the proteins by applying 150 V for 1.5 h. Turn off the voltage when the blue migration front has reached the bottom of the gel.

#### 3.2.9. Western Blotting

After protein separation, the proteins are transferred electrophonically from the resolving gel to a PVDF membrane using a Western blotting transfer machine and are subsequently detected using protein-specific antibodies according to the following protocol.

1. Disassemble the gel chamber. Remove the stacking gel. Immerse the resolving gel in transfer buffer for 10 min.
2. Cut one sheet of PVDF membrane and three gel-blotting filter papers to the same size as the gel. Wear gloves in order to avoid contaminating the membrane.
3. Soak the PVDF membrane in methanol for 30 s. Discard the methanol and immerse the membrane in $H_2O$. Do not allow the membrane to dry out during processing (see Note 10). The membrane is now ready for use. Due to the toxicity of methanol, this experiment should be performed under a hood.
4. Soak the gel-blotting filter papers in transfer buffer.
5. Assemble the "sandwich" by placing the membrane on top of two gel-blotting filter papers. Place the gel on top of the membrane and add an additional gel-blotting filter paper on top of the gel (see Note 11). Air bubbles must not be trapped between the gel and the membrane.
6. Assemble the transfer machine and transfer/blot the proteins onto the membrane by applying a current of 0.35 A for 1.5 h.
7. As soon as the transfer is ready, disconnect the power supply and disassemble the sandwich. The prestained molecular marker should have been transferred from the gel and be well-visible, indicating successful transfer.
8. Rinse the membrane once with PBS–Tween and incubate the membrane in blocking buffer for 1 h at room temperature or overnight at 4°C on a rocking platform.
9. Discard the blocking buffer and rinse the membrane twice with PBS–Tween.
10. Incubate the membrane with the primary antibody (anti-VP16 antibody, 1:1,000 in blocking buffer) for 1 h at room temperature or overnight at 4°C (see Note 12).
11. Discard the primary antibody and wash the membrane four times for 10 min each with PBS–Tween on a rocking platform in order to remove unbound primary antibody.

12. Freshly prepare the secondary antibody (anti-mouse IgG × HRP antibody, 1:5,000 in blocking buffer) and add it to the membrane. Incubate for at least 30 min on a rocking platform.
13. In order to remove the unbound antibody, wash the blot extensively by soaking the membrane in PBS–Tween four times for 10 min each on a rocking platform.
14. For detection, prepare the ECL solution according to the manufacturer's protocol and add the reagent to the membrane, making sure it covers the whole surface. Leave the reagent on the membrane surface for 1 min before discarding it and develop the membrane.

After VP16 detection, the membrane was reused for evaluation of the loading control; the membrane was dried, reactivated in methanol for 30 s, washed in PBS, and incubated with the mouse anti-beta actin antibody followed by the secondary anti-mouse HRP antibody as described in steps 10–14 in Subheading 3.2.9 (Fig. 4).

### 3.3. Construction and Validation of the Gene Network

For the construction and validation of the NOR gate, all plasmids in the gene network (pLMK105, pMK52, and pMF111) were transfected into HEK293-T cells according to the protocol described in Subheading 3.2.4, step 5. The cells were cultivated in

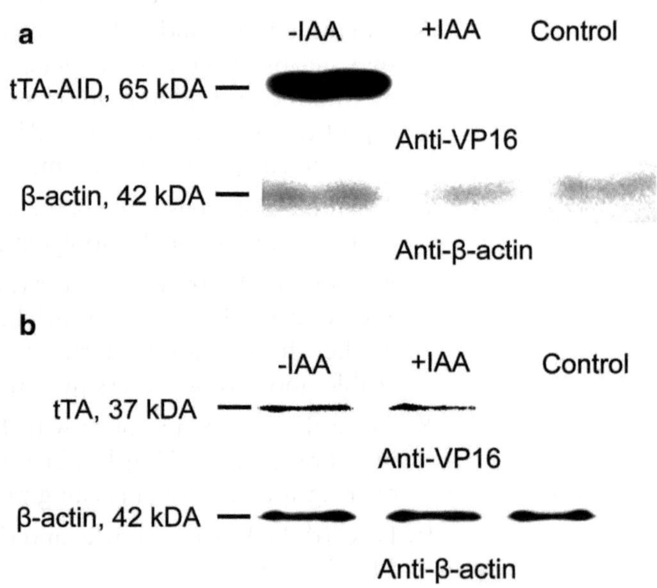


Fig. 4. Expression analysis of tTA–AID and verification of its degradation in the presence of TIR1 and IAA via Western blot using antibodies against VP16. HEK293-T cells were co-transfected with the plasmid pMK52 encoding TIR1 and either the tTA–AID-encoding plasmid pLMK105 (**a**) or the tTA-encoding plasmid pSAM200 (**b**). The cells were cultivated for 48 h in the presence or absence of 500 μM IAA. Mock-transfected HEK293-T cells were used as a negative control and an anti-beta-actin antibody was used as a loading control.

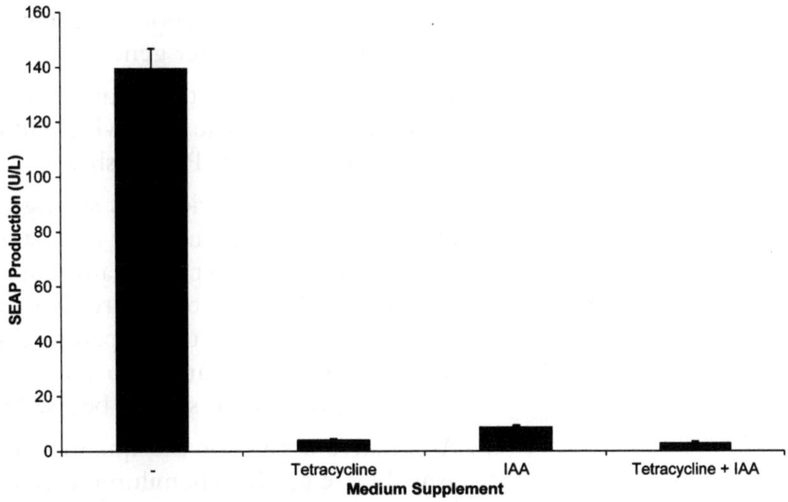


Fig. 5. Verification of the BioLogic NOR gate. SEAP production levels 36 h after transfection of HEK293-T cells co-transfected with plasmids pLMK105, pMK52, and pMF111 and cultivated in DMEM complete medium without any supplements, in medium supplemented with either 2 μg/ml tetracycline or 500 μM IAA, or in medium supplemented with both substances simultaneously.

DMEM complete medium in the absence of any inducer, in the presence of either 500 μM IAA or 2 μg tetracycline, or in the presence of both substances simultaneously. Thirty-six hours post transfection, the SEAP activity was measured as described in Subheading 3.2.6. The NOR gate output was 1 only when neither of the two substances was present. Upon addition of either substance separately or both simultaneously, the output signal was 0; thus, biological NOR gate operation was confirmed (Fig. 5).

## 4. Notes

1. Common epitope tags include HA, c-myc, and hexahistidine tags (18).
2. Common reporter genes include genes encoding for fluorescent proteins (e.g. mCherry, green and yellow fluorescent protein), the enzyme luciferase (e.g. Renilla and Firefly luciferase), and secreted enzymes (e.g. SEAP and the secreted α-amylase (SAMY)); see overview articles (19, 20).
3. The amount of expressed protein commonly depends on the number of operator sites. Therefore, plasmids containing different numbers of operator sites should be constructed and screened in cell culture in order to determine the optimal configuration.

4. If desired, the SEAP reporter gene can in principle be replaced with any other reporter gene.
5. For other plate types, the volumes given in this protocol should be multiplied by the following factors: 2 (12-well plates), 4 (6-well plate), 20 (Petri dish).
6. The transfection efficiency is strongly dependent on the total amount of DNA. Hence, for comparable results, it is not only important to transfect equal amounts of each plasmid for each experiment, but also to ensure that the total amount of DNA is constant; e.g. for two separate experiments requiring the transfection of one and two plasmids, respectively, an additional empty plasmid should be used in the former.
7. Alternative assays for the quantification of SEAP activity are available, e.g. the chemiluminescence-based SEAP reporter gene system (Roche Diagnostics AG, Mannheim, Germany, cat. no. 11 779 842 001).
8. Cell culture samples for quantification of SEAP activity can be stored at −20°C for later measurements.
9. For an optimal degree of separation between the proteins, the acrylamide concentration in the gel should be chosen depending on the size of the POI; high and low concentrations should be used for small and large proteins, respectively.
10. If the membrane has dried out, the membrane must be reactivated with methanol and rinsed in PBS.
11. The required layer configuration can vary between different machines. Since the negatively charged proteins migrate towards the anode, the membrane should always be placed between the gel and the anode.
12. Most of the primary antibodies can be reused. The number of uses depends strongly on the specific antibody, and has to be optimized for each antibody separately. After use, freeze and store the blocking buffer containing the antibody at −20°C.

## Acknowledgements

The work of MK was supported by the Swiss National Science Foundation (grant no. CR32I3_125426), the work of WW was supported by the Excellence Initiative of the German Federal and State Governments (EXC 294), and the work of MF was supported by the Swiss National Science Foundation (grant no. 31003A-126022) and in part by the EC Framework 7 (Persist).

## References


1. Weber, W., and Fussenegger, M. (2010) Synthetic gene networks in mammalian cells, *Curr Opin Biotechnol 692*, 235–249.
2. Rinaudo, K., Bleris, L., Maddamsetti, R., Subramanian, S., Weiss, R., and Benenson, Y. (2007) A universal RNAi-based logic evaluator that operates in mammalian cells, *Nat Biotechnol 25*, 795–801.
3. Kramer, B. P., Fischer, C., and Fussenegger, M. (2004) BioLogic gates enable logical transcription control in mammalian cells, *Biotechnol Bioeng 87*, 478–484.
4. Deans, T. L., Cantor, C. R., and Collins, J. J. (2007) A tunable genetic switch based on RNAi and repressor proteins for regulating gene expression in mammalian cells, *Cell 130*, 363–372.
5. Kramer, B. P., and Fussenegger, M. (2005) Hysteresis in a synthetic mammalian gene network, *Proc Natl Acad Sci USA 102*, 9517–9522.
6. Kramer, B. P., Viretta, A. U., Daoud-El-Baba, M., Aubel, D., Weber, W., and Fussenegger, M. (2004) An engineered epigenetic transgene switch in mammalian cells, *Nat Biotechnol 22*, 867–870.
7. Weber, W., Stelling, J., Rimann, M., Keller, B., Daoud-El Baba, M., Weber, C. C., Aubel, D., and Fussenegger, M. (2007) A synthetic time-delay circuit in mammalian cells and mice, *Proc Natl Acad Sci USA 104*, 2643–2648.
8. Tigges, M., Marquez-Lago, T. T., Stelling, J., and Fussenegger, M. (2009) A tunable synthetic mammalian oscillator, *Nature 457*, 309–312.
9. Swinburne, I. A., Miguez, D. G., Landgraf, D., and Silver, P. A. (2008) Intron length increases oscillatory periods of gene expression in animal cells, *Genes Dev 22*, 2342–2346.
10. Weber, W., and Fussenegger, M. (2006) Pharmacologic transgene control systems for gene therapy, *J Gene Med 8*, 535–556.
11. Malphettes, L., and Fussenegger, M. (2004) Macrolide- and tetracycline-adjustable siRNA-mediated gene silencing in mammalian cells using polymerase II-dependent promoter derivatives, *Biotechnol Bioeng 88*, 417–425.
12. Greber, D., El-Baba, M. D., and Fussenegger, M. (2008) Intronically encoded siRNAs improve dynamic range of mammalian gene regulation systems and toggle switch, *Nucleic Acids Res 36*, e101.
13. Nishimura, K., Fukagawa, T., Takisawa, H., Kakimoto, T., and Kanemaki, M. (2009) An auxin-based degron system for the rapid depletion of proteins in nonplant cells, *Nat Methods 6*, 917–922.
14. Banaszynski, L. A., Chen, L. C., Maynard-Smith, L. A., Ooi, A. G., and Wandless, T. J. (2006) A rapid, reversible, and tunable method to regulate protein function in living cells using synthetic small molecules., *Cell 126*, 995–1004.
15. Mitta, B., Rimann, M., Ehrengruber, M. U., Ehrbar, M., Djonov, V., Kelm, J., and Fussenegger, M. (2002) Advanced modular self-inactivating lentiviral expression vectors for multigene interventions in mammalian cells and in vivo transduction, *Nucleic Acids Res 30*, e113.
16. Fussenegger, M., Morris, R. P., Fux, C., Rimann, M., von Stockar, B., Thompson, C. J., and Bailey, J. E. (2000) Streptogramin-based gene regulation systems for mammalian cells, *Nat Biotechnol 18*, 1203–1208.
17. Sambrook, J., Fritsch, E. F., and Maniatis, T. (2000) *Molecular Cloning, A Laboratory Manual*, 3rd ed, Cold Spring Harbor Laboratory Press, Cold Spring Harbor, New York.
18. Fritze, C. E., and Anderson, T. R. (2000) Epitope tagging: general method for tracking recombinant proteins, *Methods Enzymol 327*, 3–16.
19. Kain, S. R., and Ganguly, S. (2001) Overview of genetic reporter systems, *Curr Protoc Mol Biol 9*, 9.6.
20. Schlatter, S., Rimann, M., Kelm, J., and Fussenegger, M. (2002) SAMY, a novel mammalian reporter gene derived from Bacillus stearothermophilus alpha-amylase, *Gene 282*, 19–31.
21. Weber, W., Rimann, M., Spielmann, M., Keller, B., Daoud-El Baba, M., Aubel, D., Weber, C. C., and Fussenegger, M. (2004) Gas-inducible transgene expression in mammalian cells and mice, *Nat Biotechnol 22*, 1440–1444.
22. Hartenbach, S., Daoud-El Baba, M., Weber, W., and Fussenegger, M. (2007) An engineered L-arginine sensor of Chlamydia pneumoniae enables arginine-adjustable transcription control in mammalian cells and mice, *Nucleic Acids Res 35*, e136.
23. Weber, W., Lienhart, C., Daoud El-Baba, M., and Fussenegger, M. (2009) A biotin-triggered genetic switch in mammalian cells and mice, *Metab Eng 11*, 117–124.
24. Mullick, A., Xu, Y., Warren, R., Koutroumanis, M., Guilbault, C., Broussau, S., Malenfant, F., Bourget, L., Lamoureux, L., Lo, R., Caron, A.

W., Pilotte, A., and Massie, B. (2006) The cumate gene-switch: a system for regulated expression in mammalian cells, *BMC Biotechnol* *6*, 43.

25. Weber, W., Marty, R. R., Link, N., Ehrbar, M., Keller, B., Weber, C. C., Zisch, A. H., Heinzen, C., Djonov, V., and Fussenegger, M. (2003) Conditional human VEGF-mediated vascularization in chicken embryos using a novel temperature-inducible gene regulation (TIGR) system, *Nucleic Acids Res 31*, e69.
26. Weber, W., Fux, C., Daoud-el Baba, M., Keller, B., Weber, C. C., Kramer, B. P., Heinzen, C., Aubel, D., Bailey, J. E., and Fussenegger, M. (2002) Macrolide-based transgene control in mammalian cells and mice, *Nat Biotechnol 20*, 901–907.
27. Malphettes, L., Weber, C. C., El-Baba, M. D., Schoenmakers, R. G., Aubel, D., Weber, W., and Fussenegger, M. (2005) A novel mammalian expression system derived from components coordinating nicotine degradation in arthrobacter nicotinovorans pAO1, *Nucleic Acids Res 33*, e107.
28. Weber, W., Link, N., and Fussenegger, M. (2006) A genetic redox sensor for mammalian cells, *Metab Eng 8*, 273–280.
29. Weber, W., Schoenmakers, R., Spielmann, M., El-Baba, M. D., Folcher, M., Keller, B., Weber, C. C., Link, N., van de Wetering, P., Heinzen, C., Jolivet, B., Sequin, U., Aubel, D., Thompson, C. J., and Fussenegger, M. (2003) Streptomyces-derived quorum-sensing systems engineered for adjustable transgene expression in mammalian cells and mice, *Nucleic Acids Res 31*, e71.
30. Gossen, M., and Bujard, H. (1992) Tight control of gene expression in mammalian cells by tetracycline-responsive promoters, *Proc Natl Acad Sci USA 89*, 5547–5551.
31. Gossen, M., Freundlieb, S., Bender, G., Muller, G., Hillen, W., and Bujard, H. (1995) Transcriptional activation by tetracyclines in mammalian cells, *Science* ***268***, 1766–1769.
32. Neddermann, P., Gargioli, C., Muraglia, E., Sambucini, S., Bonelli, F., De Francesco, R., and Cortese, R. (2003) A novel, inducible, eukaryotic gene expression system based on the quorum-sensing transcription factor TraR, *EMBO Rep 4*, 159–165.
33. Gitzinger, M., Kemmer, C., Daoud-El Baba, M., Weber, W., and Fussenegger, M. (2009) Controlling transgene expression in subcutaneous implants using a skin lotion containing the apple metabolite phloretin, *Proc Natl Acad Sci USA* ***106***, 10638–10643.
34. Malphettes, L., Schoenmakers, R. G., and Fussenegger, M. (2006) 6-hydroxy-nicotine-inducible multilevel transgene control in mammalian cells, *Metab Eng 8*, 543–553.
35. Weber, W., Malphettes, L., de Jesus, M., Schoenmakers, R., El-Baba, M. D., Spielmann, M., Keller, B., Weber, C. C., van de Wetering, P., Aubel, D., Wurm, F. M., and Fussenegger, M. (2005) Engineered Streptomyces quorum-sensing components enable inducible siRNA-mediated translation control in mammalian cells and adjustable transcription control in mice, *J Gene Med 7*, 518–525.
36. Yao, F., Svensjo, T., Winkler, T., Lu, M., Eriksson, C., and Eriksson, E. (1998) Tetracycline repressor, tetR, rather than the tetR-mammalian cell transcription factor fusion derivatives, regulates inducible gene expression in mammalian cells, *Hum Gene Ther 9*, 1939–1950.
37. Kemmer, C., Gitzinger, M., Daoud-El Baba, M., Djonov, V., Stelling, J., and Fussenegger, M. (2010) Self-sufficient control of urate homeostasis in mice by a synthetic circuit, *Nat Biotechnol 28*, 355–360.

# Chapter 23

# Synthetic Gene Networks as Blueprint for Smart Hydrogels

**Michael M. Kämpf and Wilfried Weber**


## Abstract

The rapidly emerging ability to design and construct synthetic gene networks in mammalian cells is based on the availability of mutually compatible genetic switches that enable the time-dependent induction of transgene expression in response to the dose of an externally applied stimulus. As these genetic switches are inherently compatible with mammalian cell physiology, they are as well predestined to control the functionality of cell-free synthetic devices within an overall physiologic background. In this chapter, we describe how a genetic switch that was originally designed for gene therapeutic studies can be applied in materials science to design and construct a biohybrid hydrogel that can be used to release a therapeutic growth factor in response to an externally applied stimulus for controlling cell fate and function in a time- and space-resolved manner.


**Key words:** Drug delivery, FKBP, Gene switch, Hydrogel, Inducible release

## 1. Introduction

Significant progress has been made in the past decade in developing delivery devices for bioactive compounds, such as proteins, peptides, antigens, and oligonucleotides, within the human body or in cell culture for tissue engineering. Promising materials for this purpose are hydrogels comprising hydrated, cross-linked, hydrophilic polymer networks. Loading these polymers with biopharmaceuticals and subsequent implantation to the site of administration (i.e., diseased tissue) allow the release of the cargo in a predefined manner and leads to high bioavailability at the site of action and only low side effect-prone levels in other regions of the body (1). However, once implanted in the body, the release rate of the therapeutic can hardly be adjusted to changing dosing regimes. To overcome this limitation, materials were developed to release the cargo protein in

Wilfried Weber and Martin Fussenegger (eds.), *Synthetic Gene Networks: Methods and Protocols*,
Methods in Molecular Biology, vol. 813, DOI 10.1007/978-1-61779-412-4_23, © Springer Science+Business Media, LLC 2012

response to an external trigger; however, most trigger stimuli like light (2), temperature (3), or stimulus concentrations exceeding physiological limits (4) prevented the use of such stimulus-sensitive materials in vivo. To overcome these limitations, recently, a new type of stimulus-sensitive hydrogel was developed based on a synthetic gene switch that has originally been designed for controlling transgene expression levels in gene therapy studies (5, 6) (see Fig. 1). This hydrogel consists of linear polymers like polyacrylamide that are cross-linked by homodimeric $F_M$ protein, a spontaneously dimerizing variant of human FK-binding protein 12 (5). Administration of the clinically licensed small molecule FK506 or nonimmunosuppressive analogs thereof (rapalogs) (7) leads to the monomerization of $F_M$ dimers and the subsequent dissolution of

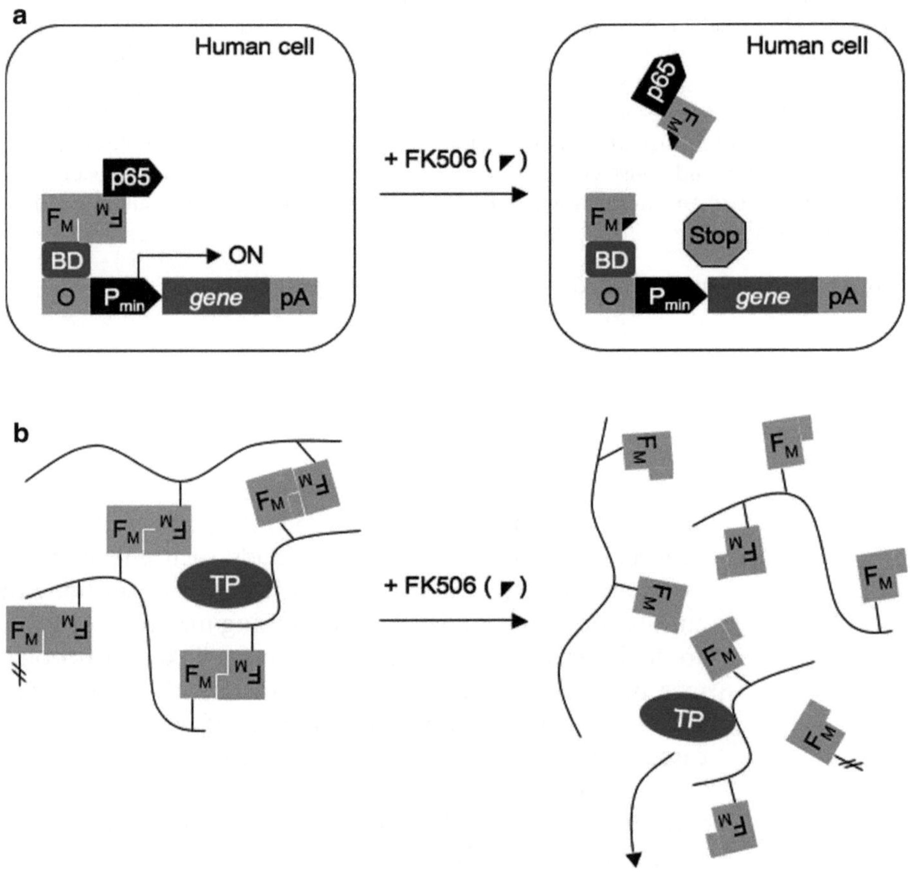


Fig. 1. Design concept of $F_M$-based inducible expression system in mammalian cells and $F_M$-based trigger-sensitive hydrogel drug depot for inducible therapeutic protein release. (**a**) By homodimerization of the chimeric $F_M$ fused to either a DNA-binding domain (DB) or a transactivator domain (p65), a split transcription factor is reconstituted and activates transgene expression in mammalian cells. Addition of the inducer molecule FK506 provokes monomerization of the $F_M$ domains and removal of the chimeric transactivator from the minimal promotor, thereby aborting transgene expression (adapted from ref. 6). (**b**) Homodimeric $F_M$ cross-links single polymer chains, thus generating a three-dimensional stable hydrogel network in which a therapeutic cargo protein (TP) can be incorporated. Disruption of the hydrogel network and accompanied release of the cargo protein are initiated by the addition of FK506 due to interruption of the $F_M$ interactions.

the hydrogel network, a process that might be used to trigger the release of a therapeutic payload protein (8).

This chapter describes in detail how the molecular components of the $F_M$-based gene switch can be applied to construct a hydrogel for the stimulus-responsive release of a therapeutic growth factor within an overall physiologic environment.

## 2. Materials

### 2.1. Protein Production and Purification

1. LB media: 10 g/l tryptone, 5 g/l yeast extract, 10 g/l NaCl, pH 7.0 (Carl Roth, Karlsruhe, Germany, cat. no. X968.3).
2. Ampicillin (Carl Roth, cat. no. K029.2).
3. *Escherichia coli* expression strain BL21 STAR™ (DE3) (Invitrogen, Carlsbad, CA, cat. no. C6010-03).
4. Expression vectors: pMK6 encoding the cross-linking protein $F_M$ (8) pRSet-$VEGF_{121}$ (9) constructed for expression of the cargo protein and vascular endothelial growth factor 121 ($VEGF_{121}$).
5. 1 M isopropyl-β-D-thiogalactopyranoside (IPTG) stock solution dissolved in water. Store in 1-ml aliquot at –20°C.
6. Photometer (GeneQuant 1300, GE Healthcare, Little Chalfont, UK).
7. French Press (APV 2000, APV Manufacturing, Bydgoszcz, Poland).
8. $Ni^{2+}$-NTA-agarose Superflow column (Qiagen, Hilden, Germany, cat. no. 30210).
9. Disposable 10 ml polypropylene columns (Pierce Rockford, IL, cat.no. 29924).
10. PBS-I; 50 mM $NaH_2PO_4$, 300 mM NaCl, 10 mM imidazole, pH 8.0.
11. IMAC wash buffer: 50 mM $NaH_2PO_4$, 300 mM NaCl, 20 mM imidazole, adjust pH to 8.0 with NaOH.
12. IMAC elution buffer: 50 mM $NaH_2PO_4$, 300 mM NaCl, 250 mM imidazole, adjust pH to 8.0 with NaOH.
13. Inclusion body dissolution buffer: 6 M urea, 500 mM NaCl, 5 mM imidazole, 20 mM Tris/HCl, adjust pH to 7.9 with HCl.
14. IMAC urea washing buffer: 6 M urea, 500 mM NaCl, 20 mM imidazole, 20 mM Tris/HCl, adjust pH to 7.9 with HCl.
15. IMAC urea elution buffer: 6 M urea, 500 mM NaCl, 250 mM imidazole, 20 mM Tris/HCl, adjust pH to 7.9 with HCl.

16. Dialysis buffer 1: 4 M urea, 1 mM EDTA, 20 mM Tris/HCl, adjust pH to 7.5 with HCl.
17. Dialysis buffer 2: 3 M urea, 1 mM EDTA, 20 mM Tris/HCl, adjust pH to 7.5 with HCl.
18. Dialysis buffer 3: 2 M urea,150 mM NaCl, 20 mM Tris/HCl, adjust pH to 7.5 with HCl.
19. Dialysis tube (3.5 kDa MWCO, Pierce Rockford, cat. no. 68035).
20. Ultrafiltration tubes (10 kDa MWCO, Corning, Amsterdam, the Netherlands, cat. no. 431483).

### 2.2. Polymer Synthesis

1. Vacuo Evaporator (Rotavapor R-210, Büchi, Flawil, Switzerland).
2. Acryloylchloride (ABCR, Karlsruhe, Germany, cat.no. AB172729).
3. *N,N*-bis(carboxymehyl)-L-lysine (Fluka, St. Louis, MO, USA, cat.no. 14580).
4. Acrylamide (AAm, Pharmacia Biotech, Uppsala, Sweden, cat. no. 17-1300-01).
5. Ammonium peroxodisulfate 10% (w/v) in water. The stock solution should be stored in aliquots at –20°C. Avoid repeated freezing and thawing cycles. Thawed APS can be stored at 4°C for up to 1 week.
6. *N,N,N′,N′*-tetramethylethylenediamine (TEMED) (AppliChem, Darmstadt, Germany, cat. no. A1148,0250).
7. Polymerization buffer: 50 mM Tris/HCl, pH 8.5.
8. Argon or nitrogen gas.
9. Dialysis tube (3.5 kDa MWCO, Pierce Rockford, cat. no.68035).
10. $NiSO_4 \times 6\ H_2O$ (AppliChem, cat.no. A0827,0250).
11. Sigmacote® (Sigma, St. Louis, MO, cat. no. SL-2).
12. Microscope slides (Carl Roth, cat. no. 0656).

### 2.3. Analytics

#### 2.3.1. Sodium Dodecylsulfate-PAGE

1. Sodium dodecylsulfate (SDS)-gel electrophoresis chamber, gel casting chambers, and accessories (e.g., Bio-Rad, Hercules, CA).
2. 5× SDS loading buffer: 10% (v/v) glycerol, 50 mM Tris/HCl, 20 g/l SDS, 2 g/l bromphenol blue, 10% (v/v) β-mercapto-ethanol, pH 6.8. Store solution at –20°C.
3. TEMED (AppliChem, cat. no. A1148,0250). Store at 4°C.
4. Ammonium persulfate (APS) stock solution of 10% (w/v) in water.
5. 4× resolving gel buffer: 1.5 M Tris/HCl, pH 8.8, 0.4% SDS. Store at room temperature.

6. 4× stacking gel buffer: 0.5 M Tris/HCl, pH 6.8, 0.4% SDS. Store at room temperature.
7. 30% acrylamide/bis solution: 29:1 (3.3% C) (Bio-Rad, cat. no. 161-0156). Due to the neurotoxicity of acrylamide, all experiments involving the substance should be performed with gloves under a ventilated hood.
8. Isopropanol.
9. Prestained protein molecular weight marker (e.g., Fermentas GmbH, St. Leon-Rot, Germany, cat. no. SM0671).
10. Running buffer: 25 mM Tris, 0.192 mM glycine, 0.1% SDS.
11. Coomassie staining solution: 0.1% (w/v) Coomassie Brilliant Blue R250, 25% (v/v) isopropanol, 10% (v/v) acetic acid.
12. Coomassie destaining solution: 20% (v/v) EtOH, 10% (v/v) acetic acid.

*2.3.2. Bradford Protein Assay*

1. Bradford protein assay reagent (Bio-Rad, cat. no. 500-0006).
2. Bovine serum albumin (BSA) (Fluka, cat.no. 05479-50g).

*2.3.3. $VEGF_{121}$ – ELISA Kit*

Peprotech, Hamburg, Germany, cat. no. 900-K10.

*2.3.4. Inducer Molecule FK506*

LC Laboratories cat.no. F-4900 dissolved as 25 mM stock solution in DMSO and stored in aliquots at –20°C.

## 3. Methods

### *3.1. Expression and Purification of the Cross-Linking Protein $F_M$ in E. coli*

The following protocol describes the expression and purification of the cross-linking protein $F_M$ using the phage promoter T7 system and purification via immobilized-metal affinity chromatography (IMAC). Other bacterial expression systems and purification methods might also be appropriate (10, 11). In order to construct the expression plasmid pMK6 (8) encoding the open reading frame of the $F_M$ protein (PDB ID code 1EYM), a hexahistidine and an HA-tag were fused N-terminally of the protein whose transcription is placed under control of the phage T7 promotor. The coding region of the $F_M$ protein was furthermore codon optimized for expression in *E. coli* to guarantee high yield of the expressed protein (Geneart, Regensburg, Germany). The N-terminal hexahistidine tag serves as affinity tag for purification purposes and coincidentally for coupling the protein to the polymer for hydrogel assembly.

1. Transform the expression vector pMK6 into chemical competent *E. coli* expression strain BL21 STAR™ (DE3) (see Note 1) using standard heat shock technique and incubate inoculum with corresponding antibiotic at 37°C and 200 rpm overnight.

2. Dilute the overnight culture 1:25 with fresh LB media supplemented with corresponding antibiotics and incubate for approximately 1–2 h at 37°C and 200 rpm until the culture reaches an $OD_{600}$ of 0.5.
3. Induce the logarithmically growing *E. coli* culture with 1 mM IPTG to switch on protein expression and incubate for 4 h at 37°C while shaking.
4. After a 4-h induction period, harvest the cells by centrifugation (10 min, 6,000 × *g*, 4°C) and resuspend cell pellet in 50 ml PBS-I per 1,000-ml initial culture volume. The resuspended cells can be shock frozen with liquid nitrogen and stored at −80°C.
5. If cell suspension was stored at −80°C, thaw at 37°C in a water bath and place subsequently on ice.
6. For efficiently disrupting, pass cells through a French press at 1,000 bar and 4°C and repeat cycles several times.
7. To remove cell debris, centrifuge the crude cell lysate 30,000 × *g* and 4°C for 20 min. Transfer supernatant (cleared cell lysate) into a fresh tube and take a 10 μl aliquot.
8. Pour $Ni^{2+}$-NTA-agarose superflow column according to the manufacturer's protocol. Use a column bed volume of 2 ml.
9. Apply the cleared cell lysate (from step 7) onto the $Ni^{2+}$-NTA-agarose column and let it pass through by gravity, and subsequently wash the column with 10-column volumes PBS-I and 10-column volumes IMAC wash buffer (see Note 2). For each step (flow through and washing fractions), take a 10 μl aliquot to track purification process by SDS-PAGE.
10. Elute $F_M$ protein from the column with 5-column volumes IMAC elution buffer. Take a 10 μl aliquot of the elution fraction to check purity of the isolated protein. The yield of purified $F_M$ protein is 50–70 mg per liter initial bacteria culture. Store at 4°C until further use.

### 3.2. Expression and Purification of the Cargo Protein $VEGF_{121}$ in *E. coli*

Since the cargo protein human $VEGF_{121}$ is processed into inclusion bodies during expression in *E. coli* and has to be isolated under denaturing conditions following refolding in its native state, there is a modified protein purification protocol compared to the one described above in Subheading 3.1. The expression plasmid pRSet-$VEGF_{121}$ was generated by PCR-mediated amplification of the cDNA of human mature $VEGF_{121}$ and ligated into the bacterial expression vector pRSet (Novagen, Madison, WI), thereby fusing it to an N-terminal hexahistidine.

1. Transform the expression vector pRSet-$VEGF_{121}$ into chemical competent *E. coli* expression strain BL21 STAR™ (DE3) (see Note 1) using standard heat shock technique and incubate

inoculum with corresponding antibiotics at 37°C and 200 rpm overnight.

2. Dilute the overnight culture 1:25 with fresh LB media supplemented with corresponding antibiotics and incubate for approximately 1–2 h at 37°C and 200 rpm until the culture reaches an $OD_{600}$ of 0.8.
3. Induce the logarithmically growing *E. coli* culture with 1 mM IPTG to switch on protein expression and incubate for 4 h at 37°C under shaking.
4. After a 4-h induction period, harvest the cells by centrifugation (10 min, 6,000 × *g*, 4°C) and resuspend cell pellet in 50 ml PBS-I per 1,000-ml initial culture volume. The resuspended cells can be shock frozen with liquid nitrogen and stored at −80°C at this step or immediately processed with the purification.
5. If cell suspension was stored at −80°C, thaw at 37°C in a water bath and place subsequently on ice.
6. Disrupt the cells by passing cell suspension repeatedly through a French press at 1,000 bar and 4°C.
7. In order to pellet the inclusion bodies including the $VEGF_{121}$, centrifuge crude cell lysate at 15,000 × *g* for 20 min at 4°C. Discard supernatant and resuspend cell pellet in 20 ml inclusion body dissolution buffer per 1,000 ml initial culture and incubate under continuous gentle stirring overnight at 4°C to dissolve the inclusion bodies.
8. Pour $Ni^{2+}$-NTA-agarose superflow column according to the manufacturer's protocol. Use a column bed volume of 2 ml.
9. Centrifuge the dissolved pellet for 1 h at 50,000 × *g* at 4°C to pellet cell debris and load $VEGF_{121}$-containing supernatant on a $Ni^{2+}$-NTA-agarose superflow column equilibrated with inclusion body dissolution buffer and let it pass through by gravity. Take a 10 μl aliquot of the flow through for analysis of the purification steps.
10. To wash the $Ni^{2+}$-NTA-agarose column and remove unspecific bound proteins, add 10-column volumes pellet dissolution buffer, and then apply 10-column volumes urea wash buffer. Keep a 10 μl aliquot for each washing step.
11. After eliminating unspecifically bound proteins (see Note 2), elute the $VEGF_{121}$ from the column with 5-column volumes urea elution buffer.
12. Add 2 mM EDTA final concentration to complex remaining divalent cations and 2 mM DTT to reduce disulfide bonds; incubate for 1 h at 4°C.

13. A three-step dialysis with decreasing amounts of urea is carried out at 4°C for refolding the protein. Transfer eluate (from step 12) in a 3.5 kDa MWCO dialysis tube and dialyze 2× for 1 h against 1 l dialysis buffer 1 (4 M urea), 2× for 1 h in 1 l dialysis buffer 2 (3 M urea), 1 h in 1 l dialysis buffer 3 (2 M urea), and finally overnight against 1 l dialysis buffer 3. Store purified and refolded $VEGF_{121}$ at 4°C.

### 3.3. Analytics of the Protein Purification Process by SDS-PAGE

During the above-described protein purification processes, it is critically necessary to collect aliquots from each single step for subsequent analysis by SDS-PAGE. This allows not only to verify the purity of the isolated protein in the eluate, but also to track all steps and ensure no undesired loss of the protein of interest.

This protocol is adapted to the Bio-Rad system, but can easily be adapted to other SDS-gel apparatus with minor changes.

1. Ensure that the glass cassettes, combs, casting stand, and casting frames are clean and dry.
2. Set up the casting stand assembly. Make sure that the caster is set up properly in order to prevent the gel solution from leaking out when casting the gel.
3. Prepare the 12% resolving gel by mixing 1.8 ml $dH_2O$, 1.3 ml 4× resolving gel buffer, and 2 ml 30% acrylamide/Bis solution in a falcon tube. In order to start the polymerization reaction, add 50 μl APS and 5 μl TEMED. Mix well and pour the gel into the cassettes so that approximately two-thirds are filled. Immediately cover the gel with isopropanol and allow the gel to solidify for 30 min.
4. Once the gel is polymerized, pour out the isopropanol and rinse the gel surface twice with water. Dry the area above the resolving gel with a tissue, taking care not to touch the surface of the gel.
5. Prepare the stacking gel by mixing 2.33 ml $dH_2O$, 1 ml 4× stacking gel buffer, and 0.66 ml 30% acrylamide/Bis solution in a falcon tube. Add 50 μl APS and 2.5 μl TEMED, mix well, and pour the gel onto the surface of the resolving gel. Insert the comb and allow the gel to polymerize for 1 h.
6. Once polymerized, gently remove the comb and rinse the wells twice with running buffer.
7. Place the gel cassettes in the electrode assembly and fill the inner and outer chambers with running buffer.
8. Preparation of the samples: Add 2 μl of 5× SDS sample buffer to each 10-μl aliquots collected during the purification progress and denature proteins for 5 min at 95°C.
9. Load the gel using a Hamilton syringe. Include one well for a prestained molecular weight standard.

10. Assemble the chamber and connect it to a power supply. Separate the proteins by applying 150 V for 1.5 h. Turn off the voltage when the blue migration front has reached the bottom of the gel.
11. Disassemble the gel chamber and transfer SDS-polyacrylamide gel into Coomassie staining solution and incubate for 1 h under gentle rotation.
12. After 1 h, discard Coomassie staining solution and rinse gel several times with $dH_2O$ before transferring it into Coomassie destaining solution; incubate for 1 h until only protein bands become visible (see Note 3).

### 3.4. Synthesis of the Polymer Poly(AAm-co-$Ni^{2+}$-NTA-AAm)

1. Dissolve 3 mmol (900 mg) acryloylchloride in 15 ml toluene and add dropwise while stirring during 4 h to an ice-cooled solution of 3 mmol *N,N*-bis(carboxymehyl)-L-lysine dissolved in 27 ml 0.44 M NaOH.
2. Evaporate toluene *in vacuo* and eliminate potential precipitates by filtration through a 0.22-um filter device. The yield of the synthesized 2,2′-(5-acrylamido-1-carboxypentylazanedidyl) diacetic acid (NTA-AAm) is approximately 50% = 1.5 mmol.
3. Dissolve 1.5 mmol NTA-AAm (from step 2) and 6.4 mmol acrylamide (AAm) in 48 ml 50 mM Tris/HCl, pH 8.5, and transfer solution in a glass flask with an air-tight valve.
4. For efficient polymerization, displace oxygen by nitrogen or argon; repeat 3–5 times to ensure that no remaining oxygen is left.
5. Add 150 μl ammonium peroxodisulfate (10% w/v) and 24 μl TEMED to initiate polymerization reaction and incubate overnight at 22°C (see Note 4).
6. Concentrate polymer poly(AAm-co-NTA-AAm) to 20 ml in vacuo and transfer to a 3.5 kDa MWCO dialyze tube; since the polymer is swelling and increases its volume dramatically during dialysis steps, close tube with 2 dialyze brackets at each end.
7. Dialyze twice against 2 l $H_2O$ for 12 h to eliminate salts and low-molecular-weight compounds, such as unpolymerized acrylamide.
8. After dialysis, supplement dialysate with 3.5 mmol $NiSO_4$ to charge NTA groups with $Ni^{2+}$.
9. Dialyze twice against 0.5× PBS-I for 12 h and twice against 0.1× PBS for 12 h to eliminate unbound $Ni^{2+}$ and concentrate tenfold in vacuo to result in an approx. 8% (w/v) solution. Store at 4°C until hydrogel formation.

### 3.5. Hydrogel Formation, Dissolution, and Analytics of Released Complexes

The coupling of the cross-linking protein $F_M$ occurs via electrostatic interaction between the hexahistidine tag and the NTA group of the polymer. Similarly, a hexahistidine tag-containing cargo protein can as well be embedded into the hydrogel. Upon addition of the inducer molecule FK506, the gel network dissolves and the cargo protein is released in a dose-dependent manner.

1. Decrease the amount of imidazole from 250 mM in the IMAC elution buffer to 10 mM by ultracentrifugation (10 kDa MWCO ultracentrifugation columns) of the $F_M$ eluate (see Subheading 3.1, step 10). Therefore, centrifuge the eluate using a 10 kDa MWCO ultracentrifugation column at 6,000 × *g* for 30 min. Refill the remaining eluate (approximately 500 μl) with 6 ml PBS-I. Repeat this step three times in order to ensure a final imidazole concentration of 10 mM (see Note 5). Finally, concentrate $F_M$ up to 50 mg/ml (see Note 6).
2. Concentrate $VEGF_{121}$ eluate (see Subheading 3.2, step 13) to 20 mg/ml by ultracentrifugation using a 10 kDa MWCO ultracentrifugation column (6,000 × *g*, 30 min) (see Note 6).
3. Prepare passivated microscope slides by dipping in Sigmacote® following rinsing with 70% EtOH.
4. For microdome hydrogel formation, provide for each hydrogel a 10 μl polymer poly(AAm-co-$Ni^{2+}$-NTA-AAm) drop on a passivated microscope slide.
5. Prepare protein solution by spiking 750 μg $F_M$ (50 mg/ml) with 37 μg $VEGF_{121}$ (20 mg/ml) per hydrogel; prepare master mix protein solution for desired number of hydrogels (see Note 7).
6. Inject carefully 15 μl of the $VEGF_{121}$-spiked $F_M$ protein solution into the 10 μl polymer drop on the passivated microscope slide; pay attention that the polymer encloses the protein solution; total volume of hydrogels 25 μl.
7. Place the glass slide with the prepared hydrogels in a PBS-I-humidified petri dish (see Note 8) and incubate for 24 h at 4°C to allow gelation of the hydrogel.
8. After gelation, the hydrogels form small disks which can easily be detached with a spatula and placed in 1 ml PBS-I in a 24-well plate.
9. Incubate for 8–24 h in 1 ml PBS-I at 22°C to eliminate unbound and not properly incorporated proteins and remove unpolymerized polymer.
10. Transfer hydrogel in 1 ml fresh PBS in a 24-well plate.
11. Add different concentrations (6–100 μM) of the inducer molecule FK506 to the hydrogels to control dose-dependent cargo protein release and incubate at 22°C under gentle rotation.
12. Take samples of the supernatant at different time points to determine time-dependent dissolution characteristics of the hydrogel.
13. Quantify released $F_M$ and $VEGF_{121}$ in the supernatant by the Bradford protein quantification method. Use 5 μl sample and

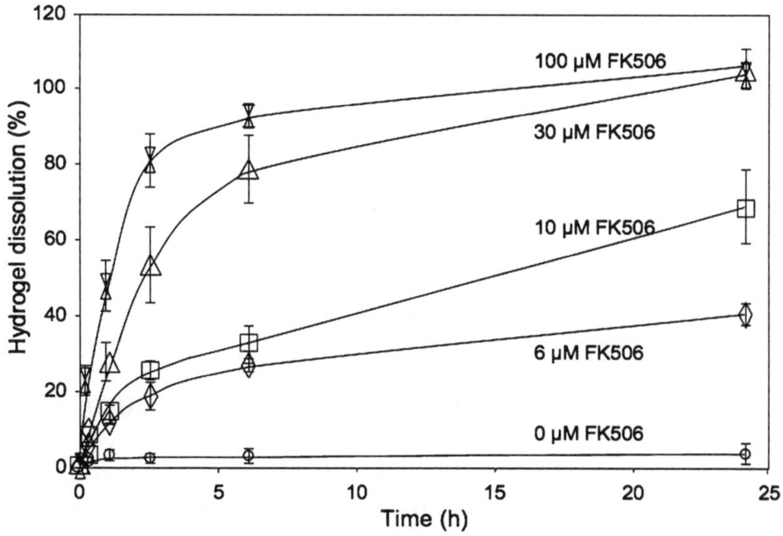


Fig. 2. Time course of dose-dependent hydrogel dissolution. Hydrogels were incubated for 24 h in PBS in the presence of increasing FK506 concentrations. Samples of the supernatant were taken at indicated time points and hydrogel dissolution was monitored by measuring total protein release via the Bradford method. *Error bars* represent the standard deviation from five experiments.

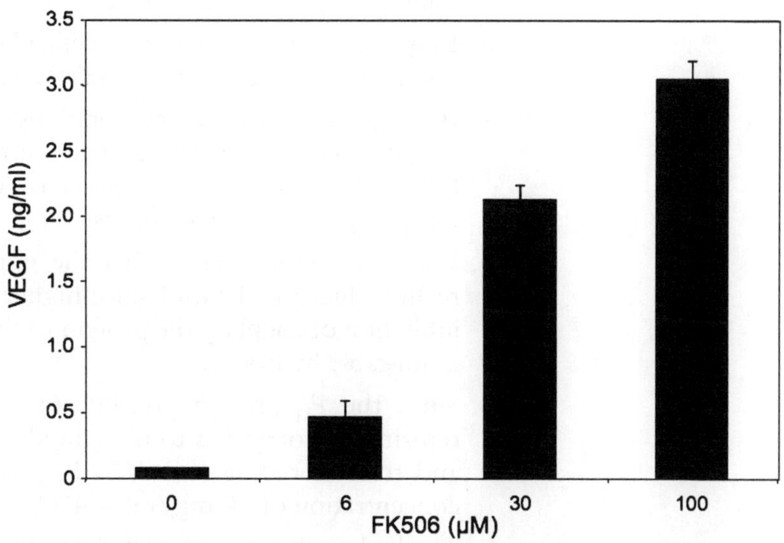


Fig. 3. Dose-dependent release of the cargo protein $VEGF_{121}$. $VEGF_{121}$-containing hydrogels were incubated for 3.5 h in PBS-I supplemented with different concentrations of FK506. $VEGF_{121}$ release in the supernatant was determined by a VEGF-specific ELISA. *Error bars* represent the standard deviation from three experiments.

fill up to 100 μl with a 1:5 dilution of the Bradford reagent. Utilize a 96-well plate and measure absorbance in a microplate reader at 595 nm (see Fig. 2). Include also a standard curve with BSA. Ensure that the samples are in the linear range; otherwise, use less sample volume.

14. To determine $VEGF_{121}$ cargo protein release, perform a VEGF ELISA according to the manufacturer's protocol (see Fig. 3).

## 4. Notes

1. For expression of some proteins, it is of advantage to use the *E. coli* expression strain BL21 STAR™ (DE3) pLysS (Invitrogen, Carlsbad, CA, cat. no. C6020-03). This strain constitutively expresses the T7 lysozyme which reduces basal expression of the protein of interest by inhibiting basal levels of T7 RNA polymerase. Therefore, it is suitable for the expression of proteins with toxic side effects for *E. coli*.
2. If you face problems with copurifying prominent nondesired proteins, wash the column with gradually increasing amounts of imidazole in the IMAC wash buffer ranging from 10 to 250 mM in 10 mM steps. Thereby, collect each washing fraction as it is not known a priori at which imidazole concentration the protein of interest is eluted.
3. In speed up the destaining process, place a tissue at the edge of the glass bin which absorbs the released blue dye rapidly.
4. It is important to avoid any vibrations or shaking as this disturbs the polymerization reaction. The success of the reaction can be followed the next day by gently rotating the glass flask and observing an increase of viscosity.
5. It is crucially important that the imidazole concentration is really reduced to 10 mM since higher concentrations result in inhibition of coupling the protein to the polymer and therefore in unstable hydrogels.
6. Since the $F_M$ protein precipitates with 50 mg/ml at 4°C, remaining protein has to be diluted to 10 mg/ml with PBS-I and then stored at 4°C. $VEGF_{121}$ can stably be stored in a concentration of 20 mg/ml at 4°C.
7. The hydrogels can optionally be stabilized by the introduction of additional cross-links using amine-specific homobifuntional linkers, such as dimethylsuberimidate (DMS). Therefore, add DMS in a molar ratio of 4:1 to the $F_M$ protein prior to hydrogel formation and incubate for 30 min at 22°C.
8. Soak a tissue with PBS-I prior to placing the glass slide with the hydrogels in the petri dish to avoid dehydration of the hydrogels. Never use $dH_2O$ as the water leads to swelling of the polymer and inhibits hydrogel formation.

## Acknowledgments

The work of MMK was supported by the GEBERT RÜF STIFTUNG (Grant No. GRS-042/07) and the work of WW was supported by the Excellence Initiative of the German Federal and State Governments (EXC 294).

### References

1. Lutolf, M. P., and Hubbell, J. A. (2005) Synthetic biomaterials as instructive extracellular microenvironments for morphogenesis in tissue engineering, *Nat Biotechnol* ***23***, 47–55.
2. Lendlein, A., Jiang, H., Junger, O., and Langer, R. (2005) Light-induced shape-memory polymers, *Nature* ***434***, 879–882.
3. Wang, C., Stewart, R. J., and Kopecek, J. (1999) Hybrid hydrogels assembled from synthetic polymers and coiled-coil protein domains, *Nature* ***397***, 417–420.
4. Miyata, T., Asami, N., and Uragami, T. (1999) A reversibly antigen-responsive hydrogel, *Nature* ***399***, 766–769.
5. Rivera, V. M., Wang, X., Wardwell, S., Courage, N. L., Volchuk, A., Keenan, T., Holt, D. A., Gilman, M., Orci, L., Cerasoli, F., Jr., Rothman, J. E., and Clackson, T. (2000) Regulation of protein secretion through controlled aggregation in the endoplasmic reticulum, *Science* ***287***, 826–830.
6. Rollins, C. T., Rivera, V. M., Woolfson, D. N., Keenan, T., Hatada, M., Adams, S. E., Andrade, L. J., Yaeger, D., van Schravendijk, M. R., Holt, D. A., Gilman, M., and Clackson, T. (2000) A ligand-reversible dimerization system for controlling protein-protein interactions, *Proc Natl Acad Sci USA* ***97***, 7096–7101.
7. Bayle, J. H., Grimley, J. S., Stankunas, K., Gestwicki, J. E., Wandless, T. J., and Crabtree, G. R. (2006) Rapamycin analogs with differential binding specificity permit orthogonal control of protein activity, *Chem Biol* ***13***, 99–107.
8. Kämpf, M.M., Christen, E.H., Ehrbar, M., Daoud-El Baba, M., Charpin-El Hamri, G., Fussenegger, M., and Weber, W. (2010) A Gene Therapy Technology-Based Biomaterial for the Trigger-Inducible Release of Biopharmaceuticals in Mice, *Adv Funct Mater* ***20***, 2534–2538.
9. Ehrbar, M., Djonov, V. G., Schnell, C., Tschanz, S. A., Martiny-Baron, G., Schenk, U., Wood, J., Burri, P. H., Hubbell, J. A., and Zisch, A. H. (2004) Cell-demanded liberation of VEGF121 from fibrin implants induces local and controlled blood vessel growth, *Circ Res* ***94***, 1124–1132.
10. Maina, C. V., Riggs, P. D., Grandea, A. G., 3rd, Slatko, B. E., Moran, L. S., Tagliamonte, J. A., McReynolds, L. A., and Guan, C. D. (1988) An Escherichia coli vector to express and purify foreign proteins by fusion to and separation from maltose-binding protein, *Gene* *74*, 365–373.
11. Smith, D. B., and Johnson, K. S. (1988) Single-step purification of polypeptides expressed in Escherichia coli as fusions with glutathione S-transferase, *Gene* *67*, 31–40.

# INDEX




Wilfried Weber and Martin Fussenegger (eds.), *Synthetic Gene Networks: Methods and Protocols*, Methods in Molecular Biology, vol. 813, DOI 10.1007/978-1-61779-412-4, © Springer Science+Business Media, LLC 2012

MIX
Papier aus verantwortungsvollen Quellen
Paper from responsible sources
FSC® C105338

If you have any concerns about our products,
you can contact us on
**ProductSafety@springernature.com**

In case Publisher is established outside the EU,
the EU authorized representative is:
**Springer Nature Customer Service Center GmbH**
**Europaplatz 3, 69115 Heidelberg, Germany**

Printed by Libri Plureos GmbH
in Hamburg, Germany